Electricity Markets

Sayyad Nojavan • Kazem Zare
Editors

Electricity Markets

New Players and Pricing Uncertainties

Springer

Editors
Sayyad Nojavan
Department of Electrical Engineering
University of Bonab
Bonab, Iran

Kazem Zare
Department of Electrical
and Computer Engineering
University of Tabriz
Tabriz, Iran

ISBN 978-3-030-36978-1 ISBN 978-3-030-36979-8 (eBook)
https://doi.org/10.1007/978-3-030-36979-8

This Springer imprint is published by the registered company Springer Nature Switzerland AG.
The registered company address is: Gewerbestrasse 11, 6330 Cham, Switzerland

Preface

Electricity generation, transmission and distribution were once primarily controlled by private companies. However, after a huge restructuring in the industry we find that retailers, large consumers, virtual power plants, demand response aggregators, electric vehicle aggregators, smart homes, energy hubs, microgrids, etc. have emerged to be key within the industry as well. Nowadays, there is a severe uncertainty of electricity market price, it is a challenge which the major electricity market players are facing. It should be noted that the electricity market price uncertainty is increasing mainly due to behaviours of market players, renewable distributed generation and responsive loads. This uncertainty poses challenges on grid operations and control, distorts the electricity market outcomes and complicates the short-run and long-run decision-making processes of electricity market players. By defining aforementioned problems and challenges of the industry, the optimal solutions will be discussed in this book. Furthermore, it can be studied in order to investigate how electricity is purchased or sold in the presence of electricity market price uncertainty. Therefore, uncertainty modelling of market price is necessary to take a decision with less risk. According to the importance of continuous and secure operation of electricity market players, further and complete studies are required in this concept. Finally, this book seeks to find, analyse and introduce features and problems of electricity market players in uncertain environment.

Bonab, Iran | Sayyad Nojavan
Tabriz, Iran | Kazem Zare

Contents

About the Editors

Sayyad Nojavan, PhD, is an Assistant Professor at the Department of Electrical Engineering, University of Bonab, Bonab, Iran. His research areas include distribution networks operation, power system operation and economics, electricity market, demand response applications, hybrid energy system, uncertainty modelling and risk management. He has also edited several books in the electrical and energy fields in Springer and Elsevier Publications.

Kazem Zare, PhD, is an Associate Professor of the Faculty of Electrical and Computer Engineering, University of Tabriz, Tabriz, Iran. His research areas include distribution networks operation and planning, power system economics, microgrids, demand response applications and energy management. He has also edited several books in electrical and energy fields in Springer and Elsevier Publications.

Chapter 1
Energy Harvesting Technologies and Market Opportunities

Farzad H. Panahi and Fereidoun H. Panahi

Nomenclature

AP	Access point
BS	Base station
CH	Cluster head
D2D	Device-to-device
EE	Energy efficiency
EH	Energy harvesting
HER	Energy harvesting rate
EHT	Energy harvesting technology
FIS	Fuzzy inference system
FQLA	Fuzzy *Q*-learning algorithm
HetNet	Heterogeneous network
ICT	Information and communication technology
IoT	Internet of Things
GHG	Greenhouse gas
M2M	Machine-to-machine
PS	Power station
QLA	*Q*-learning algorithm
QoS	Quality of service
RF	Radio frequency
RL	Reinforcement learning
RPS	Renewable power supplier
RES	Renewable energy source

F. H. Panahi (✉) · F. H. Panahi
Department of Electrical Engineering, University of Kurdistan, Sanandaj, Kurdistan, Iran
e-mail: f.hpanahi@uok.ac.ir

S. Nojavan, K. Zare (eds.), *Electricity Markets*,
https://doi.org/10.1007/978-3-030-36979-8_1

SG	Smart grid
WPT	Wireless power transfer
WSN	Wireless sensor network
UDN	Ultradense network
UE	User equipment
UWB	Ultrawide band

1.1 Introduction

Enhanced management of urban communities brings another concept, known as smart cities [1, 2], which enables environmental data gathering and better usage of city resources. Specific application areas of smart cities are intelligent transport systems, smart grids, home automation, smart agriculture, and structural health monitoring [3]. The realization of them strictly requires considerable advancements over edge systems and devices, for example, the Internet of Things (IoT). Indeed, detecting and controlling highlights of the IoT are essential empowering agents of this acknowledgment. Utilizing IoT and Information and Communication Technology (ICT) features, we can access a digitized world by means of the Internet connections, and draw one stage nearer to the Smart City idea [4, 5]. Obviously, in order to obtain nonstop tracking and control, an auxiliary or maybe a totally particular power supply needs to be prepared to the sensors. However, this strategy might also or would not be practical in some cases in general because of size constraints or environmental restrictions. Hence, energy harvesting (EH) strategies come into prominence to relieve the troubles of energy restrained networks via utilizing a stray supply or converting power from one shape to every other [6, 7].

Generally, EH, also known as energy scavenging, is the action by which energy is extracted from available external sources, commonly named as “ambient energy sources.” There is a wide assortment of ambient energy sources and relating EH technologies (EHTs) with various specialized applications. The degree of deliverable energy from every technology differs also in the range of low micro-Watt to milli-Watt. Indeed, this variety is a chance to design an appropriate EHT for IoT-gadgets according to the use cases. Some EHTs effectively offer the fundamental power yield to drive IoT gadgets. However, they may not be applicable. The energy harvester must fit in with the utilization case and the related energy forms, subjects which must be viewed when planning a gadget. In this chapter, a review of various sources, technologies, intelligent mechanisms, and also market opportunities for EH is presented.

1.2 Energy Harvesting Technologies and Challenges

Giving EH ability to smart systems and networks empowers the devices to consistently obtain their power from natural or man-made phenomena. Therefore, this gives promising features to wireless networks: self-sustainability and reliability

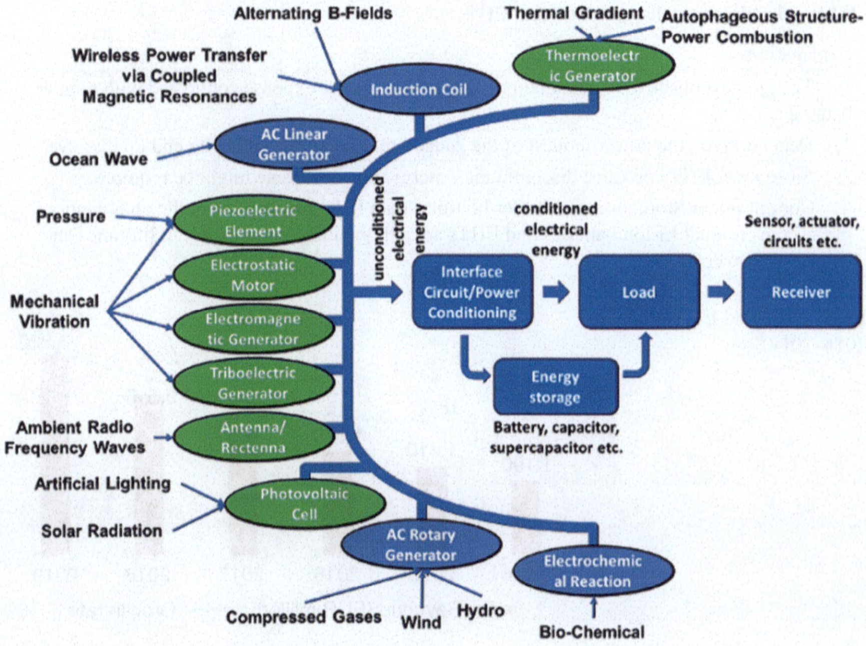

Fig. 1.1 Different forms of ambient energy sources and corresponding EHTs [15]

with network lifetimes constrained by an internal equipment instead of the energy storage. Thus, EH-enabled wireless networks will make it feasible to grow new medicinal, surveillance, and security applications which are not generally practical with ordinary battery-controlled nodes. There are a few distinctive characteristic sources and related technologies for EH: electromagnetic, solar, indoor lighting, vibrational, thermal, biological, chemical, etc. [6, 8–14] (as can be observed in Fig. 1.1). However, energy might be obtained from man-made sources by means of wireless power transfer (WPT), in a controlled way.

It is clear that the development of an EHT takes time since it needs often profound researches. Thus, there is the risk that industry surrenders the technology. For instance, small manufactures working on some energy harvesters are not fundamentally developing them with their defined utilization as harvesters any more, but as temperature sensor nodes. Generally, the major challenges of current EHT can then be summarized as in Table 1.1.

1.3 Energy Harvesting Markets and Key Players

Generally, the EH has been utilized for quite a long time for bike dynamos or solar panels. Today, it is widely applied to application fields, such as smart cities, automotive vehicles, and security systems. Development inside the areas of ICT and IoT

Table 1.1 Major challenges of current EHTs

Explanations
1. The greatest obstacle of actualizing EHT is their greater expense contrasted with regular batteries.
2. Depending on the power request of the gadget, the size of the EHT can end up massive.
3. Since most EHTs produce discontinuous energy, power storage might be required.
4. Current power storages have shorter lifetimes than EHTs. It seems that the combination of the conventional Li-ion batteries and EHTs are still the ideal with respect to lifetimes and supported power.

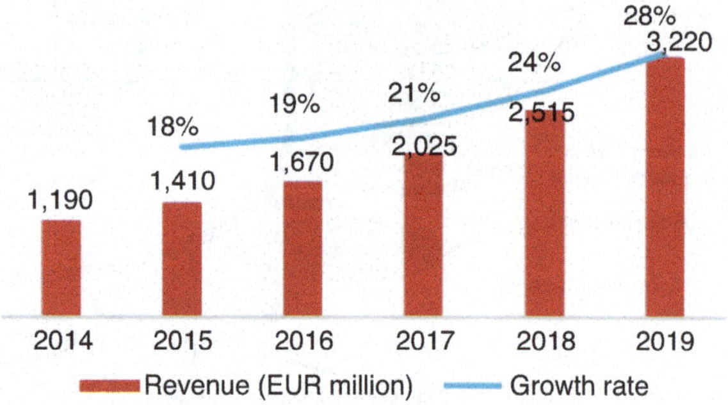

Fig. 1.2 Global EH market 2014–2019

and thus the spread of battery-based sensor systems are real power driving advances in EH and self-powered systems [16, 17]. The most well-known power sources utilized for EH are mechanical, thermal energy, and sunlight-based radiations. Recent advances in ultralow power technologies have accelerated the improvement of self-powered monitoring gadgets for a wide scope of utilizations consisting of smart grids (SGs), structural health monitoring, and biomedical telemetry [18–21]. A self-powered wireless sensor, that gains surrounding energy for driving its hardware, is among promising techniques for supporting a maintenance-free sensor network in smart networks.

The worldwide EH market demonstrates a stunning development: somewhere in the range of 2015 and 2019 it could sum at 21.9% and peak at 28% in 2019 (Fig. 1.2). Governments and public initiatives are the main drivers for EH market development. Public actors utilize EH as a key apparatus for gathering the rising energy request and saving power. In fact, EH supports SGs and IoT by powering wireless sensor networks (WSNs) that are fundamental to provide connectivity between devices. A huge number of sensors is required to monitor and manage network processes and the sensors should be powered. Ordinarily, batteries were utilized to enable the sensing nodes but they have a restricted lifetime and in a network with a huge number of wireless sensors, replacement of the batteries will not be applicable. It should be noted that EH-powered sensors need less maintenance and are easier to arrange than batteries and also more comfortable to manage in mobile-sensing strategies. To sum up, development of IoT and energy-efficient communication infrastructures for sensor networks is driving interest for wireless and battery-less sensors which will be increasingly more powered by EH. Indeed, EH wireless solutions find increasing applications in smart networks due to their low-cost installation and maintenance. In addition, EH-based wireless technology is the reliable communication strategy to provide connectivity among thousands of nodes in smart networks.

European Commission supports the business with motivation and interests in R&D of EH and storage gadgets. This is rational with European priorities as the Commission distinguishes the feasible supply of energy as the "grand challenges" confronting human communities. Indeed, green power trend is advancing the market as a lot of sustainable and renewable power sources offer an appropriate platform for the EH process. As mentioned before, EH empowers IoT and ICT by driving sensors arranges that are fundamental to associate and organize devices. Countless sensors are expected to make network procedures work and sensors should be powered. Normally, batteries were utilized to enable the sensing hardwires but they have a restricted lifetime and in a framework with a huge number of remote gadgets, the replacement of batteries is really impossible. On the other side, EH-powered sensors are self-sufficient, require less support, and are simpler to set up than batteries. Totally, advancement of IoT and ICT is driving interest for remote and battery-less sensors which will be increasingly more supported by EH.

The market for home automation is booming today and it is estimated to have 5% growth in the range of 2015–2019. Indeed, EHTs find increasing applications in this area thanks to their high cost-saving potential in setup and maintenance. Compared to copper wiring or battery, EHT is the ideal communication standards to interconnect a huge number of devices with various applications. Based on the report "EH Market Size, Share, Growth Forecast 2019 To 2027," published by Market Research Future (MRFR), EH market is increasingly influenced by growing demands for renewable power sources (RPSs). The worldwide EH market is additionally studied in detail in the report, which aims to find out how the market is likely to progress over the forecast period and what are the significant drivers affecting the market's direction over that period. In fact, EH is a basic term for the way toward catching the energy from a specific power source and storing it for later use. In spite of its simple definition, the EH market growth has not been rapid to develop because of specialized challenges in designing power storage systems. The EH market is accordingly to grow at a steady rate over the forecast period, driven by the growing interests for a considerable progress in the field of EHTs. However, organizations and new companies are expected to appear despite the heavy investments required to enter it, given the attractive growth opportunities. Generally, the global EH market is segmented on the basis of energy source, technology, application, and region (Table 1.2).

The developing interest for renewable powers such as solar- and wind-oriented energies is additionally to be a noteworthy driver for the worldwide EH market. As the generation of electric power through wind- and sunlight-based mechanism is transient and temperamental for steady power delivery, EHTs become hugely important in this area. Expanding endeavors are probably going to be taken in these fields to create successful EH frameworks over the coming years, prompting

Table 1.2 Main segments for global EH market

Type	Segments
Energy source	Chemicals, mechanical, electrical, nuclear, thermal, and gravitational
Technology	Electrodynamics, photovoltaic, thermoelectric, and others
Applications	Electronics, industrial, aerospace and defense, automotive, healthcare, and others

consistent development of the worldwide EH market over the forecast period. Key players in the worldwide EH market are probably going to concentrate on research endeavors to think of strong answers for the main consumers. Teaming up with outer research foundations is also likely to be the main way for players in the EH market, as many promising advancements in energy storage happen in research organizations. Europe, Asia Pacific, and North America presently dominate the worldwide EH market and are probably going to remain as the key players over the forecast period because of the growing government supports to renewable energy initiatives and also the growing presence of leading players in the region, which has prompted the improvement of a solid research segment in the field.

1.4 Intelligent Mechanisms for Energy Harvesting

The Internet of Things (IoT) and intelligent wireless sensor networks as a promising improvement for future telecommunications will make everything smart and empower them associating with one another instantly and transferring data pervasively [22, 23]. However, there are different challenges toward communication networks deployment consisting of security, quality of service (QoS), reliability, energy affairs, and technologies [16, 24–34]. Obviously, the IoT connected things will be more than the human population in near future. For this tremendous system of interconnected sensors, batteries are the best sources to give the ability to work the services. On the other side, IoT modules and devices require longer lifetime and supplanting the batteries oftentimes is impractical [26]. Consequently, as referenced previously, EH strategy is one of the responses to this issue. The EH-aided WSNs can work for a considerable length of time and years with the minimum level of human interventions [16].

Regarding the energy efficiency (EE) as a key basis in the designing of communication systems, some feasible approaches should be taken into account to overcome the restrictions of energy sources and network lifetime [35]. In this way, EH strategy in intelligent networks [17] has been proposed as a promising technique for expanding the lifetime of mobile and low-power nodes. Recently, RF-based EH strategies have been presented to enable portable sensors and devices to gather energy from radiated radio frequency signals in the form of ambient or devoted RF sources [36, 37]. In addition, regarding the impressive benefits of wireless powering energy-constrained sensors and maximizing the lifetime of sensor networks, EH techniques have been extensively investigated in different scenarios for WSNs [38].

Inspired by the effective performance of intelligent mechanisms and learning-based algorithms to design practical scenarios over smart networks, many researches have been focused on the acceleration of the battery charging process [39, 40]. In some other works, intelligent algorithms are utilized to guide sensor movements toward allocated power stations (PSs) as wireless battery charging points, in order to define an energy-efficient EH-enabled network. The approach depends on finding the areas of the PSs during successive movements of the mobile sensors. Thus, as stated, every sensor freely utilizes the intelligent algorithms to gradually discover a PS in the system. In a general model, considering downlink transmission, a two-tier HetNet for energy-based cooperation scenarios among sensor nodes is investigated. More explicitly, there exist N mobile sensors along with three kinds of BSs, consisting of a central BS and Q cluster heads (CHs) and furthermore M PSs, randomly distributed over the network, powered by both electrical grid and renewable power sources (RPSs) (as can be observed in Fig. 1.3).

As mentioned before, to make the EH process smart, we may have to utilize intelligent or learning-based mechanisms. So far, various kinds of reinforcement learning (RL) algorithms have been introduced [39]. Indeed, the Q-learning algorithm (QLA), as one of the most popular RL algorithms, computes the table of all values $Q(s, a)$ using continuous estimation, to form a Q-table. It should be noted that $Q(s, a)$ represents the expected value or the quality factor for a specific problem,

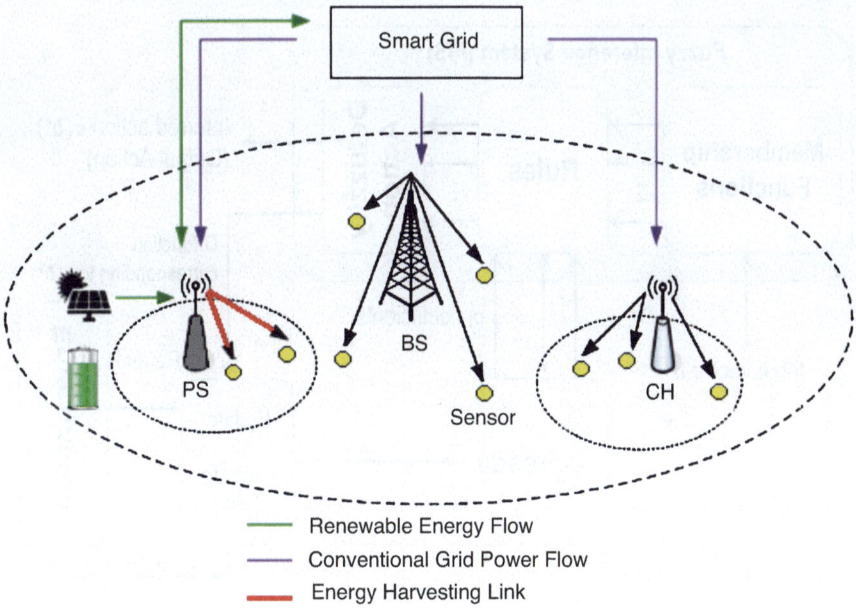

Fig. 1.3 EH model consisting of PSs and mobile sensors

which can be acquired in state $S = \{s_1, s_2, \ldots s_N\}$ for the sensor action a and the corresponding reward value r. Consequently, the related Q-table is formed according to the iterative equation as follows:

$$Q(s, a) \leftarrow Q(s, a) + \beta \Delta Q(s, a) \tag{1.1}$$

where $\beta \in (0, 1]$ is the learning rate and,

$$\Delta Q(s, a) = r + \lambda \max_{a'} Q(s', a') - Q(s, a) \tag{1.2}$$

Here, the maximization operator indicates the greatest value obtained by a mobile sensor for the action a' that may be done in next state s'. In fact, the basic QLA performance can be effectively improved by means of accurate tracking of the state-action history. This is defined by the competency factor $\lambda \in [0, 1]$, thus the enhanced learning strategy is called $Q(\lambda)$-learning [40]. The parameter λ for each state increases after the state-action process, and then exponentially decreases until the state is not checked again [39].

In general, QLA enables the mobile sensors to learn from interaction with the network, where a reward mechanism is defined for the learning process of EH. However, combining a fuzzy-control strategy with the QLA (i.e., FQLA) leads to an enhanced self-adaptive algorithm for pragmatic applications (Fig. 1.4). Indeed, the way of knowledge representations can be expressed as the primary contrast in

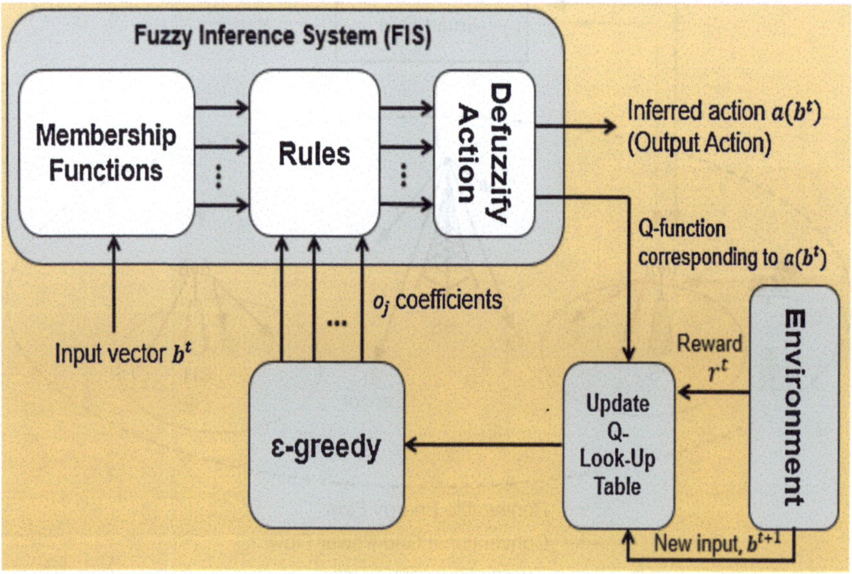

Fig. 1.4 Block diagram of FQLA to control movements of a mobile sensor

the model between the QLA and FQLA. In other words, fuzzy rules for evaluating the explored knowledge are exploited in the FQLA, while a basic look-up table (i.e., Q-table) is used in the QLA [40]. The fuzzy inference system (FIS) that plays a key role in making final decisions for FQLA includes a set of rules R and competing actions for each rule. Accordingly, the mobile sensor (i.e., the learning agent) needs to detect the best conclusion based on the related rule. It implies that an action with the highest Q-value between the feasible actions for a rule is selected. Clearly, the mobile sensors (i.e., the learning agents) should advance toward power stations (PSs) according to FQLA's decisions to detect the nearest PS. In order to estimate the optimal policy, the state-action value function $Q^{\pi}(s, a)$ is approximated in case of taking action $a \in A_s$ in state s.

Here, a practical smart scenario for mobile sensors is considered to investigate the power-saving impacts of the EH process on sensor networks. In addition, in order to evaluate whether this methodology can quicken the EH procedure in mobile sensors, a series of simulations is performed. It should be noted that the simulated environment is defined based on conventional WSN configurations, then simulation results corresponding to the conventional (Fig. 1.5a) and a FQLA-based EH models (Fig. 1.5b) are presented, respectively. Here, a basic definition to assess the EH efficiency over the system is introduced. In this way, the EH rate (EHR) is defined as $k_{\mathrm{EH}}.\left(d_{\mathrm{EH}}^{0}/d_{\mathrm{EH}}^{t}\right)^{2}$ to show the effectiveness of EH process for all mobile sensors distributed over the network, where k_{EH} is a positive coefficient and d_{EH}^{t} indicates average distance between sensors and PSs at the time of t. According to the results (Fig. 1.6), a performance degradation can be observed when the exact knowledge of PSs' locations is not available. In this case, unlike the perfect case, mobile sensors are exploring PSs using partial location information prepared by M2M communications.

Now, as a special case, another intelligent methodology is investigated in which a centralized image-processing (IP) technique is utilized to monitor the estimated network coverage and then to identify the potential locations for EH, i.e., red regions which indicate high level of RF ambient energy. The mobile sensors at that point endeavor to get to these areas during smart motions, in order to improve the network lifetime. This process is modeled based on the simulated HetNet coverage as plotted in Fig. 1.7. As stated before, smart sensors around the HetNet will try to move toward these areas to harvest energy and therefore to accelerate the battery charging process over some potential areas for the EH. As a result, the accuracy of the estimation of network coverage plays a great role to provide an efficient EH. In this technique, red areas are called peak energy points, and are assumed as the peak points of mountains. To continue, the IP-based algorithm, called peak detection algorithm can be exploited to find the peak points, which is the process of exploring for the mountains' peaks, i.e., red areas. In fact, a peak can be deduced as a higher location (i.e., higher energy for the EH model) compared with surrounding regions. Note that, in general, the concept of peaks in a specific geographical space has various interpretations [41].

Here, the peak energy regions, i.e., the red areas in the network (Fig. 1.8), are determined. As can be seen in Fig. 1.9, the peak detection is done using a mean-shift

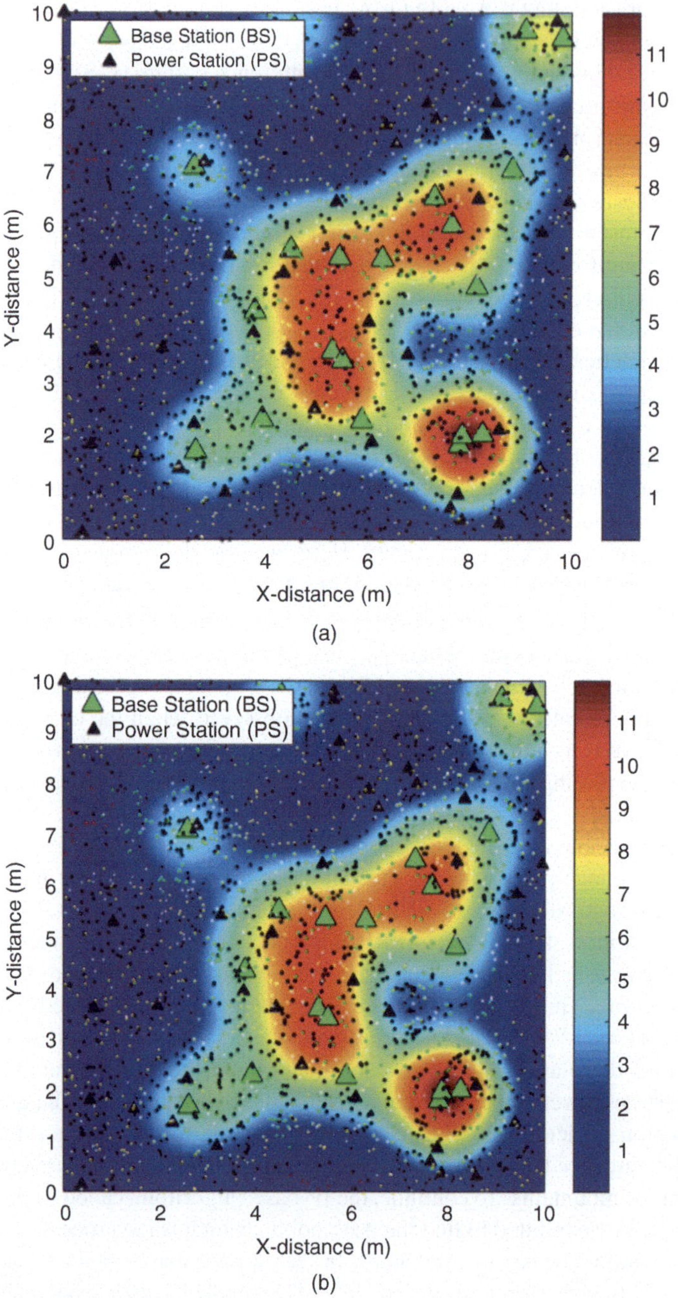

Fig. 1.5 Coverage map for conventional and intelligent EH-enabled WSNs. (**a**) Conventional EH model. (**b**) FQLA-based EH model

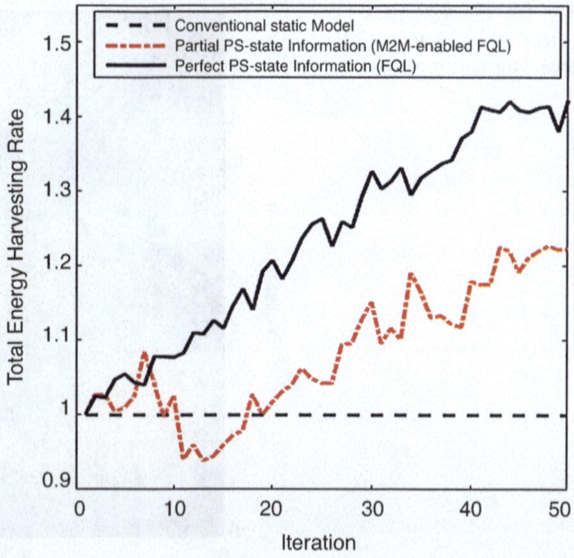

Fig. 1.6 EH rate for static and dynamic sensor scenarios

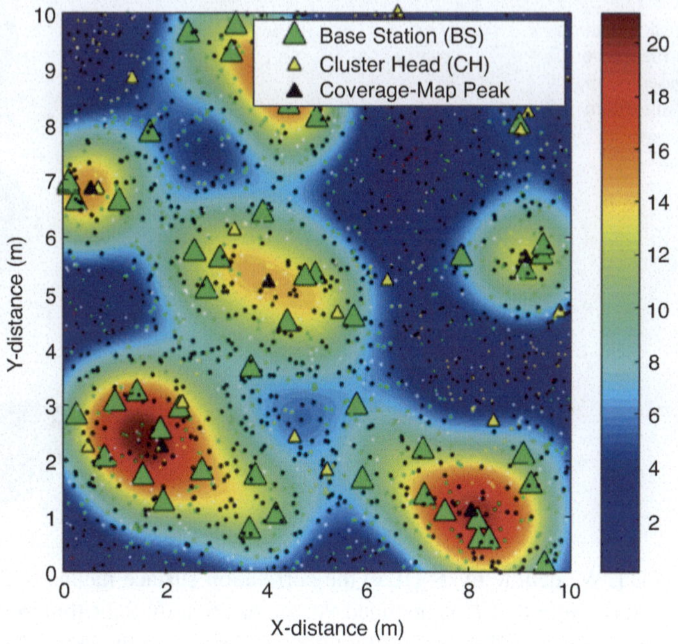

Fig. 1.7 Coverage of a sensor network in coexistence with a two-tier cellular HetNet

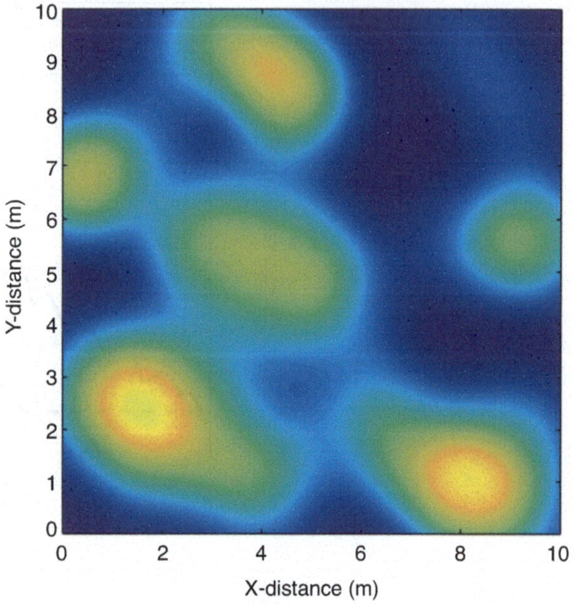

Fig. 1.8 Coverage peak detection (i.e., potential EH areas) for sensor networks

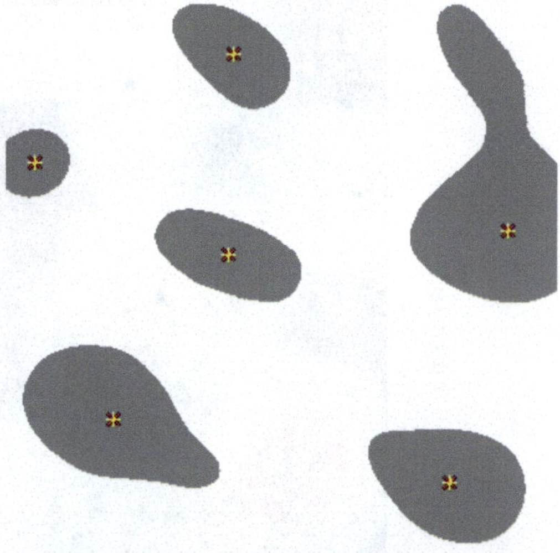

Fig. 1.9 Peak detection process for the network coverage based on the mean-shift algorithm

algorithm [41]. We denote by Γ (μ, v) the correlation surface having the origin at the highest peak. Note that Γ is obtained via the mean-shift algorithm in which the estimated points for the EH peaks are iteratively updated over the network coverage. Therefore, in an iterative way and according to mean-shift rule, each point (x, y) is updated until a stable state is reached.

The simulation results given will show how the FQLA can control the sensors' mobility to reach those potential EH areas based on smart movements (Fig. 1.10). Consequently we can understand how the FQLA can speed up the process of wire-

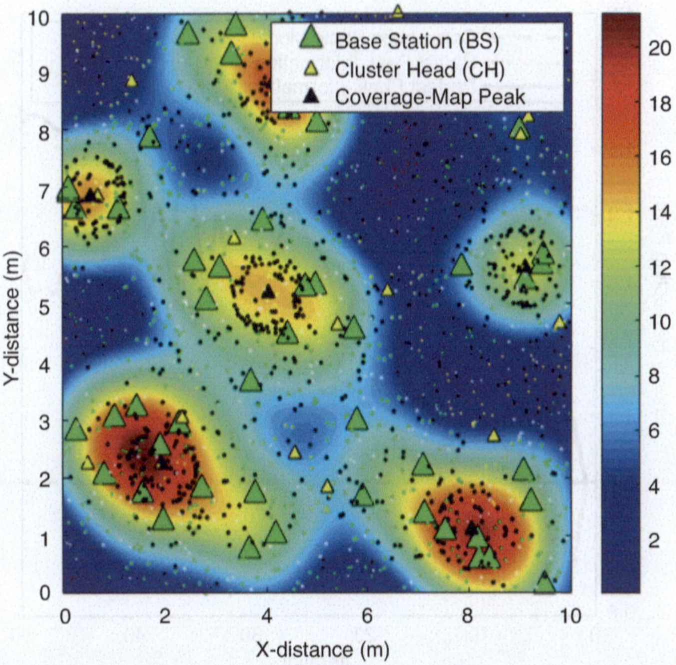

Fig. 1.10 Mobility of sensors to detect potential EH areas (i.e., peaks) over the HetNet

less EH, i.e., battery charging, for randomly distributed sensors over the network. As mentioned, the essential methodology is built on a centralized IP approach. More specifically, a centralized IP unit is deployed to monitor and estimate the instant coverage map of the HetNet, and then to detect the red areas, according to the described peak detection process. By checking the simulation results (Fig. 1.11), as expected, one can similarly observe the performance degradation in the case of partial peak information for the average EHR. The precise information of peak locations are assumed to be accessible to all sensors when the perfect case is considered.

1.5 Conclusions and Suggested Readings

The ever-growing market of communication devices leads to increase in demands for efficient power solutions. In addition, the developing interest for green powered systems and smart networks is additionally to be a noteworthy driver for the worldwide EH market. The problem of battery lifetime and EHTs is as relevant to research organizations as it is to companies and consumers. Currently, a few standalone EHTs can help to significantly alleviate this issue but researchers and companies should collaborate together to come up with intelligent EH mechanisms

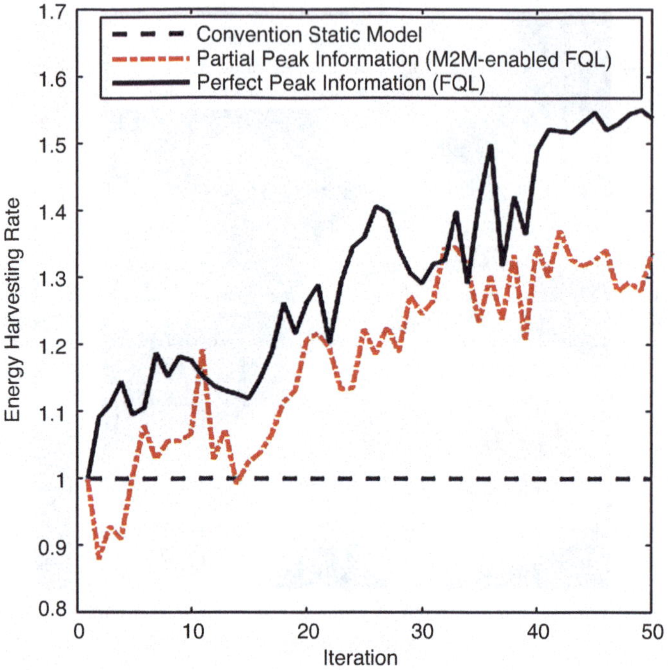

Fig. 1.11 EH rate for static and dynamic sensor scenarios

Table 1.3 Major topics and the latest research article on energy harvesting technologies

Topics	References
EH optimization and approximation algorithms	[42–44]
Joint operation scheduling (routing, data gathering, ambient harvesting)	[45–47]
WPT devices deployment	[48–50]
Electromagnetic energy harvesting and circuits and systems	[51–53]
Simultaneous wireless information and power transfer (SWIPT)	[54–56]
Green powered infrastructures and markets	[57–59]
Battery and energy storage elements	[60–62]
RFID-related electronics, self-powered sensors, and wearable devices	[63–65]
Biomass, biofuel, and bioenergy	[66–68]
Intelligent and nature-inspired optimal algorithms for EH	[69–71]
Energy-efficient and D2D-based wireless communications	[72–74]
Rechargeable sensor networks	[75–77]

to impact upon the design and appearance of future smart networks. This chapter reviewed profound analyses and contributions in the board area of EHTs and EH market along with some intelligent EH scenarios through computer simulation. Table 1.3 summarizes the general topics to prepare a comprehensive perspective on the recent studies in this area.

References

1. H. Schaffers et al., Smart cities and the future internet: towards cooperation frameworks for open innovation, in *The Future Internet, Lect. Notes Comput. Sci.*, vol. 6656 (2011), pp. 431–446
2. S. Chen et al., A vision of IoT: applications, challenges, and opportunities with China perspective. IEEE Internet Things J. **1**(4), 349–359 (2014)
3. O. Vermesan, P. Friess, Internet of Things strategic research and innovation agenda, in *Internet of Things—Converging Technologies for Smart Environments and Integrated Ecosystems*, (River Publishers, Denmark, 2013)
4. L. Atzori et al., The Internet of Things: a survey. Comput. Netw. **54**(15), 2787–2805 (2010)
5. A. Zanella et al., Internet of Things for smart cities. IEEE Internet Things J. **1**(1), 22–32 (2014)
6. S. Sudevalayam, P. Kulkarni, Energy harvesting sensor nodes: survey and implications. IEEE Commun. Surv. Tutorials **13**(3), 443–461 (2011)
7. J.W. Matiko et al., Review of the application of energy harvesting in buildings. Meas. Sci. Technol. **25**(1), 1–25 (2013)
8. J.A. Paradiso, T. Starner, Energy scavenging for mobile and wireless electronics. IEEE Perv. Comput. **4**, 18–27 (2005)
9. V. Raghunathan, C. Schurgers, P. Sung, M.B. Srivastava, Energy-aware wireless microsensor networks. IEEE Signal Process. Mag. **19**, 40–50 (2002)
10. S. Roundy, Energy Scavenging for Wireless Sensor Nodes with a Focus on Vibration to Electricity Conversion, Ph.D. Dissertation, Dept. of EECS, UC Berkeley, May 2003
11. J. Rabaey, M.J. Ammer, J.L. Silva, D. Patel, S. Roundy, Picaradio supports ad hoc ultra-low power wireless networking. IEEE Comput. **33**, 42–48 (2000)
12. J.H. Kansal, S. Zahedi, M.B. Srivastava, Power management in energy harvesting sensor networks. ACM Trans. Embed. Comput. Syst. **6**(4), 32 (2007)
13. Microstrain Press Release, Microstrain wins navy contract for developing energy-harvesting wireless sensors (July 2006), http://www.microstrain.com/news/article-65.aspx
14. J. Ieropoulos, C.M. Greenman, J. Hart, Comparative study of three types of microbial fuel cells. Enzym. Microb. Technol. **37**, 238–245 (2005)
15. Y.K. Tan, LinkedIn SlideShare (2009), [Online], https://de.slideshare.net/tanyenkheng/energyharvesting-presentation-rjc-tyk-2
16. A.S. Adila, A. Husam, G. Husi, Towards the self-powered Internet of Things (IoT) by energy harvesting: trends and technologies for green IoT, in *2nd IEEE International Symposium on Small-scale Intelligent Manufacturing Systems (SIMS)* (2018) pp. 1–5
17. P. Kamalinejad et al., Wireless energy harvesting for the Internet of Things. IEEE Commun. Mag. **53**(6), 102–108 (2015)
18. J. Han, J. Hu, Y. Yang, Z. Wang, S.X. Wang, J. He, A nonintrusive power supply design for self-powered sensor networks in the smart grid by scavenging energy from ac power line. IEEE Trans. Ind. Electron. **62**(7), 4398–4407 (2015)
19. B. Fateh, M. Govindarasu, V. Ajjarapu, Wireless network design for transmission line monitoring in smart grid. IEEE Trans. Smart Grid **4**(2), 1076–1086 (2013)
20. J.L. Wardlaw, I. Karaman, A.I. Karsilayan, Low-power circuits and energy harvesting for structural health monitoring of bridges. IEEE Sensors J. **13**(2), 709–722 (2013). L. Zhou, A.C. Abraham
21. S.Y. Tang, S. Chakrabartty, A 5nw quasi-linear CMOS hot-electron injector for self-powered monitoring of biomechanical strain variations. IEEE Trans. Biomed. Circ. Syst. **10**(6), 1143–1151 (2016)
22. C. Zhu, V.C.M. Leung, L. Shu, E.C.-H. Ngai, Green Internet of Things for smart world. IEEE Access **3**, 2151–2162 (2015)
23. A. Al-Fuqaha, M. Guizani, M. Mohammadi, M. Aledhari, M. Ayyash, Internet of Things: a survey on enabling technologies, protocols, and applications. IEEE Commun. Surv. Tutorials **17**(4), 2347–2376 (2015)

24. Z. Wang, X. Qin, B. Liu, An energy-efficient clustering routing algorithm for WSN-assisted IoT, in *IEEE Wireless* Communications *and Networking Conference (WCNC)* (2018), pp. 1–6
25. H. Hejazi, H. Rajab, T. Cinkler, L. Lengyel, Survey of platforms for massive IOT, in *2018 IEEE International Conference on Future IoT Technologies (Future IoT)*, IEEE, pp. 1–8
26. T. Ruan, Z.J. Chew, M. Zhu, Energy-aware approaches for energy harvesting powered wireless sensor nodes. IEEE Sensors J. **17**(7), 2165–2173 (2017)
27. F.H. Panahi, A. Falahati, Spectral efficient impulse radio-ultra-wideband transmission model in presence of pulse attenuation and timing jitter. IET Commun. **6**(11), 1544–1554 (2012)
28. F.H. Panahi, F.H. Panahi, M. Ghaderzadeh, A. Mohammadisarab, M2M communications as a promising technique to support green powered base stations, in *2019 27th Iranian Conference on Electrical Engineering (ICEE)*, Yazd, Iran (2019), pp. 1654–1658
29. F.H. Panahi, F.H. Panahi, S. Heshmati, T. Ohtsuki, Optimal sleep & wakeup mechanism for green Internet of Things, in *2019 27th Iranian Conference on Electrical Engineering (ICEE)*, Yazd, Iran (2019), pp. 1659–1663
30. F.H. Panahi, P. Hajimirzaee, S. Erfanpoor, F.H. Panahi, T. Ohtsuki, Smart image-processing based energy harvesting for green Internet of Things, in *2018 Smart Grid Conference (SGC)*, Sanandaj, Iran (2018), pp. 1–5
31. F.H. Panahi, S. Moshirvaziri, A. Momene, F.H. Panahi, T. Ohtsuki, D2D-aided optimal utilization of renewable energies for green powered base stations, in *2018 Smart Grid Conference (SGC)*, Sanandaj, Iran (2018), pp. 1–6
32. F.H. Panahi, P.F.i.Z.H. Panahi, Spectral-efficient green wireless communications via cognitive UWB signal model. Automatika **57**(3), 793–809 (2016)
33. F.H. Panahi, A. Falahati, Joint IR-UWB power spectral lines and interference suppression based on coded auxiliary independent signaling in presence of pulse attenuation and timing jitter. Wirel. Pers. Commun. **69**(4), 1241–1260 (2013)
34. A. Falahati, F.H. Panahi, Spectrally efficient ultra wideband transmission over wireless sensor networks, in *IET Conference on Wireless Sensor Systems (WSS 2012)* (2012)
35. W. Ejaz, M. Naeem, A. Shahid, A. Anpalagan, M. Jo, Efficient energy management for the Internet of Things in smart cities. IEEE Commun. Mag. **55**(1), 84–91 (2017)
36. G. Yang, C.K. Ho, Y.L. Guan, Dynamic resource allocation for multiple-antenna wireless power transfer. IEEE Trans. Signal Process. **62**(14), 3565–3577 (2014)
37. X. Lu, P. Wang, D. Niyato, D. In Kim, Z. Han, Wireless networks with RF energy harvesting: a contemporary survey. IEEE Commun. Surv. Tutorials **17**(2), 757–789 (2015)
38. P. Kamalinejad, C. Mahapatra, Z. Sheng, S. Mirabbasi, V.C.M. Leung, Y.L. Guan, Wireless energy harvesting for the Internet of Things. IEEE Commun. Mag. **53**(6), 102–108 (2015)
39. D. Vincze, Fuzzy rule interpolation and reinforcement learning, in *IEEE 15th International Symposium on Applied Machine Intelligence and Informatics (SAMI)* (2017), pp. 000173–000178
40. F.H. Panahi, F.H. Panahi, G. Hattab, T. Ohtsuki, D. Cabric, Green heterogeneous networks via an intelligent sleep/wake-up mechanism and D2D communications. IEEE Trans. Green Commun. Netw. **2**(4), 915–931 (2018)
41. T. Podobnikar, Detecting mountain peaks and delineating their shapes using digital elevation models, remote sensing and geographic information systems using autometric methodological procedures. Remote Sens. **4**(3), 784–809 (2012)
42. G. Han, J. Zhang, X. Mu, Joint optimization of energy harvesting and detection threshold for energy harvesting cognitive radio networks. IEEE Access **4**, 7212–7222 (2016)
43. C. Qiu, Y. Hu, Y. Chen, B. Zeng, Lyapunov optimization for energy harvesting wireless sensor communications. IEEE Internet Things J. **5**(3), 1947–1956 (2018)
44. Z. Wang, V. Aggarwal, X. Wang, Joint energy-bandwidth allocation in multiple broadcast channels with energy harvesting. IEEE Trans. Commun. **63**(10), 3842–3855 (2015)
45. Z. Ullah, I. Ahmed, T. Ali, N. Ahmad, F. Niaz, Y. Cao, Robust and efficient energy harvested-aware routing protocol with clustering approach in body area networks. IEEE Access **7**, 33906–33921 (2019)

46. J. Liu, C.R. Lin, J. Tsai, Delay and energy tradeoff in energy harvesting multi-hop wireless networks with inter-session network coding and successive interference cancellation. IEEE Access **5**, 544–564 (2017)
47. K. Chin, L. Wang, S. Soh, Joint routing and links scheduling in two-tier multi-hop RF-energy harvesting networks. IEEE Commun. Lett. **20**(9), 1864–1867 (2016)
48. H. Zhang, Y. Guo, Z. Zhong, W. Wu, Cooperative integration of RF energy harvesting and dedicated WPT for wireless sensor networks. IEEE Microwave Wireless Compon. Lett. **29**(4), 291–293 (2019)
49. S.H. Kang, J.H. Choi, F.J. Harackiewicz, C.W. Jung, Magnetic resonant three-coil WPT system between off/in-body for remote energy harvest. IEEE Microwave Wireless Compon. Lett. **26**(9), 741–743 (2016)
50. S. Wang, M. Xia, K. Huang, Y. Wu, Wirelessly powered two-way communication with nonlinear energy harvesting model: rate regions under fixed and Mobile relay. IEEE Trans. Wireless Commun. **16**(12), 8190–8204 (2017)
51. Y. Tang, A. Khaligh, A multi-input bridgeless resonant AC–DC converter for electromagnetic energy harvesting. IEEE Trans. Power Electron. **31**(3), 2254–2263 (2016)
52. F.A. Samad, M.F. Karim, V. Paulose, L.C. Ong, A curved electromagnetic energy harvesting system for wearable electronics. IEEE Sensors J. **16**(7), 1969–1974 (2016)
53. Y. Chen, F. Lai, J. You, Analysis of antenna radiation characteristics using a hybrid ray tracing algorithm for indoor WiFi energy-harvesting rectennas. IEEE Access **7**, 38833–38846 (2019)
54. G. Pan, H. Lei, Y. Yuan, Z. Ding, Performance analysis and optimization for SWIPT wireless sensor networks. IEEE Trans. Commun. **65**(5), 2291–2302 (2017)
55. T. Liu, X. Wang, L. Zheng, A cooperative SWIPT scheme for wirelessly powered sensor networks. IEEE Trans. Commun. **65**(6), 2740–2752 (2017)
56. S. Gong, S. Ma, C. Xing, G. Yang, Optimal beam-forming and time allocation for partially wireless powered sensor networks with downlink SWIPT. IEEE Trans. Signal Process. **67**(12), 3197–3212 (2019)
57. Q. Zhao, Y. Shen, M. Li, Control and bidding strategy for virtual power plants with renewable generation and inelastic demand in electricity markets. IEEE Trans. Sustainable Energy **7**(2), 562–575 (2016)
58. J. Knudsen, J. Hansen, A.M. Annaswamy, A dynamic market mechanism for the integration of renewables and demand response. IEEE Trans. Control Syst. Technol. **24**(3), 940–955 (2016)
59. D. Li, W. Saad, I. Guvenc, A. Mehbodniya, F. Adachi, Decentralized energy allocation for wireless networks with renewable energy powered base stations. IEEE Trans. Commun. **63**(6), 2126–2142 (2015)
60. J. Shen, A. Khaligh, A supervisory energy management control strategy in a battery/ultracapacitor hybrid energy storage system. IEEE Trans. Transport. Electr. **1**(3), 223–231 (2015)
61. R.H. Byrne, T.A. Nguyen, D.A. Copp, B.R. Chalamala, I. Gyuk, Energy management and optimization methods for grid energy storage systems. IEEE Access **6**, 13231–13260 (2018)
62. I.N. Moghaddam, B.H. Chowdhury, S. Mohajeryami, Predictive operation and optimal sizing of battery energy storage with high wind energy penetration. IEEE Trans. Ind. Electron. **65**(8), 6686–6695 (2018)
63. A.E. Abdulhadi, T.A. Denidni, Self-powered multi-port UHF RFID tag-based-sensor. IEEE J. Radio Freq. Identif. **1**(2), 115–123 (2017)
64. S. Seneviratne et al., A survey of wearable devices and challenges. IEEE Commun. Surv. Tutorials **19**(4), 2573–2620 (2017)
65. C. Dong, S. Li, R. Han, Q. He, X. Li, D. Xu, Self-powered wireless sensor network using event-triggered energy harvesters for monitoring and identifying intrusion activities. IET Power Electron. **12**(8), 2079–2085 (2019)
66. J. Ikäheimo, E. Pursiheimo, J. Kiviluoma, H. Holttinen, Role of power to liquids and biomass to liquids in a nearly renewable energy system. IET Renew. Power Gener. **13**(7), 1179–1189 (2019)

67. X. Chen, Y. Wang, Q. Wu, A bio-fuel power generation system with hybrid energy storage under a dynamic programming operation strategy. IEEE Access **7**, 64966–64977 (2019)
68. J. Katic, S. Rodriguez, A. Rusu, A high-efficiency energy harvesting interface for implanted biofuel cell and thermal harvesters. IEEE Trans. Power Electron. **33**(5), 4125–4134 (2018)
69. H. Liu, X. Zhao, H. Liang, Z. Li, POMDP-based energy cooperative transmission policy for multiple access model powered by energy harvesting. IEEE Trans. Veh. Technol. **68**(6), 5747–5757 (2019)
70. S. Kosunalp, A new energy prediction algorithm for energy-harvesting wireless sensor networks with Q-learning. IEEE Access **4**, 5755–5763 (2016)
71. X. He, H. Jiang, Y. Song, C. He, H. Xiao, Routing selection with reinforcement learning for energy harvesting multi-hop CRN. IEEE Access **7**, 54435–54448 (2019)
72. S. Xiao, X. Zhou, D. Feng, Y. Yuan-Wu, G.Y. Li, W. Guo, Energy-efficient mobile association in heterogeneous networks with device-to-device communications. IEEE Trans. Wirel. Commun. **8**, 5260–5271 (2016)
73. R. Atat, L. Liu, N. Mastronarde, Y. Yi, Energy harvesting-based D2D-assisted machine-type communications. IEEE Trans. Commun. **65**(3), 1289–1302 (2017)
74. F.H. Panahi, F.H. Panahi, G. Hattab, T. Ohtsuki, D. Cabrici, Green heterogeneous networks via an intelligent power control strategy and D2D communications, in *2017 IEEE 28th Annual International Symposium on Personal, Indoor, and Mobile Radio Communications (PIMRC)*, Montreal, QC (2017), pp. 1–8
75. Y. Ma, W. Liang, W. Xu, Charging utility maximization in wireless rechargeable sensor networks by charging multiple sensors simultaneously. IEEE/ACM Trans. Netw. **26**(4), 1591–1604 (2018)
76. R. Deng, S. He, P. Cheng, Y. Sun, Towards balanced energy charging and transmission collision in wireless rechargeable sensor networks. J. Commun. Netw. **19**(4), 341–350 (2017)
77. Y. Shu et al., Near-optimal velocity control for mobile charging in wireless rechargeable sensor networks. IEEE Trans. Mobile Comput. **15**(7), 1699–1713 (2016)

Chapter 2
Electricity Market Pricing: Uniform Pricing vs. Pay-as-Bid Pricing

Alireza Akbari-Dibavar, Behnam Mohammadi-Ivatloo, and Kazem Zare

2.1 Introduction

As practical economic approaches, the auctions are used in electricity markets trading. The auction leads to awarding some participants attend a financial program and compete according to market rules. Any auction is based on three fundamentals, bidding, clearing, and pricing rules. The bidding rules determine the quality and quantity of submitted bids. The clearing rules determine how to settle the market, choose the awarded participants and determine the amount of traded product. The pricing rules determine at what price the production will be traded. The electricity market auctions are sealed-bid auctions, in which pre-qualified bidders submit their bids to the market simultaneously so that they are not announced about each other submission. In these kinds of auctions, each participant submits only one price offer for each segment of production. The main disadvantage with the sealed-bid auction is that winner participants do not have information about rivals' revenue and feel that they are experiencing losses, since they have not estimated the cost accurately, which is called "winner's curse" [1]. In fact, they believe that they were able to gain more profit, if they had forecasted the prices accurately.

There are many kinds of auction designs used in electricity markets around the world. The first group is named sealed-bid auctions, including first sealed-bid auctions, uniform price auctions (UPA), and pay-as-bid auctions (PABA). The second group is dynamic pricing auctions, in particular, descending clock auctions for electricity markets. Third, the hybrid auctions contain a combination of the mentioned auctions, e.g., descending clock auction followed by PABA, or first-price sealed-bid stage followed by an iterative descending auction, and sequential

A. Akbari-Dibavar (✉) · B. Mohammadi-Ivatloo · K. Zare
Faculty of Electrical and Computer Engineering, University of Tabriz, Tabriz, Iran
e-mail: a.dibavar96@ms.tabrizu.ac.ir; bmohammadi@tabrizu.a.c.ir; kazem.zare@tabrizu.ac.ir

S. Nojavan, K. Zare (eds.), *Electricity Markets*,
https://doi.org/10.1007/978-3-030-36979-8_2

auctions [1]. We will focus on the most used auction types in electricity markets, i.e., the UPA and PABA.

In electricity markets, there is another subcategory regarding the auctions, namely, one-sided and two-sided auctions. The first one refers to some kinds of auctions, in which only one side participates strategically and submits its offers. On the other hand, in two-sided auctions, both sides are asked to bid. Electricity markets can be traded in both forms; however, the one-side auctions are dominant, since the electricity is somewhat an inelastic good, moreover, usually, the demand-side does not tend to encounter with the rules of the power market. In an ideal restructured power system, the generation companies (GenCos) are independent of the government and transmission systems. They submit their selling offers separately to the market as sealed price-quantity pairs, which are arranged according to their production costs, i.e., the variable cost associated with the amount of production. System operator as an auctioneer aggregates the submitted offers and arranges them considering their economic merits and by estimating load demand, clears the market, dispatches generators and determines the market clearing price (MCP). A feasible solution should meet loads, economic viability, and system securities [2].

In competitive electricity markets, producers take bidding strategies to make their profit as much as possible. The behavior modeling of power producers leads to the subject of power transaction games. These games are known as non-cooperative games, in which all participants try to maximize their profit without considering others' welfare. It should be pointed out that the power transactions are known as static auctions since all participants submit their bids at the same time. The electricity market is run repeatedly, which is known as the repeated game, in which participants learn the strategies progressively. These games are held as imperfect and complete information games; the first one refers to a set of games, in which the participants are not announced with the structure of system and with other participants' strategy, while the second one refers to a set of games, in which power producers are informed about the strategies of other participants made recently [3].

Considering the capacity of each participant, it belongs to sets of price-taker or price-maker group. The price-taker participants engage in the power market and have not enough authority to influence the market situation by their actions, while the price-maker participant can pose market power—as they can change the situation of the market in a way that it is profitable for themselves. This condition holds in an oligopolist market, in which there are few power plants, and their actions influence the market's price. The market power is defined as the ability of participants in changing prices to their benefit. Two types of market power have been introduced by the literature, i.e., economic withholding and physical withholding. In economic withholding, generators bid highly above their actual marginal cost; hence, the average market price goes up. In physical withholding, generators hesitate in providing full capacity, so by this mean, the active production is reduced in the market, and the market price increase occurs [4, 5]. The market power is measured by concentration ratios such as Herfindahl–Hirschman Index (HHI) [6], generalized HHI, or Lerner index [7].

It is worth mentioning at the end that the primary mission of deregulated markets is to breaking the traditional monopolistic environment and adding efficiency to the market clearing process. Hence a good auction selection is of great importance in the power market operation [8].

2.2 Contributions

Despite existing works that test and advocate one pricing scheme using evolutionary algorithms, this chapter looks on the published works and highlights the critical deductions from each one. Selecting a proper pricing mechanism for electricity markets is a sophisticated job and depends on each market's attributes. Hence, a generalized prescription cannot be extracted from the works, but reviewing these kinds of literature can provide valuable insights to the market auctioneer and operators to operate the markets with higher efficiency.

2.3 Electricity Market Pricing

There are three main auction types utilized in power markets, namely, UPA (or non-discriminatory), PABA (or discriminatory), and Vickrey auction, which is known as Vickrey–Clarke–Groves (VCG) pricing in multi-unit power system economic [9].

2.3.1 Vickrey–Clarke–Groves (VCG)

In Vickrey auction, the winner GenCos will be paid at the price of second-highest bid unlike standard sealed-bid tenders; hence, it is reputed to second-price sealed-bid auction, and Vickrey believes that this makes participants' bid more truthfully [9]. About the Vickrey auction, it should be noted that this kind of action may reduce the collusion and enhance the willingness of the bidders to bid at their real cost, but has severe problems, which is not suitable for electricity as a non-storable and inelastic good, and it creates a complex framework for bidders. Furthermore, the auctioneer must solve a substantial mixed-integer problem in large-scale power systems. So, there are few works that investigate the multi-unit Vickrey auctions for power markets [10]. In [11], some reasons are expressed regarding the weakness of Vickrey's auction for real electric markets.

2.3.2 Uniform Price Auction

The uniform pricing auction (UPA) or non-discriminatory pricing or competitive pricing or system marginal pricing is a kind of sealed-bid auction. In UPA, all of the winners would be paid at a single pre-determined price, i.e., MCP, irrespective of what they had bided, and the last winner company, which is called "*marginal unit,*" will get the bided price and have not extra surplus. This auction is a generalized form of Vickrey auction [12] and is introduced by Friedman in the 1960s [13]. The process of clearing market regardless of pricing rule is straightforward. This process is based on the equilibrium of generation and demand considering economic indices. The independent system operator (ISO) aggregates the supply offers and demand bids in a spot market, and arranges them in the form of ascending and descending curves, as shown in Fig. 2.1. The intersection of these two curves will determine the market price and consequently, the cleared quantity, which satisfies the aggregated cost of winner sellers and demand of winner buyers, respectively [2]. All sales occur at market price. The shaded area in Fig. 2.1 shows the total revenue of winner producers. The upper area of the shaded area represents the surplus of buyers. In the

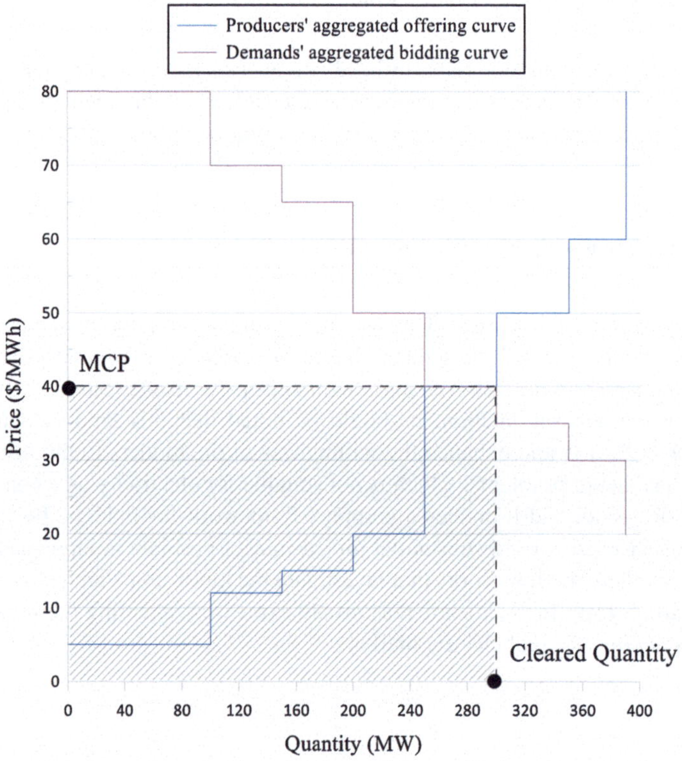

Fig. 2.1 The mechanism of UP auction

spot markets, there is only one MCP; however, there might be more than a single particular clearing price (or quantity) in private bilateral contracts, according to the agreements between buyers and sellers. Even between various zones under zonal pricing, the MCP can be different. The zonal pricing scheme divides an electrical network into several zones, where in each zone, the probability of congestion is ignored and the price is unique.

The UPA is in-use pricing mechanism for forward transactions of the energy in Europe and the USA and is the simplest pricing mechanism from the viewpoint of the suppliers and consumers. Considering the concept of UPA, the winner sellers will award the highest accepted price automatically, so there is not a loss for none of the sellers if they had bided their marginal cost [14]. On the other hand, the buyers will pay the least possible value for their traded good. Both of one-sided and two-sided auctions can be implemented through UPA [15].

2.3.2.1 The Advantages with the UPA

In the UPA, the electricity price regardless of the offered prices is the same for all winners whose offers are less than the MCP. It seems that these make the power plants to be more honest regarding price bidding. If a producer bids at a higher price than its actual marginal price, it might be not committed by the ISO and miss out the opportunity of participation in the market; on the flip side, if it bids at a lower price than its actual marginal cost, it may be committed for full capacity generation and will incur loss.

Consequently, offering marginal cost is the best strategy for the power plants to assure that they will not suffer economically, and also the ISO finds the least price, in which the demand is satisfied. This is the reason why the UPA is well-known as an efficient pricing mechanism. It is stated that the UPA is a transparent pricing mechanism which leads to the efficient selection of the least-cost producers and social welfare maximization. Considering the expected revenue equivalence, under no market power condition, the UPA is superior to the PABA in terms of dispatch efficiency and economic merits [16].

Moreover, in a single-unit case, the UPA theoretically is a cost revealing auction; however, in multi-unit cases, the probability of gaming still exists, which can reduce the overall efficiency [17]. It is stated by Nazemi and Mashayekhi [18] that only producers that their cost is near to the MCP will exercise market power and have the incentive to bid dishonestly, yet others will not risk losing the market participation opportunity by offering at above of their actual cost. Furthermore, the UPA creates opportunity and incentivized small producers to take part and take the advantage of higher prices determined by large-scale producers [1]. Since the paid price to all winner producers is the same, there is a single probability distribution for market prices [19], and this may lead to being helpful to create risk-averse decisions. The authors of [20] summarize the benefits of the UPA as the following points:

- Bidding the actual cost from supplier side.
- Minimization of consumer and social costs.
- Equality between all participants.
- Incentives for innovation of technology and efficiency for being more efficient player.
- Ease of market power mitigation and monitoring.
- At last, the UPA has been tested and well functioned over years in different markets.

2.3.2.2 The Disadvantages with the UPA

In 2001, it was declared that despite the progress of power market operation and fuel cost reduction under the UPA, the electricity prices have been increased in comparison with 1990. The reason was the domination of large-scale power plants. This was the start of the recasting of pricing rules in balancing markets in England and Wales, which was known as New Electricity Trading Arrangements (NETA) [21]. After that, the California Power Exchange reported that pricing rule transforming to PAB pricing could probably decrease the abuse of market power exercising and reduce price spikes in spot markets [22].

On the other hand, it is shown by [9] that when the market share of the producer is increased, without any communication, the other small producers collaborate and collusion appears among small producers, in a way, they bid at lower price and get higher cleared price determined by the tremendous power plants despite their small share. The authors of [20] have shown that this kind of auction experiences market power exercise easily if no mitigation and monitoring is considered; moreover, the "hockey-stick" shaped curves constructed from supply curves can lead to price volatility year by year. The market power exercising through capacity withholding is discussed by [14], which leads to inefficient dispatch and higher cost. Viehmann et al. [23] using a Q-learning approach found that UPA leads to higher prices, when there is an asymmetry sharing between the size of utilities, and if the number of suppliers is limited.

2.3.3 Pay-as-Bid Auction

PABA is a sealed-bid auction, which discriminates between winners. In this auction, each buyer will pay his bided price, and each seller will be paid at the price it had offered [15]. This auction is used when different units of a particular commodity are sold by different prices and is implemented by governments and central banks to allocate treasury bonds, for allocation of carbon credits and for electricity generation payments. Considering the rule of gaining profit, each producer offers higher than her actual price to gain more profit, and by forecasting MCP and by offering near it, each player refuses to reveal her actual cost. In other words, the participants forecast

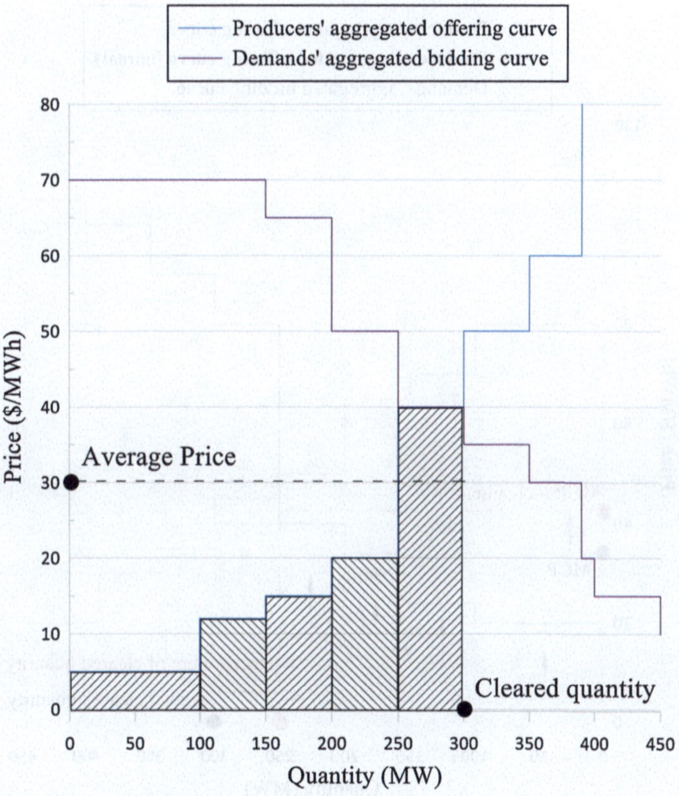

Fig. 2.2 The mechanism of PABA

the MCP or at least estimate the cost function of rivals. In this regard, the Wolak's cost function estimating tool, which is designed for the UPA, has been modified by [24] for Iranian PAB-based market. The estimation is based on the weighted average prices through defining an expected profit of the firm and the first-order conditions.

Figure 2.2 shows the market-clearing under the PABA. The market clearing process is similar to the UPA. The revenue paid to generators is highlighted by the shaded area. As a well-established tradition, the buyers will pay the average price instead of market clearing price. All players submit their offers higher than their actual cost to gain more revenue. This leads to higher prices and lowers the cleared quantities, which reduces market efficiency, as shown in Fig. 2.3.

On the other hand, there is a converse relation between the price offered and the probability of acceptance in the market and the expected profit gained so the decision-maker should compromise between two mentioned factors [25]. The PABA has attracted attention because most markets in England and Wales are traded through bilateral agreements, which are settled by PABA [26]. Also, PAB pricing is

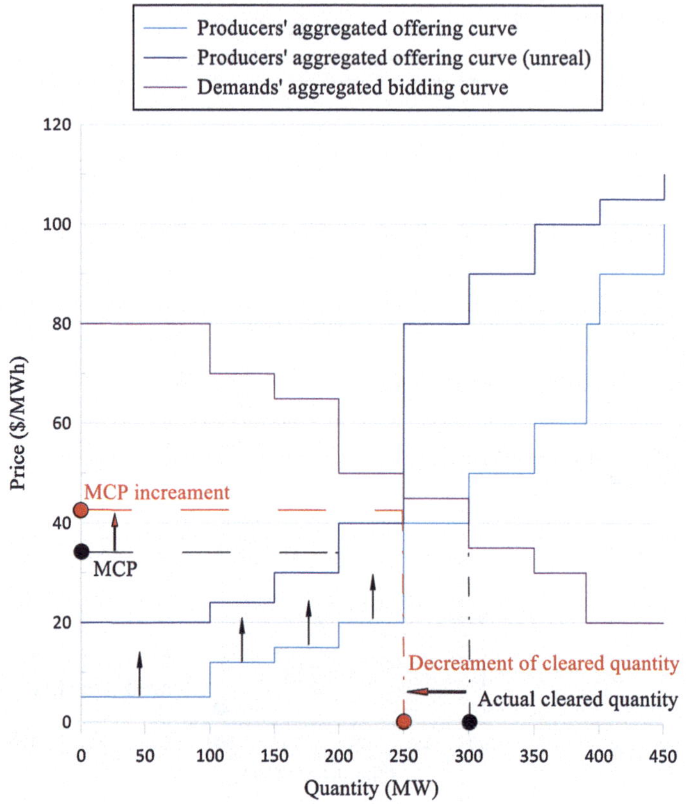

Fig. 2.3 Effect of unrealistic bidding behavior

the dominant pricing mechanism for renewable-based generation sector, e.g., solar generation and wind turbines, in most European countries [27].

Besides, the balancing markets in some European countries such as Germany and Italy are settled under PAB pricing [28]. In this regard, a multi-leader-common-follower structure is proposed by [29], which is a bilevel optimization problem. At the upper level, the producers try to find optimal offering strategy with given demand, and at the lower level, the regulator runs the market and dispatches the generation. The same authors in [30] stated that in PAB-based markets, the linear bidding is more profitable than quadratic bidding for producers. The optimal linear bidding curve for each producer is called "*best response,*" and under circumstances where no best response is found, the optimal offering strategy can be achieved by a sequence of quadratic bidding converging to "*limiting best response*" point [30]. However, the authors of [31] believe that UP pricing reduces the strategic behaviors of the participants, so it is superior to the PAB pricing for balancing markets. Hu et al. [32] state that PAB pricing mechanism of balancing markets of Germany or Italy leads to low liquidity of the intra-day markets and consequently causes higher transaction costs of the market participants. Finally, considering the European Commission proposes in creating an integrated balancing market for whole Europe, using game-theory approach, the authors of [33] stated that both German balancing market with PABA and future Europe market design with UPA have satisfying results in terms of price competitiveness and market outputs. Furthermore, switching to the UP market will not persuade participants to bid honestly and reveal actual costs but leads to underbidding.

2.3.3.1 The Advantages with PABA

The PAB pricing mechanism is a price setting; hence, the accepted bids have less flexibility in choosing the shape of their offering curve and reduce the possibility of multiple equilibria point. Besides, the PABA will result in flatten bidding, which improves competitiveness. In PAB governed market, the participants only rely on their private cost function and capacity to provide optimal bids. Also, they have no information about others' bidding and cost functions, so they would bid near their production cost truthfully. Furthermore, the increased cost in unequal sharing of capacity and volatility in bidding behavior is less than the UPA [9]. The authors of [20] stated that these auctions could reduce the price volatility and market power if there is a sufficient number of producers, and none of them have not complete data about others' offers and final cleared price. According to [34], the strategic bidding of different sized producers will not take place in discriminatory auctions, where all winners are paid based on the output of each of their plants. It is proved in [35], the consumers will prefer PABA when the demand is perfectly inelastic. Moreover, the risk of tacit collusion under PABA is relatively lower than that under UPA [36].

2.3.3.2 The Disadvantages with PABA

Discriminatory pricing is used in Iran and some Latin American countries, and the assumption is that the large-scale power plants with lower cost will bid lower prices, and consequently, the overall consumer surplus and social welfare will be high. However, this assumption seems very optimistic and should be debated. Since the payments are based on the offers of sellers, the discriminatory pricing cannot make an incentive for power plants to offer at the actual price [17]. From Ref. [20], this auction abets lower-cost producers to offer at higher prices, and this makes them not committed in the market, while high-cost resources are committed, as shown in Fig. 2.4. The unrealistic bidding leads to insufficient dispatch and unreal MCP, consequently rises the consumer payments and destroys the social welfare. It creates economic advantages for incumbents and large-scale power plants while it violates small and new entrants since they are not acquainted with the market structure and cleared prices to forecast MCP accurately. Also, the effort for reducing MCP will lead to other non-optimal dispatches, which spoils economic operation. Under this pricing scheme, the equilibrium depends on probability distribution of demand

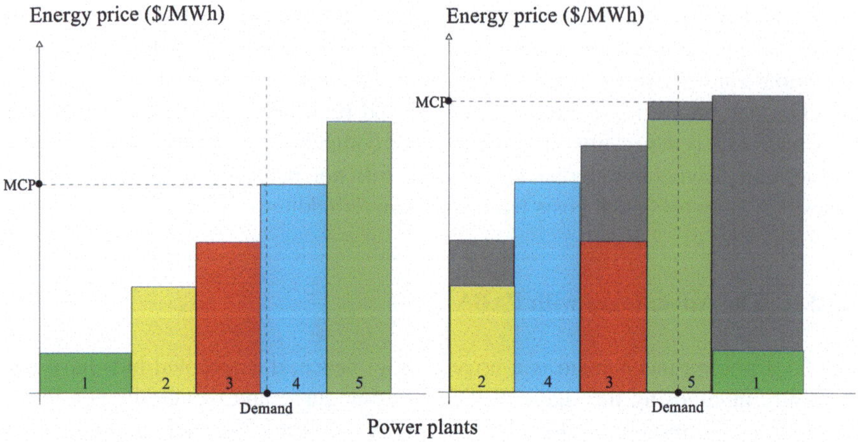

Fig. 2.4 Inefficient dispatch under PAB auction

shock which makes producers to take randomized strategies instead of deterministic ones, known as mixed strategy Nash equilibrium. Noted by [37], some barriers exist for small producers in the PABA, which can be destructive in long-term runs. It is shown by [38] that the PABA lead to lower costs for consumers in short term; however, it will result in incorrect pricing signals for investment and consequently increases the long-term costs.

2.4 Switching from UPA to PABA?

While economists claim that the UPA and PABA will lead to the same expected revenue and consumer surplus, this is conditional on the assumption of accessing to complete information of players cost function. For instance, it is shown by [39] that both auctions will lead to an equivalent revenue when the sellers are announced by the value of buyers and set their optimal supply and reserve prices, and when the information of demand is incomplete, the optimal design PABA weakly dominates the UPA. Figure 2.5 shows the total payment equivalence under both UP and PAB pricing mechanisms for perfect information environment, as the shaded area below the MCP is equal for both auctions. In the PABA, under complete information environment, the players know the MCP and offer at the maximum acceptable price to get maximum profit, and under UP, all of the generators offer at their true marginal cost and they will automatically receive MCP (i.e., the maximum acceptable price).

However, the perfect information assumption does not always happen, and if the demand is uncertain, the UPA will be vulnerable and results in price volatility, while under the PABA, the markup function will compensate some part of the demand uncertainty [40]. On the advocation of PABA, Oren stated that this auction

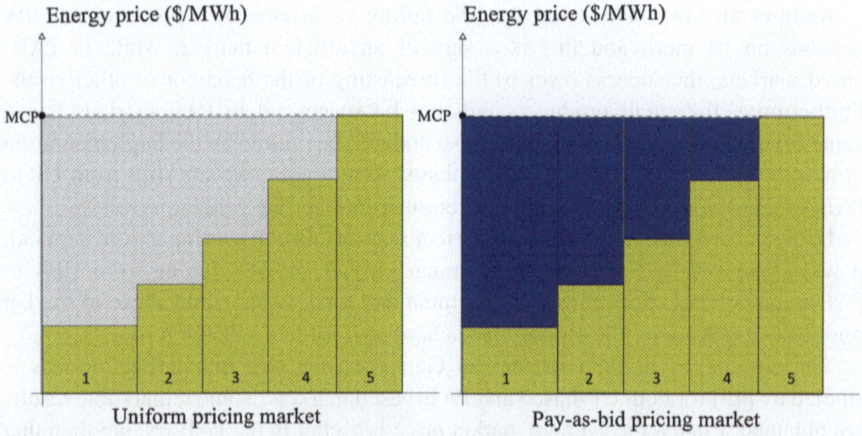

Fig. 2.5 The equivalence of UP and PAB pricing payments under perfect competition

is prioritized when there is a need for fragmentation of product, e.g., in reserve markets, where the performance (in terms of spinning reserve, non-spinning reserve, regulation, etc.) is criterion for payments [40].

There is an argue for UPA or non-discriminatory pricing among winners since the payments are the same regardless of merits and performance; hence, this pricing mechanism should not be chosen when some of the generators are under the control of the government. Meanwhile, the UPA can increase the possibility of collusion in the market [15]. When some influential players try to exercise market power to increase the MCP, other feeble players will automatically complicit without any action, and the overall efficiency will go down. In line with the previous expression, the regulator, Ofgem, (or members of its supervising authority, GEMA) expressed that the UPA can increase the market power exercising of large-scale producers [21]. However, it is proved that the collusion in terms of capacity withholding still exists under PAB pricing somewhat; moreover, proof of abuse under PAB pricing is harder since the suppliers can claim they used "guess the clearing price" principle and forecasted prices wrongly [6].

In an oligopoly situation, it is stated by [41] that the average prices are higher in the UPA than the PABA, but it is difficult to deduce what pricing rule is more efficient.

By analyzing bidding behavior of large-scale power plants, Yamamoto and Tezuka [42] found that in PABAs, the price bids of high-cost generators reach to their upper limit and this makes an opportunity for fringe power plants to win. In PABA, the power plants' price curve is flatter, and they offer the same prices for each quantity they sell [42]. But when the scheme is changed from the UPA to PABA in the monopoly, the power plants with market power behave insufficiently, which reduces social welfare [42].

Kahn et al. [43] have stated that the failure or success of a producer in UPA depends on its merit and this is a sign of an efficient market, while in PAB-based markets, the success owes to the forecasting of the behavior of other rivals. Furthermore, the small producers will not be successful in PAB markets, since generally they have higher forecasted cost and will be injured by the implementation of market power by large-scale power plants. Also, the immediate shift from UP to PABA may decrease the efficiency and competition among generators.

Using an analytical evolution in a perfect market and considering elastic demand, it was expressed by Federico and Rahman [44] that the switching from UPA to PABA can decrease both consumer payment and total welfare, and abuse of market power under PABA by the monopolist is harder than that in the UPA.

The problem of bidding strategy of GenCos using experimental economics is studied by [45] for both UP-based and PAB-based markets. Some remarkable results are obtained. Firstly, the average market price is higher in the PAB pricing than that in the UPA pricing and the prices are more stable, which leads to further payments to GenCos. Secondly, the prices are getting higher under both settlements when loads are becoming less elastic. Finally, under the UPA, the competition among GenCos is more intense.

For reactive power procurement, it is mentioned by [46] that the UPA is not preferable since it pays all participants the same price without considering their performance and merit regarding system condition and their location, so the PAB-based market is recommended for these kinds of power delivery.

In markets with high penetration of renewable resources, due to the variability of the production of these resources, they offer zero price to the market, when the settlement is UPA and receive the MCP. However, this reduces the income and may lead to discouragement where the large portion of the energy is provided by this kind of resources. The authors of [47] tested changing from UP pricing to PAB pricing, by two energy share scenarios with different renewable penetration rate. The results indicate that the UPA is superior for the fluctuating renewable resources. The authors of [48] compare the bidding behavior of producers in Iranian (PAB-based) and Danish markets (UPA-based) with respectively low and high penetrated wind turbines. The results indicate that in Iran's market, with system capacity margin increment, the trends for higher price bidding are disappeared and peak load condition leads to lower variability in bidding behavior compared with the Danish market. Also, it is concluded by [49] that changing the pricing rule from UP to PAB will affect the bidding behavior of wind farm, as they show a non-conservative behavior but also leads to lower expected revenue. The photovoltaic systems sizing in spot markets under both pricing schemes is addressed by [50] employing a genetic algorithm. It is found that the net present worth of installed solar systems for the owners of these systems under UP pricing is a little more than that in the PAB-based market.

For day-ahead markets, it was shown by [51] that the changing from UPA to PABA increases the prices because of the collective learning models. Producers tend to find MCP, and this leads to bidding up. However, the referred phenomena might be mitigated in real-time markets.

Based on two market benchmarks, namely, the monopoly and the perfect competition, the authors of [44] have deduced that demand-weighted average prices will increase under PABA compared with the UPA if demand is highly uncertain. Also, players with market power will react inefficiently under PABA, may lead to lower output and welfare deterioration and probably increases the market power exercising in the medium run. Finally, the UPA structure incentivizes players to compete by their supply function considering demand uncertainty in repeated interactions; however, forcing them to compete by pricing under PABA by using continuous bilateral contracts may remove these equilibria.

Ren and Galiana [52, 53] theoretically and by simulation have been found that the expected profit of generators and consumers' payments are equal in both UPA and PABA; but, the variation of these expected profits and payment are more under UPA. Moreover, the risk of not realization of these expected values is higher in the UP.

Xiong et al. [54] compared the UPA and PABA using a multi-agent framework where producers develop bids using Q-learning algorithm. The authors have found that PAB pricing can reduce the market price and price volatility. Furthermore, in the PABA, the aggregated supply curve is more flattened than that in UPA.

The authors of [55] analyzed the strategic behaviors of a tremendous player and a small player in a short-term run under both UPA and PABA using game theory and auction theory. They found that under a static game considering two players and inelastic demands, the Nash equilibrium point under PABA yields less revenue than UPA. Also, under the elastic demand assumption, the PABA leads to great expected served demand.

Using an adaptive game, it is shown by [56] that when the market is approaching in perfect competition, the market prices go down under the UP mechanism. However, when the market structure is likely to oligopoly, the PAB pricing is more efficient. Moreover, the PAB pricing leads to a uniform sharing of generators, while under the UPA producers have various market shares.

Holmberg [57] suggests switching from the UPA to PABA, when the risk of power shortage is not too high, and the market is tended to be imperfect. Also, from the viewpoint of the auctioneer, in the presence of risk-averse producers, the PAB pricing has priority.

On the contrary, Genc [58] stated that there is higher risk in the presence of chaotic bidding or mixed strategy of producers under PABA, which makes prices unpredictable and this is harmful to both consumers and generators.

The pricing behavior of producers is modeled in Vickrey, UP, and discriminatory auctions by [59]. It is found that firstly, the UPA always outperforms by the discriminatory auction in terms of consumer surplus and secondly, the Vickrey auction is the best auction considering productivity efficiency. However, another comparison between three auctions is ambiguous. Furthermore, the decision about selecting UPA or PABA requires a compromise between productive efficiency and prices amount.

Finally, the best strategy seems to be a hybrid auction combining PABA as the last stage of payment and the UPA is used in the allocation of generation and classification part. This scheme is called "*simultaneous descending clock auction plus PAB negotiation stage,*" which reduces the collusion and inefficient dispatch problems in one go, according to [60].

2.5 Conclusion

In this chapter, a brief introduction to the electricity market has been provided. Various auctions are outlined. Uniform, discriminatory, and Vickrey auctions are suggested for electricity markets. Each auction and a comparison between them are presented. Uniform and discriminatory pricing are compared to each other according to existing literatures. However, the various authors have their own opinions, the selection between two pricing settlements is still ambiguous, and as a general statement, it depends on the effects of each pricing rule on the market performance. As a general conclusion, in the following, the definite attributes and results of each pricing mechanism are summarized:

- UPA should not be chosen in government-oriented power markets.
- UPA may lead to tacit collusion between producers, especially small generators.
- UPA is vulnerable against demand uncertainty.
- UPA can encourage producers to reveal their actual cost, which is beneficial for long-term planning.
- The UPA may lead to innovation and improvement of system investments.
- There is the winners' curse under UPA between winner sellers.
- PABA will insult small producers unless they are large enough.
- PABA does not reveal the actual cost of producers.
- The probability of collusion is lower under PABA.
- Under PABA condition, the cheap plants are concerned regarding their bidding since they can gain more profit, but there is a considerable risk.
- The PAB pricing is useful in the case of frequency regulating markets or reactive power compensation services, where the performance is important.
- The PABA may lead to consolidation of the industry and concentration of the market.
- Seems the UPA outperforms by the PABA in terms of renewable-based future markets.
- The best auction design can be formed as a combination of both UP and PAB pricing schemes.

References

1. L. Maurer, L. Barroso, *Electricity Auctions: An Overview of Efficient Practices* (The World Bank, Washington, DC, 2011)
2. M. Shahidehpour, H. Yamin, Z. Li, *Market Operations in Electric Power Systems: Forecasting, Scheduling, and Risk Management* (John Wiley & Sons, 2003)
3. P. Bajpai, S.N. Singh, Bidding and Gaming in Electricity Market: An Overview and Key Issues, in *Proceedings of National Power System Conference (NPSC)*, Chennai (2004), pp. 338–346
4. A.K. David, F. Wen, Market power in electricity supply. IEEE Trans. Energy Convers. **16**(4), 352–360 (2001)
5. A.F. Rahimi, A.Y. Sheffrin, Deregulated electricity markets. IEEE Trans. Power Syst. **18**, 486 (2003)
6. S. Heim, G. Götz, Do Pay-as-Bid Auctions Favor Collusion? Evidence from Germany's Market for Reserve Power (2013). ZEW - Centre for European Economic Research Discussion Paper No. 35. Available at SSRN: https://ssrn.com/abstract=2278873 or https://doi.org/10.2139/ssrn.2278873
7. A. Nazemi, A. Farsaei, S. Khalil Moghaddam, Estimating market power by introducing a new Lerner index in the Iranian electricity market. Energy Sources Part B Econ. Plan. Policy **11**, 882 (2016)
8. D. Fang, J. Wu, Optimal pricing mechanism for generators and large consumers considering power transmission cost, in *Asia-Pacific Power and Energy Engineering Conference, APPEEC* (2010)
9. L.F. Sugianto, K.Z. Liao, Comparison of different auction pricing rules in the electricity market. Mod. Appl. Sci. **8** (2014)
10. L.M. Ausubel, P. Milgrom, The lovely but lonely Vickrey auction" in *Combinatorial Auctions* (2013)
11. M.H. Rothkopf, Thirteen reasons why the Vickrey-Clarke-groves process is not practical. Oper. Res. **55**(2), 191–197 (2007)
12. W. Vickrey, COUNTERSPECULATION, auctions, and competitive sealed tenders. J. Finance **16**, 8 (1961)
13. M. Friedman, A program for monetary stability. NY: Fordham University Press, 63–65 (1960)
14. A. Ockenfels, V. Grimm, G. Zoettl, The Pricing Mechanism of the Day Ahead Electricity Spot Market Auction on the EEX, European Energy Exchange EEX (2008)
15. M. Morey, Power Market Auction Design: Rules and Lessons in Market Based Control for the New Electricity Industry, Prep. Edison Electr. Inst. (2001)
16. J. Lin, F. H. Magnago, Electricity Markets, Theories and Applications. John Wiley & Sons, (2017)
17. R. Ethier, R. Zimmerman, T. Mount, W. Schulze, R. Thomas, A uniform price auction with locational price adjustments for competitive electricity markets. Int. J. Electr. Power Energy Syst. **21**, 103 (1999)
18. A. Nazemi, M. Mashayekhi, Competitiveness assessment of Iran's restructured electricity market. Energy Econ. **49**, 308 (2015)
19. D.J. Swider, C. Weber, Bidding under price uncertainty in multi-unit pay-as-bid procurement auctions for power systems reserve. Eur. J. Oper. Res. **181**, 1297 (2007)
20. J.P. Pfeifenberger, R. Lueken, K. Spees, L. Mwalenga, Uniform Price vs. Differentiated Payment Auctions, a Discussion of Advantages and Disadvantages, June 2017. Website: www.ieso.ca/-/media/Files/IESO/Document-Library/engage/ica/ICA-20170615-Brattle-Uniform-Price-Auctions.pdf?la=en
21. G. Shuttleworth, I. MCKENZIE, A Comparative Study of the Electricity Markets in UK, Spain and Nord Pool, Report Prepared by NERA Confin. Rome (2002)
22. N. Fabra, N.H. Von Der Fehr, D. Harbord, Modeling electricity auctions. Electr. J. **15**, 72 (2002)

23. J. Viehmann, S. Lorenczik, R. Malischek, Multi-Unit Multiple Bid Auctions in Balancing Markets: An Agent-Based Q-Learning Approach, Energiewirtschaftliches Institut an der Universitaet zu Koeln (EWI) (2018)
24. E. Maghool, A. Naghavi, S.F. Ghaderi, Estimating cost function in power markets under pay-as-bid pricing rules using observed bid data, in *2008 IEEE Electrical Power and Energy Conference - Energy Innovation* (2008)
25. J. Sadeh, H.R. Mashhadi, M.A. Latifi, A risk-based approach for bidding strategy in an electricity pay-as-bid auction. Eur. T. Electr. Power **19**, 39 (2009)
26. F. Stacke, P. Cuervo, A combined pool/bilateral/reserve electricity market operating under pay-as-bid pricing. IEEE Trans. Power Syst. **23**, 1601 (2008)
27. CEER report on tendering procedures for RES in Europe - ceer.eu. [Online], Available: https://www.ceer.eu/ceer-tendering-procedure?inheritRedirect=true, Accessed 18 May 2019
28. N. Mazzi, J. Kazempour, P. Pinson, Price-taker offering strategy in electricity pay-as-bid markets. IEEE Trans. Power Syst. **33**, 2175 (2018)
29. D. Aussel, P. Bendotti, M. Pištěk, Nash equilibrium in a pay-as-bid electricity market: part 1–existence and characterization. Optimization (2017)
30. D. Aussel, P. Bendotti, M. Pištěk, Nash equilibrium in a pay-as-bid electricity market part 2 - best response of a producer. Optimization (2017)
31. F. Müsgens, A. Ockenfels, M. Peek, Economics and design of balancing power markets in Germany. Int. J. Electr. Power Energy Syst. **55**, 392 (2014)
32. J. Hu, R. Harmsen, W. Crijns-Graus, E. Worrell, M. van den Broek, Identifying barriers to large-scale integration of variable renewable electricity into the electricity market: a literature review of market design. Renew. Sust. Energ. Rev. **81**, 2181 (2018)
33. F. Ocker, K.M. Ehrhart, M. Belica, Harmonization of the European balancing power auction: a game-theoretical and empirical investigation. Energy Econ. **73**, 194 (2018)
34. C.D. Wolfram, Strategic bidding in a multiunit auction: An empirical analysis of bids to supply electricity in England and Wales. Rand J. Econ., 703–725 (1998)
35. P. Hästö, P. Holmberg, Some inequalities related to the analysis of electricity auctions. Appl. Math. Lett. **19**, 814 (2006)
36. P. Klemperer, What really matters in auction design. J. Econ. Perspect. **16**(1), 169–189 (2002)
37. C. Vázquez, M. Rivier, I.J. Pérez-Arriaga, On the use of pay-as-bid auctions in California, some criticisms and an alternative proposal, Inst. Investig. Tecnológica, Tech. Rep. IIT-00-077A (2000)
38. F.P. Sioshansi, *Competitive Electricity Markets: Design, Implementation, Performance* (Elsevier, Amsterdam, 2011)
39. M. Pycia, K. Woodward, Pay-as-bid: Selling divisible goods, Available SSRN 2417512 (2016)
40. S. Oren, When is a pay-as bid preferable to uniform price in electricity markets, in *IEEE PES Power Systems Conference and Exposition*, 2004., Pp. 1618–1620. IEEE, (2004)
41. N. Fabra, N.H. Von Der Fehr, D. Harbord, Designing electricity auctions. RAND J. Econ. **37**, 23 (2006)
42. Y. Yamamoto, T. Tezuka, A large generator's bids in an electricity supply auction: uniform pricing versus pay-as-bid pricing, in *Electr. Eng. Japan (English Transl. Denki Gakkai Ronbunshi)*, vol. 160, (2007), p. 41
43. A.E. Kahn, P.C. Cramton, R.H. Porter, R.D. Tabors, Uniform pricing or pay-as-bid pricing: A dilemma for California and beyond. Electr. J. *14*(6), 70–79 (2001)
44. G. Federico, D. Rahman, Bidding in an electricity pay-as-bid auction. J. Regul. Econ. **24**, 175 (2003)
45. H. Chen, Experimental Analysis of Uniform Price and PAB Auctions in Electricity Markets, in *2007 International Power Engineering Conference (IPEC 2007)* (2007), pp. 24–29
46. N. Amjady, A. Rabiee, H.A. Shayanfar, Pay-as-bid based reactive power market. Energy Convers. Manag. **51**, 376 (2010)
47. S. Nielsen, P. Sorknæs, P.A. Østergaard, Electricity market auction settings in a future Danish electricity system with a high penetration of renewable energy sources - a comparison of marginal pricing and pay-as-bid. Energy **36**, 4434 (2011)

48. M.H. Bashi, G. Yousefi, H. Gharagozloo, H. Khazraj, C.L. Bak, F.F. Silva, A Comparative Study on the Bidding Behaviour of Pay as Bid and Uniform Price Electricity Market Players, in *Proceedings - 2018 IEEE International Conference on Environment and Electrical Engineering and 2018 IEEE Industrial and Commercial Power Systems Europe, EEEIC/I and CPS Europe 2018* (2018)
49. M.H. Bashi, G. Yousefi, C.L. Bak, J.R. Pillai, Long term expected revenue of wind farms considering the bidding admission uncertainty. Energies (2016)
50. M. Mohammadi, GA-based optimal sizing of PV under pay as bid and uniform power market pricing considering uncertainty of solar radiation, Teh. Vjesn. (2013)
51. E. Guerci, M.A. Rastegar, Comparing system-marginal-price versus pay-as-bid auctions in a realistic electricity market scenario, in *Managing Market Complexity*, (Springer, Berlin, Heidelberg, 2012), pp. 141–153
52. Y. Ren, F.D. Galiana, Pay-as-bid versus marginal pricing - part I: strategic generator offers, IEEE Trans. Power Syst. (2004)
53. Y.U. Ren, F.D. Galiana, Pay-as-bid versus marginal pricing - part II: Market behavior under strategic generator offers, IEEE Trans. Power Syst. (2004)
54. G. Xiong, S. Okuma, H. Fujita, Multi-agent based experiments on uniform price and pay-as-bid electricity auction markets, in *2004 IEEE International Conference on Electric Utility Deregulation, Restructuring and Power Technologies. Proceedings*, Vol. 1, pp. 72–76. IEEE, (2004)
55. Y.S. Son, R. Baldick, K.H. Lee, S. Siddiqi, Short-term electricity market auction game analysis: Uniform and pay-as-bid pricing. IEEE Trans. Power Syst. **19**, 1990 (2004)
56. C.C. Skoulidas, C.D. Vournas, G.P. Papavassilopoulos, An Adaptive Game for Pay-as-Bid and Uniform Pricing Power Pools Comparison, in *Athens, the 3rd Mediterranean Conference and Exhibition on Power Generation, Transmission T Distribution and Energy Conversion* (2002)
57. P. Holmberg, Supply function equilibria of pay-as-bid auctions. J. Regul. Econ. **36**, 154 (2009)
58. T.S. Genc, Discriminatory versus uniform-price electricity auctions with supply function equilibrium. J. Optim. Theory Appl. **140**(1), 9–31 (2009)
59. N. Fabra, N.-H. Von Der Fehr, D. Harbord, Designing Electricity Auctions: Uniform, Discriminatory and Vickrey," in *First Conference on Wholesale Markets for Electricity* (2002)
60. R. Moreno, B. Bezerra, L. A. Barroso, S. Mocarquer, H. Rudnick, Auctioning Adequacy in South America Through Long-Term Contracts and Options: From Classic Pay-as-Bid to Multi-Item Dynamic Auctions, in *2009 IEEE Power and Energy Society General Meeting, PES '09* (2009)

Chapter 3
Integrated Gas and Power Networks

Alireza SoltaniNejad, Ramin Bahmani, and Heidarali Shayanfar

Nomenclature

Sets

h	Index for energy hubs
l	Index for transmission line
η	Index of natural gas supply contract
GU	Set of gas-fired generating units

Parameters

$\overline{F_k}$	Maximum capacity of transmission line k
Z_{ki}	Gas compressibility factor at compressor inlet
$\alpha_k, \beta_k, \gamma_k$	Gas consumption coefficients of compressor k
$R_k^{\max}$	Compression ratio of compressor k
$\pi_i^{\max}, \pi_i^{\min}$	Max and min pressure at node i
$\mathrm{WS}_i^{\max}, \mathrm{WS}_i^{\min}$	Max and min amount of gas supply at node i
A	Pipe-nodal incidence matrix
NC	Number of candidate compressors and existing compressors
NCG	Number of coal-fired generators
NWS	Number of gas suppliers

A. SoltaniNejad · R. Bahmani · H. Shayanfar (✉)
Center of Excellence for Power Systems Automation and Operation, School of Electrical Engineering Iran University of Science and Technology, Tehran, Iran
e-mail: alireza_soltani@elec.iust.ac.ir; r_bahmani@elec.iust.ac.ir; hashayanfar@iust.ac.ir

S. Nojavan, K. Zare (eds.), *Electricity Markets*,
https://doi.org/10.1007/978-3-030-36979-8_3

NWL	Number of gas loads
WL_i	Natural gas load at node i
CarbonCost	Carbon emission price
$P\mathrm{lineCost}_i$	Investment cost of installing pipeline i
$C\mathrm{lineCost}_i$	Investment cost of installing compressor i
$E\mathrm{lineCost}_i$	Investment cost of installing electricity line
ξ_1, ξ_2	Carbon emission coefficient of coal-fired generator and gas-fired generator, respectively
M_k	Large enough input value
$P_{gk}^{\max}, P_{gk}^{\min}$	Max and min capacity of generator k
$\mathrm{Cost}_{\mathrm{gas}i}$	Gas purchase cost of supplier i
a_i, b_i, c_i	Coefficients of the operation cost of generator i
PL_i	Real power load at node k
μ_1, μ_2, μ_3	Gas fuel rate coefficients of generator i
GHV	Gas gross heating value
M_{ij}	Gas pipeline constant depending on diameter, length, temperature, friction, and gas composition
ς	Coefficient of converting net present value to annualized investment cost
$\mathrm{pf}_g^{\mathrm{Gas}}$	Participation factor of gas supply facilities g [p.u]
$\mathrm{CF}_{i,t}$	Capacity factor of electricity unit i during time period t [p.u]
$\varnothing$	Energy conversion factor
HHV	High heating value
e_a^{ptg}	Efficiency of PtG facility a
σ	Discretized storage and inflow/outflow rate used to linearize the properties of the NG storage
$\rho^{\mathrm{in}}, \rho^{\mathrm{out}}$	Inflow and outflow rate of storage
NT	Number of periods in the duration time
L_t^{E}	Electricity power output within energy hub
W_{o}	Cost of firm natural gas contract
SU, SD	Startup and shutdown cost of a unit
ρ_{LS}	Penalty price of electricity load shedding
P_{LS}	Electricity load shedding
VOLL	Penalty price of shed load
$\rho_{\mathrm{gas}}, \mathrm{gs}_{\mathrm{s}}$	Price of natural gas and operation cost of gas storage s

Variables

z_i	Binary decision variable, 1 if electricity line i is installed, and 0 otherwise
fP_k	Natural gas flow of pipeline
H_k	Power for compressor k

σ	Specific heat ratio
fc_k	Gas flow rate at compressor k
π_i, π_j	Pressures at node I and j, respectively
τ_k	Amount of gas tapped by compressor k
x_i	Binary decision variable, 1 if pipeline i is installed, and 0 otherwise
y_i	Binary decision variable, 1 if compressor i is installed, and 0 otherwise
z_i	Binary decision variable, 1 if electricity line i is installed, and 0 otherwise
WS_i	Natural gas injection of gas supplier i
fp_i	Natural gas flow of pipeline
fc_k	Gas flow rate at compressor k
fl_k	Power flow on transmission line k
B_k	Electrical susceptance of transmission line k
$\theta_{\mathrm{fr}(k)}, \theta_{\mathrm{to}(k)}$	Voltage angle at "from" and "to" buses of transmission line k
$P_{\mathrm{g}k}$	Real power supply from generator k
$p_{g,t}^{\mathrm{G,N}}$	Gas production of new gas supply projects g in time period t [TJ/h]
$p_{g}^{\mathrm{C,N}}$	Gas supply capacity of new gas projects [TJ/year]
$p_{g,t}^{\mathrm{G,Ex}}$	Gas production of existing gas supply projects g in time period t [TJ/h]
$p_{i,t}^{\mathrm{N}}$	Electricity production of new unit i during time period t [MW]
$p_{i}^{\mathrm{C,N}}$	Power capacity to be built for new unit i [MW]
$p_{i,t}^{\mathrm{Ex}}$	Electricity production of existing unit i during time period t [MW]
$p_{i}^{\mathrm{C,Ex}}$	Power capacity of existing unit i [MW]
G_{aht}	Gas production of PtG facility a at load block h of year t
P_{aht}^{bc}	Base-case power consumption of PtG a at load block h of year t
ψ	NG flow rate between NG node s i, j in time t
AC	Total available capacity
$\mathrm{AU}_{int}, \mathrm{AL}_{lnt}$	Binary variable which is equal to 1 if unit i/line l is available, being 0 otherwise
RM	Grid resilience metric
$f_i()$	Electric load loss cost function
$\mathrm{pd}_{i,b,t}$	Load curtailment
W	Cost of natural gas contract
$P_{i,t}^0$	Generation of unit I at hour t
$\mathrm{LD}_{j,t}^0$	Preventive load shedding at bus at hour t
$v_{\mathrm{sp},t}$	Production of natural gas in well sp at hour t
$\mathrm{GC}_{\mathrm{s},t}, \mathrm{GD}_{\mathrm{s},t}$	Storing/releasing rate of storage s at hour

3.1 Introduction

Electricity is one of the best forms of energy that has features such as easy controllability, conversion to different types of energy, convenient transmission, and economical production compared to other types of energy, and such characteristics of electrical energy lead to increased consumption and the tendency of technological progress toward electrical technology [1]. Note that the emergence of smart power networks can create demand-side management to control demand, but increasing demand is proportional to population growth over time, requiring the development of resources to meet that demand [2, 3]. Two approaches have been developed in order to meet the increased electricity demand.

The first approach is the development of traditional hierarchical power systems. Traditional power systems include centralized power plants as generating units, transmission systems, subtransmission, and distribution systems that deliver electrical energy from production to end users. With the development of power plants, transmission networks, subtransmission, and distribution networks, this increase in electricity demand can be met [4]. Concentrating on electricity generation not only does not provide the opportunity for efficient power generation technologies due to the high distance from the consumer but also makes high losses for the transmission of electrical energy to the end users. Also, due to hierarchical energy transmission, the reliability of the power system is reduced, since disruption in any part of the generation, transmission, and distribution can cause a power supply failure.

The disadvantages of the traditional power systems development approach have increased the tendency toward consumer side production and the presence of distributed generations in the subtransmission and distribution systems. The presence of these low-capacity products in the power grid has also led to the emergence of structures such as micro-grids as well as concepts such as virtual power plants, which turns distribution networks active rather than reactive parts of the system [5]. The second approach can address a variety of perspectives, including the cost of electricity generation, power system efficiency, reliability, environmental issues, and other issues to the power system. In both of the above approaches to meet increased demand, distributed power generations have been added which could be gas-fired units.

In the past decade, as natural gas has less pollution, lower prices, and more abundance than other fossil fuels, its consumption in the electricity generation sector is increasing. Transmission of natural gas from wells to end users requires pipelines, storage facilities, compressors, and valves. Figure 3.1 schematically illustrates the electricity and natural gas systems. Gas flow paths from gas wells to gas units and other gas consumers are shown. With some similarities to the power system, the steady-state gas flow of a pipeline is a function of the pressure difference between its two ends, the gas properties (compressibility factor, specific gravity), and the physical properties of the pipe (diameter, length, friction coefficient). It can be concluded that the importance and role of gas pressure in the gas grid is the same as the voltage in the grid. Compressors are installed in a location where the

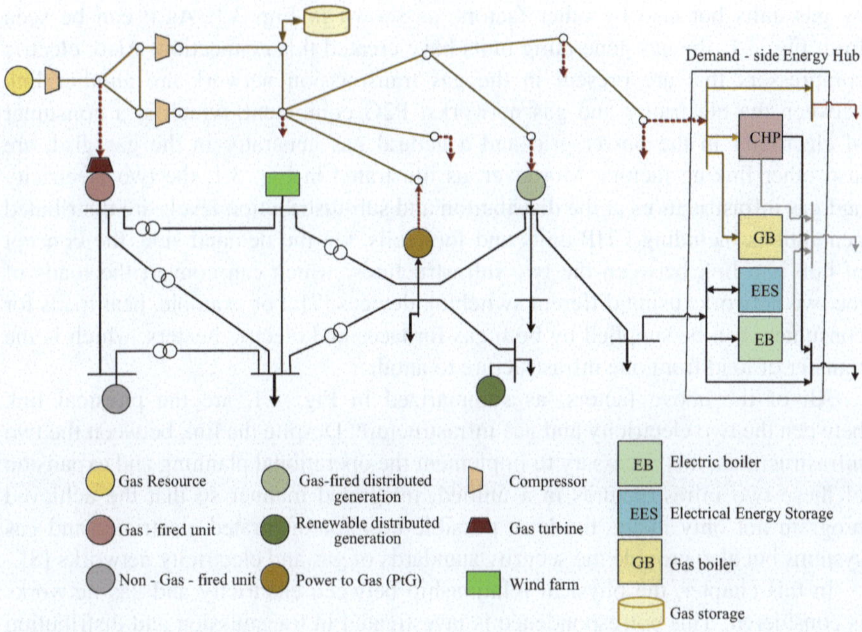

Fig. 3.1 Physical interdependency of gas and electricity networks in transmission, distribution, and demand side

transmission network needs to be increased to enhance the transmission capacity of the grid so that by increasing the gas pressure, it is possible to transfer gas to distant locations, which is very similar to the performance of the transformer in the power system. These compressors have power consumption that must be supplied. At critical points in the power grid where the compressors cannot be fed, the compressors are fed using a gas turbine. The amount of gas consumed by a compressor depends on the pressure increase. Valves and control devices in the gas grid are the same as breakers and fuses in the grid. Unlike electricity where large-scale storage is not yet technically or economically feasible, natural gas can be stored for future consumption. There are three main types of natural gas storage facilities: (a) underground storage, (b) LNG tanks, and (c) pipelines. Another important difference between electricity and natural gas systems is that electricity moves at the speed of light, while natural gas is transmitted at a maximum speed of 30 km/h.

The distinct advantages of combined cycle units such as high productivity, rapid response, and shorter installation and commissioning time have doubled the importance of gas. As a result, government and market agents have increased investment in the construction of new gas plants [6]. Furthermore, the abundance of natural gas resources in many places, such as the USA, Russia, Europe and Latin America, can be considered as the main factor in the growth of natural gas consumption. The electricity and gas infrastructures are interconnected not only by gas units but also by other factors, as shown in Fig. 3.1. As it can be seen from Fig. 3.1, the gas generating units have created this connection. Also, electric compressors that are present in the gas transmission network are another link between the electricity and gas networks. P2G equipment, which is a consumer of electricity in the power grid and a natural gas generator in the gas grid, are also other linking factors. Moreover, as illustrated in Fig. 3.1, the two electricity and gas infrastructures at the distribution and sub-distribution levels are distributed generations including CHP units and fuel cells. On the demand side, the concept of hub is a link between the two infrastructures, which can convert the loads of the two networks using different switching devices [7]. For example, heat loads for consumers can be supplied by both gas furnaces and electric heaters, which is the transfer of load from one infrastructure to another.

All of the above factors, as summarized in Fig. 3.1, are the physical link between the two electricity and gas infrastructure. Despite the link between the two infrastructures, it is necessary to implement the operational planning and expansion of these two infrastructures in a unified, integrated manner so that the achieved program not only incurs the least possible cost for integrated electricity and gas systems but also provide the security standards of gas and electricity networks [8].

In this chapter, the physical relationship between electricity and gas networks is considered. This correspondence is investigated in transmission and distribution levels. Moreover, the effect of several facilities such as PtG and gas-fired units is studied. In addition, various approaches such as resiliency and reliability are also taken into account.

The rest of the proposed chapter is listed as follows: the expansion co-planning of electricity and gas networks is discussed in Sect. 3.2. Afterward, the operation planning of integrated gas and electricity system is presented in Sect. 3.3. Finally, the conclusion and future works are described in Sect. 3.4.

3.2 Expansion Co-planning of Electricity and Gas Networks

With the increasing consumption of gas in the electrical and nonelectrical sectors, the development of corresponding infrastructure is essential to meet the demand of both power plant and non-power producer customers. Considering the benefits of electrical energy mentioned in the previous section, the development of power system is also of great importance. Since these two infrastructures are interconnected through a great number of equipment in different sectors, it therefore requires integrated planning to be able to make decisions on capacity and timing of new equipment installation based on economic considerations, environmental issues, and security factors [9]. Gas pipelines, gas import pipeline capacity, gas storage equipment, compressors, and LNG terminal capacities are among the equipment to be developed and strengthened in planning the development of the gas transmission network. The development of power transmission networks focuses more on strengthening and building electricity transmission lines.

In most integrated expansion co-planning of the electricity and gas networks, the objective function is to minimize the total cost of operation and expansion, and hence, in the modeling of this program, the constraints of the networks electricity and gas operation are also included. Since the computational burden of integrated electricity and gas planning is high, most of the studies use the simple form of operating constraints. Although some models have used the exact model, they have finally provided a method to reduce the computational burden. Here, like most studies, we use a simple model.

Equations (3.1)–(3.18), similar to Reference Modeling [10], are used toward the co-operation planning of electricity and gas infrastructure. Equations (3.1)–(3.18) model the relationships in the operation planning of the power system. Equation (3.14) shows the power equilibrium requirement per bus of the power system at each time interval. The constraints (3.10) and (3.12) represent the permissible quantity for power through existing and constructible DC power lines, respectively. The DC load distribution equations for the existing and constructible power lines are presented in Eqs. (3.9) and (3.11), respectively.

$$-z_k\overline{\mathrm{fP}_k} \le \mathrm{fP}_k \le z_k\overline{\mathrm{fP}_k} \tag{3.1}$$

$$H_k = \sigma\,\mathrm{fc}_k\left[\left(\frac{\pi_j}{\pi_i}\right)^{Z_{ki}} - 1\right] \tag{3.2}$$

$$\tau_k = \gamma_k H_k^2 + \beta_k H_k + \alpha_k \tag{3.3}$$

$$1 \le \frac{\pi_j}{\pi_i} \le R_k^{\max} \tag{3.4}$$

$$0 \le \frac{\pi_j}{\pi_i} - 1 \le y_k R_k^{\max} \tag{3.5}$$

$$\pi_i^{\min} \le \pi_i \le \pi_i^{\max} \tag{3.6}$$

$$\mathrm{WS}_i^{\min} \le \mathrm{WS}_i \le \mathrm{WS}_i^{\max} \tag{3.7}$$

$$\sum_{i=1}^{\mathrm{NP}} A_{mi}.\ \mathrm{fp}_i + \sum_{i=1}^{\mathrm{NC}} U_{mi}.\ \mathrm{fc}_i + \sum_{i=1}^{\mathrm{NWS}} V_{mi}\mathrm{WS}_i - \sum_{i=1}^{\mathrm{NWL}} C_{mi}\mathrm{WL}_i - \sum_{i=1}^{\mathrm{NC}} D_{mi}\tau_i = 0 \tag{3.8}$$

$$\mathrm{fl}_k = B_k\left(\theta_{\mathrm{fr}(k)} - \theta_{\mathrm{to}(k)}\right) \tag{3.9}$$

$$-\overline{F_k} \leq \mathrm{fl}_k \leq \overline{F_k} \tag{3.10}$$

$$-M_k\left(1-z_k\right) \leq \mathrm{fl}_k - B_k\left(\theta_{\mathrm{fr}(k)} - \theta_{\mathrm{to}(k)}\right) \leq M_k\left(1-z_k\right) \tag{3.11}$$

$$-z_k\overline{F_k} \leq \mathrm{fl}_k \leq z_k\overline{F_k} \tag{3.12}$$

$$P_{gk}^{\min} \leq P_{gk} \leq P_{gk}^{\max} \tag{3.13}$$

$$\sum_{i=1}^{\mathrm{NL}} T_{mi}\mathrm{fl}_i + \sum_{i=1}^{\mathrm{NG}} G_{mi}P_{gi} - \sum_{i=1}^{\mathrm{ND}} W_{mi}PL_i = 0 \tag{3.14}$$

$$\mathrm{WL}_k = \left(\mu_1 P_{gi}^2 + \mu_2 P_{gi} + \mu_3\right)/\mathrm{GHV} \tag{3.15}$$

$$\mathrm{fP}_k^2 = sgn_{ij}M_{ij}\left(\pi_i^2 - \pi_j^2\right) \tag{3.16}$$

$$-M_1\left(1-z_k\right) \leq \mathrm{fP}_k^2 - sgn_{ij}M_{ij}\left(\pi_i^2 - \pi_j^2\right) \leq M_1\left(1-z_k\right) \tag{3.17}$$

$$-\overline{F_k} \leq \mathrm{fl}_k \leq \overline{F_k} \tag{3.18}$$

Equation (3.7) illustrates the equilibrium condition of gas flow in each node of the gas network. Equations (3.16) and (3.17), respectively, demonstrate the gas capacity in the existing and constructible pipelines. Equations (3.2)–(3.6) illustrate the operation constraints of gas compressors in terms of gas pressure before and after the compressor as well as the physical characteristics of the compressor. Similar to the power grid, Eq. (3.6) shows the permissible gas pressure range in each node. Equation (3.15) shows the relationship between the natural gas and electricity infrastructures through power plants. Equation (3.19) illustrates the cost of expansion co-planning investment, which includes the cost of gas pipelines, the cost of installing the compressor, and the cost of developing the power lines, respectively. Equation (3.20) defines the cost of generating electricity, which includes the cost of purchasing gas as well as the cost of fuel for non-gas-fired units. Equation (3.21) is aimed toward calculation of carbon emission cost. The sum of the three cost functions in Eqs. (3.19)–(3.21) is considered as the objective function in the expansion co-planning.

$$\text{InvestmentCost} = \varsigma.\left(\sum_{i=1}^{\text{NCP}} x_i\,P\text{lineCost}_i + \sum_{i=1}^{\text{NCC}} y_i\,C\text{lineCost}_i + \sum_{i=1}^{\text{NCL}} z_i\,E\text{lineCost}_i\right) \tag{3.19}$$

$$\text{ProductionCost} = \sum_{i=1}^{\text{NWS}} 8760.\text{Cost}_{\text{gas}i} * \text{WS}_i + 8760.\sum_{i=1}^{\text{NCG}}\left(a_i P_{gi}^2 + b_i P_{gi} + c_i\right) \tag{3.20}$$

$$\text{CarbonEmissionCost} = \sum_{i=1}^{\text{NCG}} 8760.\text{CarbonCost} * \xi_1 * P_{gi} + \sum_{i=1}^{\text{NGG}} 8760.\text{CarbonCost} * \xi_2 * P_{gi} \tag{3.21}$$

A number of studies have also addressed the development of gas suppliers as well as the development of generating units in the power system. Equations (3.22) and (3.23), respectively, show the limitation of gas production in each of the existing and constructible gas producers. Equations (3.24) and (3.25) also show the capacity constraints of the existing production units, respectively. It should be noted that in all of the above modeling equation used for the purpose of integrated development, a binary variable is used which represents the state of constructing the equipment after the expansion co-planning has been implemented [11].

$$0 \le p_{g,t,b}^{\text{G,N}} \le \sum_{b=1}^{\text{B}} p_{g,b,w}^{\text{C,N}}.\text{pf}_{g,b}^{\text{Gas}} \tag{3.22}$$

$$0 \le p_{g,b,t}^{\text{G,Ex}} \le p_{g,b}^{\text{C,Ex}}.\text{pf}_{g,b}^{\text{Gas}} \tag{3.23}$$

$$0 \le p_{i,t}^{\text{N}} \le p_i^{\text{C,N}}.\text{CF}_{i,t} \tag{3.24}$$

$$0 \le p_{i,t}^{\text{Ex}} \le p_i^{\text{C,Ex}}.\text{CF}_{i,t} \tag{3.25}$$

If planning for the integrated development of electricity and gas infrastructure involves the retirement of units such as coal-fired power plants after several years, the results show that in order to compensate lack of production capacity, it is required to build more gas units. This will in turn increase the connectivity of the two infrastructures; and consequently pipelines, compressors, and other gas network equipment need to be further developed to meet the increased demand for gas in the power generation sector. As a result, the investment decisions of these two infrastructures are strongly linked. Also, considering the retirement of the units in expansion co-planning of the electricity and gas networks can significantly increase the computational complexity [12].

As illustrated in Fig. 3.1 another equipment that creates a connection between the electricity and gas infrastructure is the PtG equipment. Modeling of this equipment is presented in Eq. (3.26):

$$G_{aht} = \varnothing . P^{\mathrm{bc}}_{aht} . e^{\mathrm{ptg}}_{a} / \mathrm{HHV} \tag{3.26}$$

This equipment is in fact a consumer of electricity in the power grid and an injector in the gas grid. One of the applications of PtG equipment is the optimal use of renewable power sources, especially wind units. With the use of PtG units, the surplus power of wind units can be converted to gas, thus preventing their loss. Otherwise, the capacity of the power lines should be increased in order to transfer the excess power to the consumption areas. Reducing the amount of power cutoff by wind farms also results in a reduced use of fossil fuels, thus, reducing the environmental pollution. As a result, the use of PtG units can increase the penetration of renewable wind products, delay the development of the electricity transmission network, as well as reduce the use of fossil fuels and carbon emission [12].

Another equipment used in the gas grid is gas storage. These gas storages, which are more applicable than electric energy storage in large-scale utilization, can increase system reliability. The reliability of the system may be reduced due to a disruption in the gas grid or increased demand for gas units because of high demand for electricity. Also, it might be the reason for the sudden decline in the production of renewable units, which gas storages can significantly solve. In the equations, gas storage modeling is discussed.

Equations (3.27) and (3.28) show the limitation of the injected gas and the storage output. Equation (3.29) also states that charging and discharging operations do not occur simultaneously. Equation (3.30) shows the limitation of the level of gas stored in the storage, which can be between permissible values. Equation (3.31) illustrates that the amount of gas in storage at any time is equal to the amount of gas in the interval before input gas is added and the output gas is reduced [13].

$$\sum_{j} \psi_{j,s,t} \leq \sum_{\sigma} v^{\mathrm{in}}_{\sigma,s,t} \rho^{\mathrm{in}} \tag{3.27}$$

$$\sum_{j} \psi_{j,s,t} \leq \sum_{\sigma} v^{\mathrm{out}}_{\sigma,s,t} \rho^{\mathrm{out}} \tag{3.28}$$

$$\sum_{\sigma} v^{\mathrm{in}}_{\sigma,s,t} + \sum_{\sigma} v^{\mathrm{out}}_{\sigma,s,t} = 1 \tag{3.29}$$

$$S_s \leq x_{s,t} \leq \overline{S_s} \tag{3.30}$$

$$x_{s,t} = x_{s,t} + \sum_{j} \psi_{j,s,t} - \sum_{j} \psi_{s,j,t} \tag{3.31}$$

The use of gas storages can significantly increase the reliability of the system, so that depletion of these large-scale storages can compensate a sudden increase of gas demand of the system in short term. The location of these gas storages can have a substantial impact on reducing the gas load cutoff (power plant and non-power plant). If the location of these gas storage are defined in order to reduce the cutoff cost of the gas loads, the results show that the optimal location of these gas storage facilities can lead to a reduction in the penalties that should be paid to the consumers, which can be achieved by minimizing the number of gas cutoffs caused by the pipeline outages. In addition to reducing the number gas disruptions, the average price of gas also declines due to the imbalance in gas [14].

As shown earlier, one of the criteria that can be considered in expansion co-planning of electricity and gas infrastructure is reliability. To this end, different studies have considered different criteria of reliability in expansion planning to achieve the reliability of a single system to a certain extent. One of these criteria is the loss-of-load expectation (LOLE) matrix, which by definition is the average number of hours or days in a given time period (usually 1 year). The load is greater than the production capacity. This definition is illustrated in (3.32) [15].

$$\mathrm{LOLE}_h = \sum_{t=1}^{\mathrm{NT}} P_t \left(\mathrm{AC}_h \leq L_t^{\mathrm{E}}\right) \tag{3.32}$$

The results of applying reliability metrics to expansion co-planning of gas and electricity infrastructure indicate that in order to achieve a certain level of reliability (the numerical value of reliability indexes determine the level of reliability of the system), the system expansion cost is increased, because more elements are needed to be added to the system, which in turn increases the cost of system expansion.

Another reliability criterion is the $N - 1$ index, which is a simplified case of the $N - k$ index, where k is the number of system elements out of operation. According to this criterion, a system that normally operates should be able to continue operating at k-element exit. Adding this criterion to expansion co-planning can improve the reliability of system. This criterion is illustrated in Eq. (3.33) [12].

$$s.t \sum_{i} (1 - \mathrm{AU}_{int}) + \sum_{l} (1 - \mathrm{AL}_{lnt}) \leq 1 \tag{3.33}$$

This criterion for the gas network can also be considered in expansion co-planning of electricity and gas. The $N - 1$ criterion can be applied using the contingency matrix [16]. This matrix includes set of 0 and 1, in which 0 shows the disruption of an element and one shows the normal condition of that element. If n illustrate the number of element which are going to be considered, this matrix $n.n + 1$. In this matrix, the first column depicts the normal condition (with no disruption) and other columns only show the outage or disruption of an element.

Thus, this matrix is applied to the modeling of the constraints of the operation so that in addition to the modeling of normal conditions, the constraints of the system can also be modeled in conditions of an element outage. In this regard, the disconnected element can be a pipeline or a compressor.

To examine the impact of considering the $N-1$ criterion in the electricity and gas network, three cases are taken into account. In the first case it is assumed that the $N-1$ criterion is not considered for any infrastructure. In this case, although the investment cost may be low, the power outage can be very high in the event of a power outage. It is assumed in the second case to consider only in the $N-1$ power system rather than gas network. In this case, although the investment cost of electricity infrastructure increases due to the development of transmission lines, the amount of electricity outages is not much lower than before. In this case, it may be because of gas pipeline outflow, as a result of which gas units may not be able to generate their nominal capacity due to the weak gas supply network. Thus, the amount of interruption is still high. However, if the $N-1$ criterion is taken into account in both infrastructures, by installing more pipelines, the amount of curtailed electrical load can be reduced to zero (Figs. 3.2 and 3.3). In a system with high penetration of gas-fired units, the production capacity of such systems is a function of the rated capacity of the installed units as well as the capacity of the gas network infrastructure.

Flexibility-based expansion co-planning is another approach that can be applied to expansion planning. As defined by Presidential Policy Directive 21 (PPD-21),

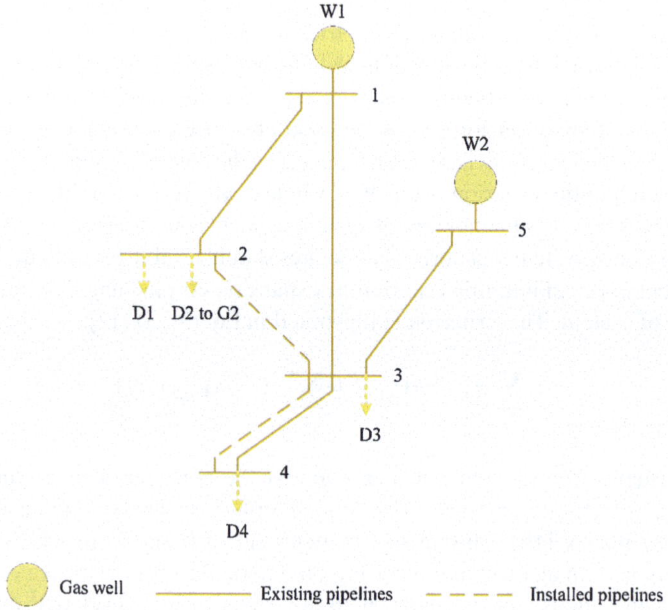

Fig. 3.2 Power system combined with gas system [8]

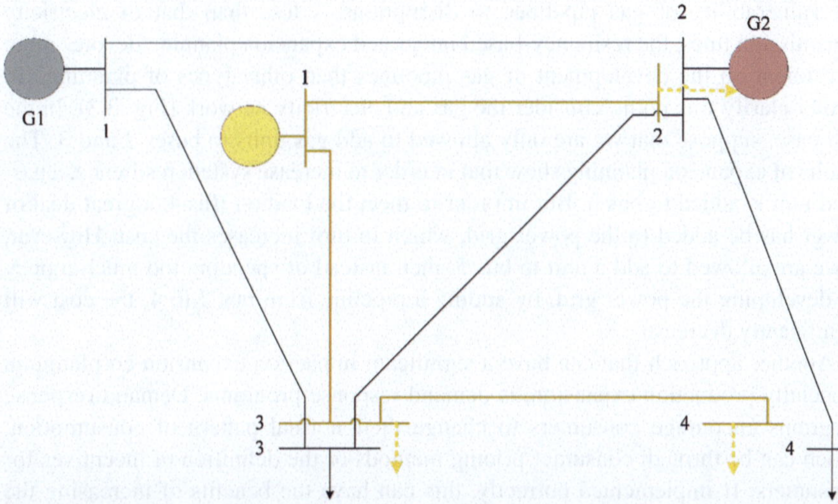

Fig. 3.3 Four-bus system [9]

resilience is the ability to adapt to changing conditions and the ability to tolerate and recover from impaired conditions. The disturbances can be deliberate attacks, major accidents, or natural disasters. The set of possible events and measures proposed to eliminate the disorder are two important parts of the above definition, the first to identify hazardous conditions in the system and the second is the planning and recovery plans for the disorder. Criteria such as length of service unavailability, cost of system recovery, or cost of preventing severe disruption are used to evaluate the resilience. One of the criteria that can be used to measure the resilience of a power system is the minimum disruption level after the most severe event that has been modeled in relation (3.35) [17].

$$\mathrm{RM} = \max\min \sum_{i,b,t} f_i\left(\mathrm{pd}_{i,b,t}\right) \tag{3.34}$$

In this equation, f is a function that calculates the disruption cost. In Resilience-Based expansion Planning, Eq. (3.34) is also added to Planning as Eq. (3.35). In this regard, the $\mathrm{RM}_{\max}$ is the maximum allowable cutoff value after the worst disturbance in the electricity and gas system, which can be approximated by sensitive loads such as hospitals and security centers.

$$\mathrm{RM} = \max\min \sum_{i,b,t} f_i\left(\mathrm{pd}_{i,b,t}\right) \leq \mathrm{RM}_{\max} \tag{3.35}$$

The results show that in order to achieve a certain level of resilience in the network, adding equipment to increase system tolerance against major disruptions is a necessity. However, in integrated electricity and gas networks, equipment can be added to the electricity or gas network to increase network resilience. Since the vulnerability of gas pipelines to disruptions is less than that of electricity transmission lines, the resiliency-based integrated expansion planning devotes more investment on the development of gas pipelines than other types of planning. To further clarify this point, consider the gas and electricity network (Fig. 3.3). In the first case, suppose that we are only allowed to add gas units to buses 2 and 3. The results of expansion planning show that in order to increase system resilience, a gas-fired unit is added to bus 3. But in order to meet the load on Bus 4, a great deal of power has be added to the power grid, which in turn increases the cost. However, if we are allowed to add a unit to bus 5, then instead of spending too much money on developing the power grid, by adding a pipeline from bus 2 to 4, the cost will significantly decrease.

Another approach that can have a significant impact on expansion co-planning, especially production expansion, is demand response programs. Demand response programs encourage consumers to change their normal pattern of consumption, which can be through consumer pricing methods or the definition of incentives for consumers. If implemented correctly, this can have the benefits of increasing the load factor and making the load curve smoother. This long-term load curve change can have a significant impact on expansion planning, including integrated expansion plans for the electricity and gas networks. As a result of the correct implementation of the electrical demand response programs, the peak load has been reduced, which can lead to a significant reduction in the capacity of the new gas-fired units. Thus, in countries such as GB where some of their gas is supplied by imports, the results show that implementation of demand response programs can reduce imports by 90 million cubic meters by 2050. The cost savings associated with implementing GB demand Response Programs over a 50-year period from 2010 are estimated at approximately $ 60 billion. Although cost savings are not large in the gas network, the use of demand response plays an important role in improving gas network security.

Even though electric charge response programs reduce the cost of developing the gas network, the cost of operating the gas network increases because in this case part of the peak load transferred to other hours is supplied by cheap gas units. This will likely increase gas consumption and thus increase the cost of operating the gas network. The simultaneous reduction of the maximum gas supply capacity (import and development of the gas network) and the increase in the amount of gas consumed and the flow of gas in the gas network increase the utilization coefficient of the gas network. Therefore, the use of load response programs in the power system can bring great benefits to both the electricity and gas infrastructure [18].

Applying the concept of energy hub has also made sense for gas response programs in the gas infrastructure. The energy hub has a number of input energy carriers and a number of output energy carriers. The input carriers either appear unchanged in the output or are converted to other carriers using converters. With converters such as electric heaters and PtG equipment in the energy hub, there is an opportunity to, for example, provide consumers with thermal energy rather than being supplied with gas, which means that there will be a reduction in gas consumption. In other words, consumers are encouraged to only change their primary energy source. Changing the source of primary energy in the electrical sector is more likely; so, its use can significantly reduce the cost of developing electricity and gas infrastructure.

As shown in Fig. 3.1, both the electricity and gas infrastructures are interconnected at the level of the electricity and gas distribution networks, which can be connected through equipment such as renewable distributed generation, PtG, as well as on the consumer side through concepts such as energy hubs. Unlike gas transmission networks, in gas distribution networks, the operating and technical loss costs of the development planning are not considered in the gas distribution network. This is because the gas is distributed in a limited area in gas distribution networks as compared to the transmission network; so, these compressors are not present in gas distribution networks. Similarly, the cost of constructing gas storage tanks is not taken into account as they are mostly installed at the transmission level of these gas reserves.

Distributed generations are low-capacity products whose presence in the distribution network has quit a few advantages for the power grid. Reducing the cost of transmission network expansion, reducing transmission losses, and increasing energy efficiency are some of the benefits of using these generations. In the expansion planning, if the development of distributed generation, including gas-fired distribution units, is not considered, then the cost of distribution network development will be high, given that the distribution network infrastructure needs to be developed to access power from the power grid. However, the cost of developing the gas network is not high as no new load is added to the network.

In the latter case, the development of gas-fired distributed generation is considered, except that the distribution network expansion planning is first implemented and the optimal location is obtained from the view of the electricity distribution network after the gas distribution capacity and location have been determined, and then the gas distribution network expansion planning is implemented. In this case, since the distributed generation is implemented in order to reduce the cost of operating power network, the cost of operating the electricity grid is at its lowest value compared to other cases. However, the expense of developing a gas network is high, because expansion planning must be in place for gas-fired distribution facilities.

In the third case, integrated planning for the development of the electricity, gas distribution network, and gas-fired distributed generations is implemented. The results show that in this case, since the location of the gas distribution needs to be determined in order to reduce the total cost of the two systems, the cost of operating the distribution network is therefore far from its optimal value (the second case). However, it should be noted that the total cost in this case is the lowest of the three preceding cases. Therefore, the expansion co-planning of the electricity and gas distribution network and the gas-fired distributed generation can provide a program with the lowest total cost and with considering the constraints of both systems. Note that since the optimal location of distributed generation is different in the second and third cases, expansion co-planning is appropriate for organizations that own both systems.

Like transmission network expansion planning, in most distribution network planning, total operating cost and network expansion are considered as the objective function. The expansion of equipment such as capacitive banks can also be considered in these plans [19]. Since the use of capacitive banks reduces the reactive power output of gas-fired distributed generation, it means increasing the capacity of distributed generation in providing active power, which in turn reduces the cost of developing distributed generation. On the other hand, reducing the development of gas-fired distributed generation means reducing the cost of expanding the gas distribution network. Therefore, in integrated energy systems such as electricity and gas infrastructure, investment decisions of one infrastructure can affect the costs of investing in another infrastructure. Therefore, in order to reach the lowest cost decisions, it is necessary to have integrated investment planning and development of such infrastructure.

3.3 Operational Co-planning of Electricity and Gas Networks

In addition to expansion planning, the relationship between the two electricity and gas infrastructures should also be considered in short-term operational planning. The most basic operational planning is the unit commitment. The system operator is responsible for the short-term planning of the system. The operator must employ the production units with regards to the anticipated load for each day and hour, so that the total operating cost is minimized. The unit commitment determines this optimal program. After the optimal sequence of units has been determined, the power output of each unit in the system is determined by the economic dispatch. Before the program can be announced to the units, the system security must also be checked in order to consider whether the units can be loaded within the permissible range of the transmission and all restrictions. This program is named Security Constrained unit commitment (SCUC). As in today's power systems we see a high penetration of gas-fired units in the manufacturing sector, so the security constraints that is needed to be considered after the unit commitment program are not just about the power grid, because the constraints of the fuel infrastructure, i.e., the Gas network, should also be considered in this planning. Figure 3.4 shows the SCUC program in integrated electricity and gas networks.

First, the unit commitment program whose modeling is shown in Eqs. (3.1)–(3.19) is fulfilled. Equation (3.35) is the operating cost that is considered as the objective function. Operating costs include, respectively, the cost of gas contracts, the cost of producing non-gas plants, the cost of startup and shutdown of the units, as well as the penalty for not supplying energy. Details of modeling are given in [20]. It should be noted that there are usually two approaches to model the cost of gas-fired units. In the first approach, the cost of these units is modeled as other units using the quadratic cost function [20]. However, in the second approach, the cost of gas units is modeled by the cost of gas purchased from the gas network using contracts.

$$\min\left\{\sum_{\eta} W_{o,\eta} + \sum_{t}\sum_{\eta} W_{\eta t} + \sum_{t}\sum_{i \notin \mathrm{GU}} \left[F_{c,i}\left(p_{it}\right).I_{it} + \mathrm{SU}_{it} + \mathrm{SD}_{it}\right] + \sum_{t} \rho_{\mathrm{LS}}.P_{\mathrm{LS}}\right\} \tag{3.36}$$

To examine the impact of considering gas network constraints on operation planning and SCUC, suppose that the planning shown in Fig. 3.4 is implemented on the gas and electricity network shown in Figs. 3.5 and 3.6 for both cases with and without the gas network constraints.

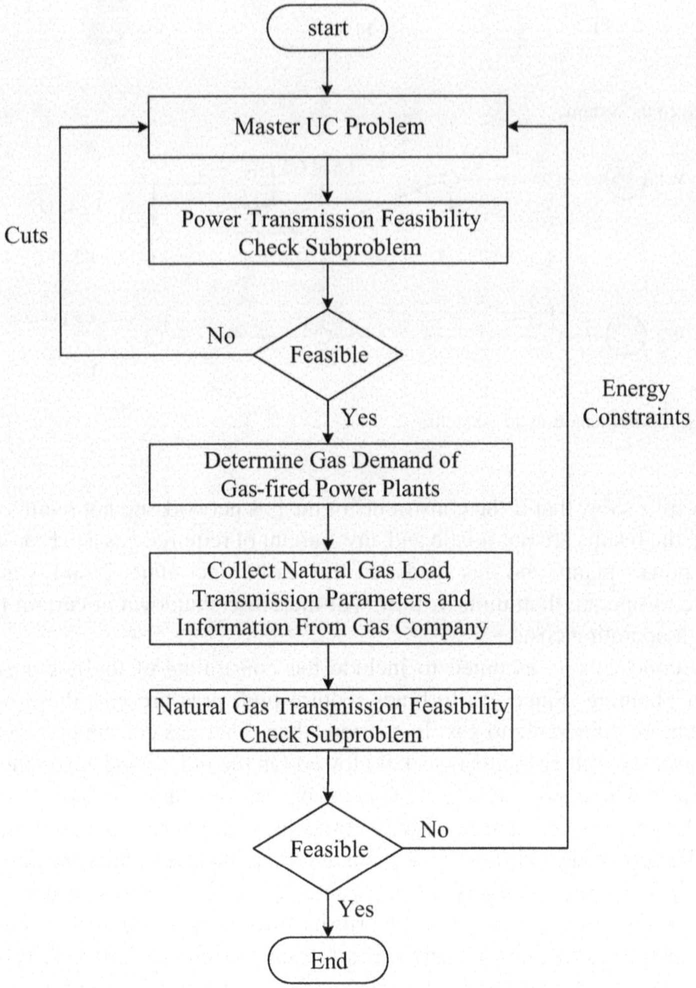

Fig. 3.4 SCUC program in integrated electricity and gas networks

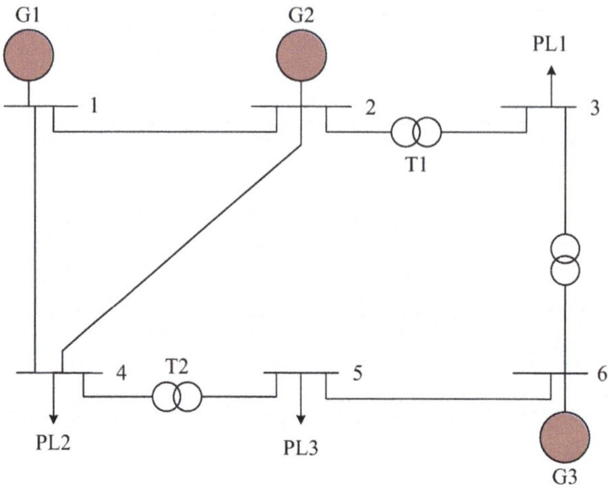

Fig. 3.5 Six-bus system

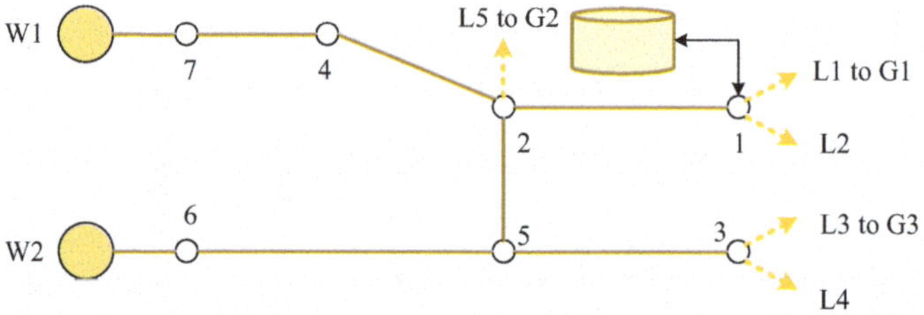

Fig. 3.6 Seven-node natural gas system

The results show that if the constraints of the gas network are not included in the planning, the results are optimistic and any amount of required gas can be consumed by both power plants and gas loads. In this case, since units 2 and 3 are more expensive to operate than units 1, they will therefore shutdown at certain times to avoid high operating costs.

The second case is assumed to include the constraints of the gas network in operation planning. Since gas turbines require high pressure gas, these units are therefore more vulnerable to gas fluctuations than other gas consumers, so the cost of gas contracts with gas units is usually lower. On the other hand, since the cost of gas contracts where gas can be cut off is lower, gas units are usually less costly in order to lower their cost and be able to compete with other units in the electricity market. Therefore, they choose the type of contract where gas units can choose low cost and low priority contracts. In this context, if a fault occurs in the gas grid, the loads of the power plant are in the priority of disconnection. The results of the SCUC implementation, considering the constraints of the gas network in Fig. 3.6, show that the operating cost has increased in this case. Unlike the previous case, in which low-cost units were operated, in this case, due to the constraint of pipeline 1 and also the higher priority of load 2, plant 1 produces less power and plants 2 and 3 compensate for this power loss, resulting in an increase in the operating costs.

Another difficulty that may occur in the electricity and gas networks is the simultaneous peak in electricity and gas consumption, which may lead to power plant outage and power disruption in the network accordingly, since the power consumers, especially residential, have higher priority than power plants. Applying demand response programs in the electrical sector can smooth the electrical load curve, and thus reduce the peak load synchronization intensity of the two infrastructures [21]. Pipeline exit can also be simulated in operational planning as a factor that could threaten system reliability. For example, in Fig. 3.6, suppose pipeline 2 between nodes 2 and 5 exits the circuit, followed by unit 1, which is a large unit. Following the exit of unit 1 from the system, units 2 and 3, although increasing their output power, but cannot compensate for the power output of a large unit such as 1, which in turn causes a large power outage in the power grid.

As can be seen, the above presented programming is very vulnerable to the exit of an element and incurs a great deal of cost. Therefore, it may be advisable to include the exit of this element in operational planning to increase system reliability against such events. As mentioned in the previous section, the k-element output can also be considered. The results show that the greater the number of outbound elements, the longer the execution time of a program that is used to evaluate such a definite number in operational planning. This type of modeling is shown in Eq. (3.37).

$$\begin{aligned} &\min f\left(x^0\right) \\ &s.t\ h^0\left(x^0\right) \le 0 \\ &\qquad \max\min\ g\left(x^k\right) \le \mathrm{RF}^{\max} \\ &h^k\left(x^k\right) \le 0 \end{aligned} \tag{3.37}$$

This type of modeling is Master–Slave modeling. The Master section executes the SCUC problem under normal conditions and calculates the optimal response under normal conditions. Then the solution obtained by the Slave section is examined in the worst case scenario, and the total gas consumption of the gas units should not exceed its permissible value. So the obtained answer should ultimately be able to comply with what is supposed in the worst of circumstances. Therefore, unlike the method shown in Fig. 3.4, in this method, the results of the operation of the normal conditions are obtained by looking at the worst outage and possible condition, which can in turn increase the efficiency of the proposed method and enhance the efficiency of the answers in real cases.

Applying gas storage is another method that can improve the performance of the gas network under abnormal conditions. The use of gas storages can make the consumption curve of the gas grid smoother and thus prevent problems such as overloading pipelines during peak hours of consumption. For this purpose, assume in Fig. 3.4 that power plant 1 has a reservoir, so by cutting off pipeline 2, the unit can remain in network with gas stored during off-peak hours. As a result, the amount of power outage and the cost of operating the power grid are significantly reduced. Gas storage devices also have an operating cost, which is shown in Eq. (3.38), which is a function of the gas charged and discharged.

$$\min f\left(x^0\right) = \sum_t \left\{ \sum_{i \notin GU} \left(\mathrm{SU}_{i,t}^0 + \mathrm{SD}_{i,t}^0\right) + \sum_{i \notin GU} F_i^{\mathrm{c}}\left(P_{i,t}^0\right) + \mathrm{VOLL}.\sum_j \mathrm{LD}_{j,t}^0 + \sum_{\mathrm{sp}} \rho_{\mathrm{gas}} v_{\mathrm{sp},t}^0 + \mathrm{gs}_{\mathrm{s}}.\sum_{\mathrm{s}} \left(\mathrm{GC}_{\mathrm{s},t}^0 + \mathrm{GD}_{\mathrm{s},t}^0\right) \right\} \quad (3.38)$$

As noted in the sections above, the gas and electricity infrastructures have a great deal of interdependence in the operation sector, thus, it has been seen that the outage of a single element can have economic and security consequences. Therefore, it may be possible to reduce the impact of each infrastructure from changes in the other infrastructure, despite the physical link between the two electricity and gas infrastructure. One of the approaches employed in a number of studies is the use of units capable of changing fuel input. That is, units that can use two fuels, for example, oil and gas. The effect that these units can have is the change in the peak hours of gas consumption, which has been found to be very problematic in both systems if they are within the same range as peak hours. Therefore, the use of these units is an appropriate strategy to reduce the gas consumption peak that can increase the security of the power system during peak hours of gas consumption.

Another product whose presence in the power grids is increasing is renewables. Global pollution and the reduction of fossil fuels are some of the factors that have led to the development of these resources. These resources, despite their benefits, are not predictable and controllable because of the natural source of energy in such products and create power fluctuations in the power grid. One way of controlling these oscillations is to use fast-reacting units, such as gas units, which can control these oscillations to some extent by rapidly changing the output power. The use of gas units to control these fluctuations may reduce the fluctuations to some extent, but by changing the output power of the gas units and thereby changing the amount of gas consumed by these units, these fluctuations are transmitted to the gas network. It should be noted, however, that due to the existence of gas pipelines that enable gas storage, its inertia to change is lower than the power grid and in fact it fluctuates and moves from zero power system with high inertia to the gas grid with low inertia. Therefore, the uncertainty of renewable DGs, especially wind turbines, leads to fluctuations in the gas grid, which can in some cases cause gas to break again many times in the gas grid. To this end, the planning of the operation of the power grid should be re-programmed to avoid interruptions [22]. In power distribution networks, the use of gas-fired units such as CHP and CCHP units increases the flexibility of the distribution network in eliminating the power fluctuations of renewable units. Flexibility is the use of existing capacity (distributed load and generation) in the distribution network to improve system performance. By eliminating fluctuations in the distribution grid by gas-fired distributed generation, the power fluctuation of the distribution grid is prevented to enter the upstream and local capacities of the grid have been used instead of being compensated by the reservation of larger units [23].

Given the benefits of renewables, especially wind turbines, it may be perceived that the greater the penetration of wind turbines on the power system, the lower the operating cost. This is not a general term, as the economic system for the development of wind turbines needs to be examined, as well as the power system in which these renewables will be installed. For example, in a power system where units such as gas-fired units are capable of significantly increasing and decreasing power, turbine power fluctuations are well controlled and power outages are avoided, resulting in decreasing operating costs. However, inadequate capability of increasing and decreasing power leads to expensive units being operated or load curtailment, which can increase the operating cost itself. Therefore, the installation and development of renewable products, especially wind turbines, should be subjected to the provision of infrastructure to control the fluctuations created by these renewable DGs [24].

It has been observed that fluctuations in the electricity and gas networks can affect the operating cost as well as the security of the integrated system. One of the factors that can increase this impact is the presence of electric compressors in the gas grid. To examine this more closely, consider two modes. In the first case, the power grid load is assumed to be constant, but the gas grid load changes and increases over a given interval. In this case, as the amount of gas consumed increases, the flow of gas in the grid increases, resulting in an increase in the power consumption of the compressors, which in turn increases the cost of operating the power grid, and the greater the number of electric compressors available in the grid, the greater the cost of operation will be [25]. Compressors delay the expansion of the network by increasing the transmission capacity of the gas network.

As the presence of renewable production units increases, due to the physical limitations of the power grid in power transmission, overcapacity or Excess Power may emerge that is not applicable. In addition to gas-fired units, PtG equipment can help make better use of renewable productions. PtG equipment, such as electrolyzers, not only make better use of renewable products like solar cells, but are also environmentally friendly for carbon dioxide absorption. In order to develop deployment of PtG equipment, assessments should be made of the amount of extra energy generated over a period of time and appropriate bids should be invested accordingly. For example, the development of PtG equipment is not economical in a way that its full capacity can be used only 1% of the time per year [26].

Given the relationship between the gas and electricity infrastructures in different sectors, the results show that the use of co-planning methods can provide more comprehensive, economical, as well as safer planning than planning separately. As shown in the sections above, by shifting planning from separate planning to integrated planning, the results are different, resulting in the benefits of integrated planning for countries and companies that own both of the electricity and gas infrastructure. While in some countries they may have separate ownership of these two infrastructures. So is it possible that the above methods lose their effectiveness under such circumstances? The answer is that in a number of studies the issue of separate ownership of these infrastructures has also been taken into account in planning. In such systems where information exchange is limited, the problem can be subdivided into two subproblems, using a decentralized solution framework, each of which is solved separately and in this case information security is also respected. Reference [27] is one of the references that fully explains the decentralized solution framework.

3.4 Conclusion

Electricity and gas are important infrastructures to supply energy, and gas is regarded as primary energy and electricity as secondary energy. These two infrastructures are interconnected through different equipment and at different levels of transmission and distribution networks. As a result, the investment and operation decisions of either of these infrastructures can have significant economic and reliability impacts on the other infrastructure. Therefore, these two infrastructures should be considered as an integrated energy infrastructure, in which the planning needs to be implemented simultaneously. This chapter examined the physical relationship between the two gas and electricity infrastructures in the transmission, distribution, and consumption sectors. Initially, factors linking the two infrastructures such as gas units and PtG equipment were mentioned. Also, in part of this chapter's introduction, some of the similarities and differences between the two infrastructures were explained. The rest of this chapter is divided into two parts: expansion co-planning and operational co-planning. In the expansion co-planning of electricity and gas infrastructure section, the first common modeling that has been considered in many studies so far has been presented. Then the equipment and approaches that could be considered in this development plan were also examined. Equipment such as gas storages, PtG equipment, as well as approaches such as reliability and flexibility are among the items described in the expansion co-planning section. In the operational co-planning section, unit commitment modeling was presented for both infrastructures. Then, the impact of $N-1$ criteria on planning and other topics was discussed. The relationship between gas and electricity infrastructures can be investigated from not only the physical perspectives but also other aspects such as market approaches in future works. Moreover, since technologies and other research areas have employed artificial intelligence in order to enhance the efficiency and productivity of such systems, machine learning and big data can be used in decision-making and data science research topics to help researchers to have better forecasting and proper co-planning of electricity and natural gas systems.

References

1. H. Zhao, Q. Wu, S. Hu, H. Xu, C.N. Rasmussen, Review of energy storage system for wind power integration support. Appl. Energy **137**, 545–553 (2015). Elsevier
2. B. Odetayo, J. MacCormack, W.D. Rosehart, H. Zareipour, A real option assessment of flexibilities in the integrated planning of natural gas distribution network and distributed natural gas-fired power generations. Energy **143**, 257–272 (2018)
3. A. Jindal, M. Singh, N. Kumar, Consumption-aware data analytical demand response scheme for peak load reduction in smart grid. IEEE Trans. Ind. Electron. **65**(11), 8993–9004 (2018). ieeexplore.ieee.org
4. H. Nemati, M.A. Latify, G.R. Yousefi, Coordinated generation and transmission expansion planning for a power system under physical deliberate attacks. Int. J. Electr. Power Energy Syst. **96**, 208–221 (2018)
5. A.G. Zamani, A. Zakariazadeh, S. Jadid, A. Kazemi, Stochastic operational scheduling of distributed energy resources in a large scale virtual power plant. Int. J. Electr. Power Energy Syst. **82**, 608–620 (2016)
6. Y. Wen, X. Qu, W. Li, X. Liu, X. Ye, Synergistic operation of electricity and natural gas networks via ADMM. IEEE Trans. Smart Grid **9**(5), 4555–4565 (2018)
7. M. Salimi, M. Adelpour, S. Vaez-ZAdeh, H. Ghasemi, Optimal planning of energy hubs in interconnected energy systems: a case study for natural gas and electricity. IET Gener. Transm. Distrib. **9**(8), 695–707 (2015)
8. B. Li, R. Roche, D. Paire, A. Miraoui, Optimal sizing of distributed generation in gas/electricity/heat supply networks. Energy **151**, 675–688 (2018)
9. M. Chaudry, N. Jenkins, M. Qadrdan, J. Wu, Combined gas and electricity network expansion planning. Appl. Energy **113**, 1171–1187 (2014)
10. Y. Hu, Z. Bie, T. Ding, Y. Lin, An NSGA-II based multi-objective optimization for combined gas and electricity network expansion planning. Appl. Energy **167**, 280–293 (2016)
11. J.B. Nunes, N. Mahmoudi, T.K. Saha, D. Chattopadhyay, A stochastic integrated planning of electricity and natural gas networks for Queensland, Australia considering high renewable penetration. Energy **153**, 539–553 (2018)
12. C. He, L. Wu, T. Liu, Z. Bie, Robust co-optimization planning of interdependent electricity and natural gas systems with a joint N-1 and probabilistic reliability criterion. IEEE Trans. Power Syst. **33**(2), 2140–2154 (2018)
13. B. Odetayo, M. Kazemi, J. MacCormack, W.D. Rosehart, H. Zareipour, A.R. Seifi, A chance constrained programming approach to the integrated planning of electric power generation, natural gas network and storage. IEEE Trans. Power Syst. **33**(6), 6883–6893 (2018)
14. T.D. Diagoupis, P.E. Andrianesis, E.N. Dialynas, A planning approach for reducing the impact of natural gas network on electricity markets. Appl. Energy **175**, 189–198 (2016)
15. X. Zhang, L. Che, M. Shahidehpour, A.S. Alabdulwahab, A. Abusorrah, Reliability-based optimal planning of electricity and natural gas interconnections for multiple energy hubs. IEEE Trans. Smart Grid **8**(4), 1658–1667 (2017)
16. Y. Zhang, Y. Hu, J. Ma, Z. Bie, A mixed-integer linear programming approach to security-constrained co-optimization expansion planning of natural gas and electricity transmission systems. IEEE Trans. Power Syst. **33**(6), 6368–6378 (2018)
17. V. Zahedi Rad, S.A. Torabi, H. Shakouri G, Joint electricity generation and transmission expansion planning under integrated gas and power system. Energy **167**, 523–537 (2019)
18. M. Qadrdan, M. Cheng, J. Wu, N. Jenkins, Benefits of demand-side response in combined gas and electricity networks. Appl. Energy **192**, 360–369 (2017)
19. B. Odetayo, J. MacCormack, W.D. Rosehart, H. Zareipour, A sequential planning approach for Distributed generation and natural gas networks. Energy **127**, 428–437 (2017)
20. M.S. Cong Liu, Y. Fu, Z. Li, Security-constrained unit commitment with natural gas transmission constraints. IEEE Trans. Power Syst. **24**(3), 1523–1536 (2009)

21. X. Zhang, M. Shahidehpour, A. Alabdulwahab, A. Abusorrah, Hourly electricity demand response in the stochastic day-ahead scheduling of coordinated electricity and natural gas networks. IEEE Trans. Power Syst. **31**(1), 592–601 (2016)
22. M. Qadrdan, J. Wu, N. Jenkins, J. Ekanayake, Operating strategies for a GB integrated gas and electricity network considering the uncertainty in wind power forecasts. IEEE Trans. Sustain. Energy **5**(1), 128–138 (2014)
23. Y. Jiang et al., Coordinated operation of gas-electricity integrated distribution system with multi-CCHP and distributed renewable energy sources. Appl. Energy **211**, 237–248 (2018)
24. A. Alabdulwahab, A. Abusorrah, X. Zhang, M. Shahidehpour, Coordination of interdependent natural gas and electricity infrastructures for firming the variability of wind energy in stochastic day-ahead scheduling. IEEE Trans. Sustain. Energy **6**(2), 606–615 (2015)
25. J.H. Zheng, Q.H. Wu, Z.X. Jing, Coordinated scheduling strategy to optimize conflicting benefits for daily operation of integrated electricity and gas networks. Appl. Energy **192**, 370–381 (2017)
26. T. Estermann, M. Newborough, M. Sterner, Power-to-gas systems for absorbing excess solar power in electricity distribution networks. Int. J. Hydrog. Energy **41**(32), 13950–13959 (2016)
27. C. He, L. Wu, T. Liu, M. Shahidehpour, Robust co-optimization scheduling of electricity and natural gas systems via ADMM. IEEE Trans. Sustain. Energy **8**(2), 658–670 (2017)

Chapter 4
Transmission Pricing: Right Insights for Suitable Cost Allocation Methods

M. A. Benetti and M. Sperandio

4.1 Introduction

The ongoing energy transition process modifies many of the electric power system fundaments and leads to a paradigm change [1–5]. New technological elements are being inserted into the networks, different generation technologies are arising, and clients are becoming more engaged, with active participation in system decisions [6–10]. The current power system is getting pretty distinct from the traditional model, where consumers were passive, disengaged, and unresponsive to the market price signals [11–13]. Several publications expose particularities and challenges that arise mainly as a consequence of new generation technologies and advances in communication, the main drivers of this process [14–19].

Facing this scenario, it is essential to analyse the main changes caused by this transition energy process, its repercussions in the power sector, and the trends that may be inferred [20, 21]. Additionally, regulatory and economic fundaments that guide the pricing of transmission systems must be explored, in order to identify which characteristics are appropriate for the development of modern TCA strategies [22, 23].

From this context, some reflections emerge and a central inquiry may be made. Are the TCA methods adequate to provide efficient tariffs in this energy transition context? To answer it some issues need to be defined, such as: which criteria to employ for evaluating the quality of TCA methods, and which changes are ongoing across the electric power systems. To investigate these issues, a review about the state-of-the-art of TCA methods is performed and two other questions arise. Is there consensus about the best TCA method? Does the literature present a significant

M. A. Benetti · M. Sperandio (✉)
Department of Electromechanical and Power Systems, Federal University of Santa Maria, Santa Maria, RS, Brazil
e-mail: marcelo.benetti@mail.ufsm.br; mauricio.sperandio@ufsm.br

S. Nojavan, K. Zare (eds.), *Electricity Markets*,
https://doi.org/10.1007/978-3-030-36979-8_4

number of recent and relevant publications about the TCA topic? The chapter answers these questions and specifies the most suitable TCA publications in terms of their adherence to the modern transmission fundaments.

Therefore, the great contribution of this chapter is to evaluate the current TCA methods and to provide appropriate insights that may be applied to the development of modern TCA algorithms. Future directions about transmission pricing in the face of energy transition process are inferred and the right strategies to obtain efficient TCA algorithms are pointed. The chapter is innovative because it considers the classical principles of transmission pricing, but also the modern requirements that arise with the energy transition. This strong set of TCA fundaments allows to evaluate methods and to point right strategies in an updated format.

The next sections of this chapter are organised as follows. Section 4.2 presents the modern electric power systems in the energy transition context and the changes that arise. Additionally, transmission pricing fundaments with focus on TCA are approached. In Sect. 4.3, a wide review about the published TCA methods is realized, and the most suitable publications are indicated. Finally, Section 4.4 brings the chapter conclusions and future directions that may be inferred.

4.2 Transmission Pricing in Modern Electric Power Systems

Several issues have recently dominated the debate over electricity market reforms around the world [24–27]. The main ones are: energy matrix with low carbon emissions, renewable energies, energy efficiency, distributed generation (DG), and demand response (DR) [28–32].

The reduction of carbon emissions has been sought by a number of countries through the adoption of goals linked to a specific year [25, 33, 34]. The insertion of renewable energies into the energy matrix, besides promoting the decreasing of emissions, aims to promote the following aspects: sustainable growth, job creation, energy security, and technological development [35–37].

Another important issue, the energy efficiency, remains a technological challenge for the power and industrial sectors [32, 34, 38]. In terms of DG, its feasibility is verified in an increasing number of regions due to technological advances in the processes of electricity generation, combined with the electricity prices raising [39–43]. Thus, the growth of prosumers, consumers who may also produce energy surplus compared to their load, is modifying the traditional dispatch paradigm [44–46]. The customer involvement is another issue. As they are becoming active participants, distinct dynamic tariff designs are been proposed. Their modelling considers the customer responses in face to the network charges [47–49].

4.2.1 *Energy Transition Ongoing*

Some direct repercussions may be identified from the energy transition process. Among them, difficulty in distinguishing the boundaries between transmission and distribution systems [50, 51]. It implies in a redefinition of the planner aims and of the operator roles in these two systems. Uncertainties associated with all segments of the power sector are increasing [52–54]. Thus, the major challenge in planning and operation areas is the creation of strategies that manage distributed and centralized generation, in an efficient way, considering the greater dynamism of agents, and the higher level of uncertainty about their behaviour [55–58].

4.2.1.1 Major Changes and Repercussions

We may observe that distinct areas are undergoing relevant changes in their framework and design due to the energy transition process that happens across the power systems. Figure 4.1 summarizes these areas in an integrated arrangement.

In the following, the major changes and repercussions that arisen from the energy transition are discussed, and contributory publications in terms of strategies to deal with the challenges are highlighted.

Market Models and Tariff Approaches

Market improvements need to be made through adapting models to the new reality that includes the presence of renewable energy sources (RES) and plug-in electric vehicles (PEVs), as suggested in [59–61]. Besides, other articles propose modifications in the approaches that generate tariffs in regulated segments, as shown in [62, 63].

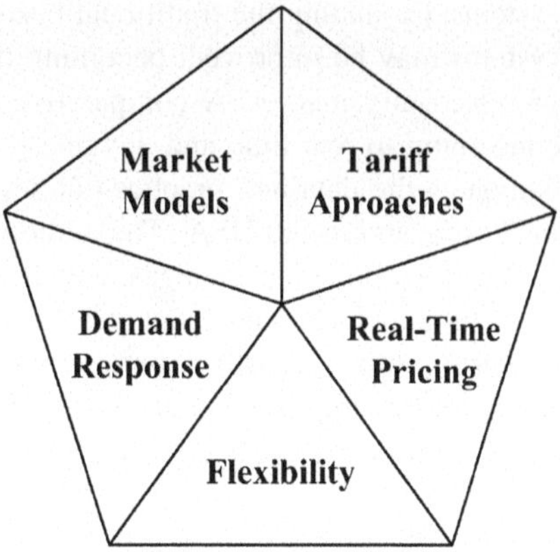

Fig. 4.1 Electricity areas with intense modifications

In [59], the authors discuss the need to adjust the market design regards to ancillary services, network charging, and balancing. According to them, flexible resources must have better prices, and locational signals need to be introduced to avoid transmission and distribution congestion. The effects of RES in markets, exploring the weather dynamics and price arbitrage concerning to wind source, is analysed in [60]. The impact of PEVs in market rules of the transmission system operator (TSO) is discussed in [61]. As the TSO priority is the security of supply, changes in its rules are challenging, because their impact are uncertain with regard to supply. However, the rules must be improved looking for sustainability and competitiveness. Different tariff candidates for residential microgrids are investigated in [62]. The evaluation is conducted analysing their effects on load and generation profiles, as well as on energy bills. Insights about different electricity network tariff designs are provided in [63]. According to it, a tariff may aggravate the regional distribution disparities, if it neglects the prosumer impacts. Furthermore, deep first connection charges and tariff corridors may increase the equity among regions.

Real-Time Pricing

Another important issue related to the modern electricity markets is the adoption of real-time pricing. It has been employed and improved in several power systems, as reported by [64–66].

A review about real-time electricity markets (RTMs) in North America, Australia and Europe, detailing their market architectures and incentive policies, is presented in [64]. RTMs maintain the flexibility and reliability of power systems. The total value of distributed energy resources (DER) must be measured besides the impact on the customer. The DER impact on the grid and surrounding market participants must be considered. A new model to measure the importance of customer peak shaving with DG technologies is presented in [65]. It compares the customer incentives provided by utility rates with the real-time prices market in New England. Policy and customer-utility interactions must integrate new technologies, such as the DG systems, expanding the traditional business models. The financial impacts of these systems may be improved, becoming the customers well informed participants of the electricity market. A comparative analysis between zonal and nodal designs concerning to real-time and day-ahead operational signals is performed in [66]. It compares the data and practices of several TSOs, placed in interconnected zones in Europe and in the USA. The authors conclude that a nodal signal is natural for RTM with topological changes, with flow control procedures included in the pricing algorithm, and a zonal design is best suited for real-time corrective actions through bilateral contracts. Further studies are needed to evaluate market design at the long-term timeframes, on the matters of grid development and locational signals.

Flexibility and Demand Response

Flexibility is a resource that gains importance in a scenario where the presence of generation sources with high uncertainty is growing. Moreover, DR is a mechanism that is becoming possible due to advances in measuring and communications. An expressive number of publications address these issues providing a set of future perspectives [67–71].

A review about flexibility products summarizes the main necessary attributes and some approaches to market designs [67]. The authors indicate the necessity of standardized and simple definitions to flexibility products, considering energy-constrained resources, such as PEV and storage. A market structure is proposed in [68] to contract jointly transmission and distribution services from load flexibility. The market presents great potential in providing economic benefits to its users and solutions to congestion management. The multi-dimensional flexibility (MDF) of energy consumers is an important issue addressed in [69], which proposes a day-ahead market design for MDF services. This market may help flatten the locational marginal prices (LMPs) across the peak load period and may help the operator to have an optimal strategy of adopting different MDF bids. In [70], expressions regard to DR are aggregated in a novel framework which unifies concepts and terminology used in the literature. The role of aggregators and the need to incentive their entry are discussed in [71], which assesses their economic features to enable the increase of DER. Different scenarios are explored to define the factors that determine their role in power systems.

4.2.1.2 How the Energy Transition Impacts on the TCA Paradigm

As the energy transition process increases the uncertainties and the speed of changes in the power sector, TCA methods must contemplate these characteristics in their modelling. Large power plants are being replaced by RES, which profoundly impact distribution networks. This segment goes through a revolution with the insertion of several new technological elements. In addition, measurement devices in real-time and communication advances allow new interactions among different agents. Therefore, TCA methods need to capture this new dynamic phenomenon. There is a wide challenge in this regard. With greater uncertainties, regulation must extend its traditional role of ensuring the predictability of investment. The strong emphasis of transmission pricing approaches on long-term investment may be revised. They must focus on pricing the system usage, in terms of costs and benefits, by the agents. Mechanisms to improve the system discipline, reducing generation uncertainties and the transmission congestion are positive. In our view, these mechanisms should be gradually adopted to improve short-term operational and market conditions without long-term contradictions.

An important aspect to note is that RES are already competitive, so their participation on power systems is increasing, which forces regulation to update more frequently. In mature electricity markets, monomial tariffs have been replaced by binomial tariffs, which separate the energy price from the transportation cost. It represents a clear withdrawal of subsidies, as RES already stand on their own. Additionally, it grows the importance of TCA methods, as they impact an increasing number of agents. About the flexibility, it has become a scarce resource that represents a challenge to the system operation, but also to the regulatory model. In our view, agents who provide this resource must be rewarded by TCA methods, because if someone is providing this resource, it means that someone is not.

4.2.1.3 Integration of Devices and Systems

Smart grid is a topic whose technology and experience have been acquired for more than 10 years, although various challenges and issues need to be properly faced, as addressed by [72, 73]. New technological devices spread throughout the systems are treated mostly in isolated approaches. However, the integration is an evident aspect that emerges from the energy transition process. Figure 4.2 exposes the main fields involved with this process.

Based on a wide set of publications, the mains fields of this integration are presented in the following.

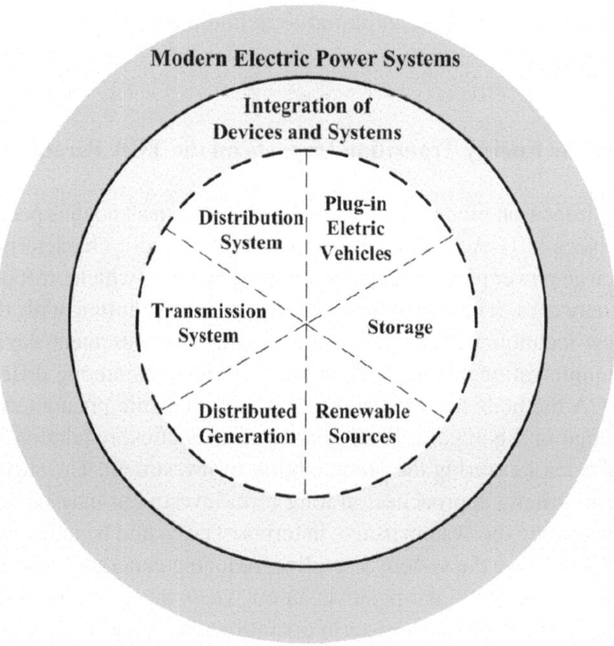

Fig. 4.2 Integrated fields of modern electric power systems

Plug-in Electric Vehicles and Storage

PEVs are a significant source of uncertainty because their departure time, location and load are unknowns. Thus, a stochastic model for day-ahead energy resource scheduling is proposed in [74], integrated with the dynamic electricity pricing for PEVs. Results show that this model is more effective than similar deterministic models. The impact of PEVs in market rules of the TSO is discussed in [61]. As the TSO priority is the security of supply, changes in its rules are challenging because the consequences are uncertain in relation to supply. However, the rules must be improved looking for sustainability and competitiveness. The energy storage development allows new applications and it naturally creates practical challenges, such as the proper usage in distribution networks to overcome voltage unbalances due to PEV and RES penetration [75]. Regulation rules are a frequent barrier to the storage development in several countries [76]. The most relevant regulatory barrier is the classification as a generation asset, and not as a complementary source that provides flexibility to the system. To decrease barriers, a market structure that values the storage as a complementary element of flexibility is necessary. Then, there is still an implementation issue in TCA design of how to minimize the unavoidable effect of distortion of the long-term charges on the reaction of these flexible assets over short-term prices. Another issue that must be faced is the development of an efficient manner of charging a storage system connected in the transmission level with tariffs for generators and loads. It remains an open issue with great importance.

Renewable Sources and Distributed Generation

The need of flexible resources, like storage and DR, arises with large amounts of RES, as the development of more integrated approaches to address the interdependence between technical and institutional elements to achieve a main policy goal for shaping the energy system of the future [77]. Concerning to the dispatchable photovoltaic units, their increase causes the lack of distribution network capacity in some systems [78]. Thus, distribution systems must be updated with the most advanced technologies as much as possible. The batteries' lifetime should be investigated for the implementation of effective operating tools [79]. Schemes that integrate RES into power sector are a controversial issue. The Feed-in Tariff (FiT) scheme needs further investigations to improve its rules in order to decrease the integration costs and to avoid the subsidy dependent pathway [80]. In [81], the authors present state-of-the-art DG deployment methodologies, discussing two specific questions: the optimal location for DG installation and its optimal capacity. Reference [82] reviews the planning of DG in distribution networks, pointing out that artificial intelligence techniques are very suitable for optimal planning of DG. In [83], a local market with agents trading electricity directly among each other is presented. From the analysis, a peer-to-peer market with intelligent agents shows to be the most advantageous.

Transmission and Distribution Systems

In [84], an integrated charging mechanism between transmission and distribution is proposed. New investments are divided in four groups: technical innovation, sustainability, services promotion, and others. The mechanism seeks to ensure sustainable modernization of the Chinese power system, which requires substantial investments. This mechanism is analysed in [85] through a dynamic approach to optimize investment policies on the electrical network of a Chinese city, practical recommendations are accomplished. In [50], a planning model for the distribution network expansion is proposed, which employs transmission system usage tariff information, in a coupled way. This integrated model minimizes system charging while expanding the distribution system. An economic and technical model that combines the transportation and distribution processes of energy is presented in [51]. This model may be employed in the wholesale or retail market, demanding complex measurement systems and high processing capacity. At last, a new method to obtain coordinated economic dispatch between transmission and distribution systems is proposed in [86]. The method employs LMP sensitivities to enhance convergence, and it may be very useful to settle problems caused by great participation of DG on networks.

4.2.1.4 How the Power Sector Integration Trend May Be Captured by TCA Methods

It is essential that economic incentives noticed by individual agents lead to the global optimum. To do this, the regulation needs to provide adequate incentives for individuals, allowing them to make their own decentralized decisions, but moving the system towards the global optimum. To achieve this purpose, an important aspect that may be improved by TCA methods is the distribution network modelling. It must be made because the major technological changes have occurred in distribution level and these changes reverberate across the transmission system. Traditionally, the greatest focus has been given to generation due to the possibility of generators plan their location depending on tariffs, which is less probable for the loads. Therefore, it is necessary to assess better the impact on distribution due to transmission tariffs, and the opposite way as well. Stability is an important attribute of TCA methods to planning activities. However, when the coupling between transmission tariffs and distribution planning is desired, tariffs must not be overly stable as they should respond to dynamic load changes in distribution.

To capture the integration trend, TCA methods must refine their approaches in other important aspects. First, the slack bus definition is critical for indicating which generators will response the load changes, whose dynamics are increasingly unpredictable and faster. In terms of the generation dispatch used to calculate transmission tariffs, it is essential to establish a clear criterion. We believe that the dispatch must prioritize the reduction of electrical losses into the transmission, which provides an overall gain to the system. Moreover, generators who contribute to this reduction must be valued through discounts on transmission tariffs. Another important aspect to capture the integration trend is the employment of quite representative models of the transmission system, which allows to obtain realistic power flows. For instance, PEVs may behave as a load or as a generation for the power system. They may still be in different geographical locations over time, connecting in distinct electrical points of the system. Therefore, it is critical for the accurate modelling of new technological elements as described.

4.2.1.5 Outlook over Selected Transmission Systems

We may notice that the energy transition process affects the transmission systems in different ways, because each one has particular characteristics. Table 4.1 approaches some selected transmission systems and their main issues, as described in distinct publications.

From the Table 4.1, it may be verified that some issues are common, even with the particularities, such as: power flow congestion, presence of RES, cost adjustments, and regulatory discussions. One general aspect is the necessary evolution of actual

Table 4.1 Main issues of the selected transmission systems

System region	Main issues
Australia	Nowadays: a restructuring process has been tackled Transmission pricing: an important theme [87]
Brazil	Hydroelectric resources: far from the load centres Transmission costs: relevant to investments [88, 89]
Central Europe	Wind and solar power growth: affects considerably the transmission networks Transmission grid: faces congestion troubles [90]
China	Long distance transmission: fundamental to pricing Trans-regional power transmission: multiple interests and controversial applied technology [91, 92]
India	Inter-state transmission system: evolution of the employed cost allocation method Targets: to reduce congestion and losses [93]
Mexico	Regulation: guarantees the efficient usage of the grid Not solved issue: how to promote effective investment in transmission expansion [94]
South America	Transmission regulatory rules are being analysed in: Argentina, Brazil, Chile, Colombia, and Peru Three topics are discussed: network expansion, transmission remuneration, and cost allocation [95]
United States of America	DG: reduces the utility revenue but it does not necessarily decrease the costs Trend: growth of the costs with *kWh* energy sales decline, if the recovery is based on *kWh* sales [96]

rules in all systems. Finally, the selected regions represent broad geographic areas with physical barriers, which increase the importance of appropriate transmission strategies.

4.2.2 *Transmission Fundaments*

The power sector is subject to a set of principles, requirements, and laws that drives both the physical features of electricity and the fulfilment of stakeholder expectations. This set, together with other practical implications, limits the number of possibilities in which this sector may be regulated. Therefore, before addressing TCA methods, the core of this chapter, a brief characterization about network economics and regulation is realized based on [97–99].

4.2.2.1 Economics and Regulation

Transmission network is the key element to guarantee competition conditions in electricity markets. The network access and usage must be regulated by an equitable and non-discriminatory set of rules. In addition, costs incurred in the construction, maintenance, and operation of the network must be shared fairly post. Regulation may be understood as a set of rules employed to control, drive, or manage an activity, institution, or system. Therefore, the main purpose is to modify the outcome that would be obtained if the human beings were authorized to interact freely. It prevents inefficient results in distinct places and time frames which otherwise may occur. In terms of performance, regulation must guide the industry towards improving the collective benefit perceived by its participants. It is important to emphasize that regulation is not the only manner to protect investors and consumers. Their interests are advocated by courts, which uses legislation and laws to judge the actions. The difference is that regulation constitutes an ex-ante action, whereas a judicial action is ex-post.

Considering the regulatory spectrum, two questions may be made to guide the actions. Which criteria must be employed to specify when one outcome is better than another? Moreover, which is the most efficient strategy to reach the aimed purpose? To answer these questions, three basic elements must guide the regulatory framework. First, the employed rules design, to drive behaviours towards the objectives determined by the regulator. Second, the power industry structure, which must have a sufficient number of similar competitors. Third, the supervision of behaviour, which involves the monitoring of possible infractions and the constant revision of rules to ensure their effectiveness. The growth of renewable sources with their intrinsic uncertainty, often located far from load centres, is taking the transmission regulation to its limits. Total transmission costs must be driven by investment costs, because maintenance and operation are nearly proportional to the volume of assets. On reinforcements, their costs must be borne by those for whom they were built. Therefore, there is a close relationship between network cost allocation and network planning.

The employment of several transmission lines in parallel to connect two distinct locations is not an economical practice. The economies of scale impute relevant implications to the grid. It is not rational for a competition among several small networks when power may be transmitted over a single large line in a cheaper way, as long as a significant part of the capacity is used. Transmission is a service that could not be effectively provided by competing firms. Line construction may be assigned via a competitive auction, but after built just the winner firm may provide the transmission service between two buses. In short, it represents a natural monopoly. Another distinguished transmission feature is that the lines are forever. Networks are extended and reinforced, however transmission lines are rarely dismantled. Finally, the burden of transmission investments and costs in the overall electricity bill may be substantially distinct, according to the particularities of a specific power system.

4.2.2.2 Principles to Allocate the Costs

Investment in a new network facility may just be justified if its cost is less than the aggregated benefits provided for the users. Transmission expansion is necessary since the generation and load are getting more connected, and it is justifiable because the benefits for all agents exceed the transmission investment cost. Thus, the cost allocation to beneficiaries may be approached as a sub-product of the network expansion planning. The allowed annual revenues that transmission firms receive are established by the regulator. So, network users are charged with a transmission tariff.

All transmission costs must be recovered from the charges paid by users. They are constituted by network investment costs, operation and maintenance costs, and other administrative costs related to transmission business. Line losses and grid constraint costs constitute in generation costs. Thus, these costs, besides system operator costs and ancillary service costs, should be collected from system users through other charges. Transmission charges may be separated into connection charges and system usage charges. The last is the focus of this chapter. Typically, each new user who is being connected pays for its access into transmission facilities, whereas system usage charges must cover the remaining costs of the network.

The process of allocating transmission network costs among users must obey a set of principles and requirements that arise as a combination of microeconomic theory, regulatory practice, power systems engineering, and technological context. In the following, four high level principles are presented, based on [97], which we consider a powerful reference in relation to this content. These principles, together with three additional requirements, form the guide used to evaluate several TCA methods proposed by different articles.

Principle 1: Costs in Proportion to Benefits

The responsibility of each user in grid investment must be the key concept of any TCA method, even if it is hard to implement. Though fairness is important, consistent incentives for investments are the main reason for embracing this principle. Additionally, transmission tariffs must play the useful role of sending proper locational signals for the agents. Thus, the total network investment cost may be minimized through optimum decisions of siting. To gain economies of scale, new facilities are built with capacities above what will be needed in their early years. make that new facilities be constructed with capacities above what is necessary in their initial years. Therefore, the total asset cost must not be charged to the current users, who may not need the entire capacity. This cost must consider the entire lifetime of the asset and the respective users over the total period.

Principle 2: Independence of Commercial Transactions

Transmission tariffs must not depend on commercial transactions that occur across the users. Tariffs must be charged to those who benefit from the presence of a facility, regardless of business relationships. The resulting bilateral trades will be all different towards distinct scenarios, since generators and loads may buy or sell from different players each time. But in the end, the loads will be supplied and the cheaper generators will be dispatched. Therefore, this context leads to the same pattern of flows through the grid. Thus, there is no justification to discriminate the users according to commercial transactions. Instead, tariffs must depend on location of users within the system topology and on the pattern of power injections over time. When tariffs fail to be independent of commercial transactions, it may cause an accumulation of transmission charges. Such practice tends to strangle trade and to prevent buyers from accessing cheaper sellers.

Principle 3: Establishment of Tariffs Ex-Ante

The transmission tariff for a new user must be specified ex-ante and not updated, at least not for a considerable period of time. Ideally, when a new agent requests connection to a specific point of the grid, the operator must provide the transmission tariffs to be charged from the agent for approximately 10 years ahead. The tariff trajectories for potential new agents applying for connection at the year X must not be modified during the next 10 years for these agents. However, with additional information over the year X, the tariff trajectories announced at the year $X + 1$ and applied for new entrant agents during this year ($X + 1$) must be different, that is, updated.

Principle 4: Adoption of Appropriate Tariff Formats

In the transmission tariff design, two issues must be clearly distinguished: the amount paid by each agent and the specific format of the tariff. The format may present a volumetric characteristic (*$/MWh*), a capacity characteristic (*$/MW*), a fixed characteristic (*$*), or to be a combination of them. The volumetric tariff to generators represents an added component of their variable production costs. So, it distorts their efficiency in the wholesale market and it increases the short-term electricity price. The capacity tariff is the appropriate format, since the generators consider it as an additional fixed cost for investors in new facilities.

4.2.2.3 Requirements to Develop Algorithms

In addition to the presented principles that must guide the development of new approaches to deal with the TCA theme, the algorithms must consider some practical requirements. They are associated to the energy transition process ongoing throughout the power sector and to general regulatory basis. Meticulous description of several issues in the current energy context may be found in [16, 17].

Requirement 1: Deployment of Effective Benefits and Costs

The algorithms must be developed in a proper way that allows the clear perception of incentives and costs by the agents. This requirement is a challenge to the algorithms, since the agents need to be induced to make individual decisions that lead the power system to its global optimum, in terms of operational resources and grid investments. That is, the agents must receive locational signals that incentive them to exploit all system gaps, conditioned to operational, market, and system security constraints. A relevant aspect that must be considered by algorithms is the flexibility, a resource that is becoming increasingly scarce with the growth of RES. The transmission tariffs must encourage the agents to contribute to increase this important operational resource.

Requirement 2: Provision of Predictable Signals

The agents must also receive price signals that allow to plan the locations of future facilities and the annualized fixed costs. Predictable signals do not necessarily mean uniform signals, but stable outcomes that permit an accurate planning. This predictability characteristic is particularly important in a context with the growth of uncertainties across the whole power sector. Predicable tariffs are essential for the decision of which places are advantageous to install wind and solar generation, which has several candidate points. Thus, algorithms need to provide tariffs that contribute to the overall uncertainty reduction in the power sector, but they must not suppress the locational characteristics of the previous requirement. Such set of requirements evidences the complexity of the TCA topic.

Requirement 3: Fulfilment of General Regulatory Basis

Finally, the main regulatory fundaments must be observed and followed by the algorithms, because the transmission system is considered a natural monopoly, and therefore it need to be designed in this regard. As a consensus, the major fundaments are:

- economic sustainability;
- economic efficiency;
- transparency;
- simplicity, as far as possible;
- equity or non-discriminatory access;
- stability to reduce the regulatory uncertainty and;
- consistency with liberalization and with the regulatory framework of each region.

4.2.2.4 The Ideal TCA Method

The employment of fundamental principles and requirements in a practical TCA method is not a trivial task. A great difficulty is to characterize the benefits of each agent associated with the transmission network. Traditionally the strategy has been to emphasize the loading of transmission lines without exploring in detail the dynamic changes at load and generation buses. The ex-ante tariff publication, which allows an agent plans its future location, is the most difficult principle to apply. Two reasons may be identified for this. First, since the theme is very complex and controversial, usually other theoretical and practical issues emerge as the focus. Second, the theme is rarely addressed together with the transmission expansion planning. It may be noted over the years an undesirable decoupling between transmission expansion and transmission pricing. These fields must follow the current integration trend that it is observed across the power sector.

4.3 Transmission Cost Allocation Methods: Review and Analysis

About modern pricing methods for transmission systems, the cost allocation may be identified as the major theme and the focus of this chapter. Many different strategies may be employed to discriminate the total cost of transmission across the agents in a fair and viable manner.

Table 4.2 Categories of TCA methods

Non-power flow based	Postage-stamp or Pro-rata	
	Contract path	
	MW-mile	
Power flow based	Distribution factors	Direct current (DC)
		Generation shift
		Generalized generation
		Generalized load
	Tracing algorithms	Bialek's method
		Kirschen's method
	Power flow comparisons	Marginal
		Incremental
		Aggregated
	Current based	Current conjugate
		Cross term
	Alternating current (AC) flow sensitivity indices	Line utilization factors
		Reactive power adjustment factor
	Full AC power flow solutions	
	Power flow decomposition	
Incremental cost	Standard long-run incremental cost	
	Long-run fully incremental	
Marginal cost	Long-run marginal cost	Investment cost-related pricing
		DC load flow pricing
	Corrected short-run marginal cost	
Alternative strategies	Participation factors	
	Benefit factors	
	Cooperative game	
	MVA-mile methodology	
	Equivalent bilateral exchange	
	Nodal method	
	Zbus method	

These TCA methods have been divided into an extensive list of distinct categories, depending on the reference adopted [14, 16, 97, 99–103]. A comprehensive set classifying the TCA methods in categories is presented in Table 4.2.

It may be noted from Table 4.2, that each category possesses various different mathematical techniques. This broad set of techniques demonstrates the specific characteristic of the TCA theme, which depends on local particularities of each power system, such as regulatory framework, political interests, economic targets, geographical barriers, and technological limitations.

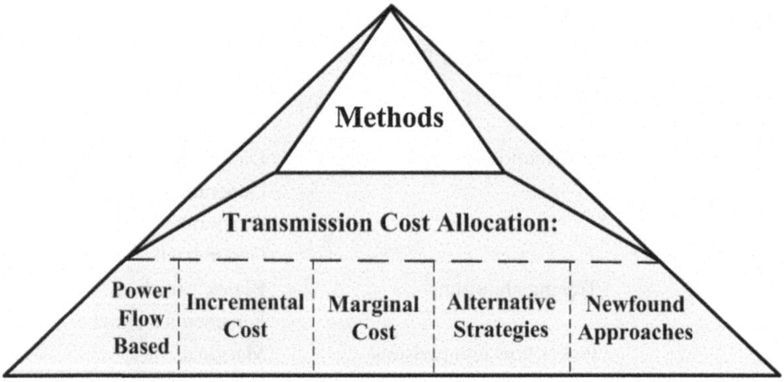

Fig. 4.3 TCA categories used to organize the relevant publications

4.3.1 Relevant Publications

In this chapter, to organize the analysis carried out on a wide range of relevant publications found in the literature, we subdivided the published TCA methods in five categories: power flow based, incremental cost, marginal cost, alternative strategies, and newfound approaches. Figure 4.3 illustrates these categories.

To classify the evaluated publications into one of the five used TCA categories, we adopted the following statements. Power flow based methods examine the transacted power quantity and the electrical distance between the source and the sink point. In incremental cost methods, agents pay the full cost of new facilities required by the transaction. The charges are evaluated considering the new transmission costs caused by them. The marginal cost category is characterized by focusing on the distinctions across nodal prices that arise due to transmission constraints in the grid. Alternative strategies form the group composed by different methods developed from other categories, modifying or mixing some characteristic. Finally, newfound approaches constitute the most innovative publications proposed to allocate the transmission costs.

4.3.1.1 Power Flow Based

A cross-regional TCA technique that considers reactive power flows is presented in [104]. It is based on the share of usage by trade, being applicable to power markets with bilateral trades and pool. In [105], a transmission pricing scheme employing a tracing method is proposed. Transmission fixed, congestion and loss cost are regarded in the scheme, and results demonstrate that the tracing method may be a fairly way to calculate them. In [106], a methodology based on power transfer distribution factors is presented. These factors provide topological sensitivity to the methodology and results show that the charges are influenced by location of buses, network connectivity, and proximity to generation.

4.3.1.2 Incremental Cost

A transmission charging methodology based on long-run incremental cost (LRIC) is presented in [107]. It recognizes the trade-offs between short-run congestion cost and future investment cost, attributing positive tariffs for charges and negative ones for rewards in congestion areas. Thus, the methodology may decrease congestion and network investment costs. The article [108] proposes an approach based on a min-max optimization technique. It provides less-dispersed tariffs seeking to contribute to a safer environment for investors in generation. A TCA method that embraces scheduled incremental cost and unscheduled interchange cost components is illustrated in [109]. The scheduled component is used to recover the incremental cost for extra MVA flow, whereas the unscheduled component is designed for monitoring, protecting, and retaining the grid discipline.

4.3.1.3 Marginal Cost

In [110], an approach for the TCA problem is developed with fundaments of long-run marginal cost (LRMC) and nodal exchange factors. A bi-level model is formulated and results indicate that the locational tariffs and the congestion level may be determined under distinct load conditions. A zoning algorithm is proposed in [111], aggregating nodal tariffs to constitute tariff zones. It ensures zonal tariff diversification and aggregates the nodal usage tariffs geographically through a weighted average approach. Reference [112] shows that variants of marginal participation approach may not be effective to allocate the costs. Modifications to the min-max fair marginal participation approach are tackled providing a satisfactory allocation. A methodology that divides a power system into a set of zones, considering transmission usage and transmission loss tariffs, is presented in [113]. It enables to illustrate the results on a unique zonal tariff map, indicating the reinforcement requirements to the network.

4.3.1.4 Alternative Strategies

A novel approach based on an economic principle to allocate fixed cost of transmission systems is presented in [114]. It uses the critical capacity concept of a line and takes into account the congestion to share the transmission revenue requirement that must be provided by each agent. This approach is most suitable with congested systems. A min-max fair tracing algorithm to allocate the transmission usage cost is introduced in [115]. The algorithm realizes a non-iterative procedure to min-max fair price determination, handling with numerical problems expected in large systems. The authors in [116] approach the TCA problem at large systems with multiple interconnected regions. A new multi-area decoupled scheme is developed, allowing that each region performs its own cost allocation from its chosen rules. An economic strategy focused on user benefits is used in [117] to allocate transmission costs. It provides a fair cost allocation to those producers and consumers that utilize the network, proportionally to transmission benefits. In [118], a method based on agent responsibilities of the network usage in terms of physical and economic resources is proposed. It employs the resolution mechanisms of cooperative game theory to obtain a fair and effective cost allocation. Its weakness is the game dimension, which increases exponentially with the number of agents. Game-theoretic models taking into account the Nucleolus and Shapley value techniques are discussed in [119] to allocate transmission costs between generators and loads. The employed techniques provide stable and unbiased outcomes to participants. A min-max fairness policy is applied to choose an appropriate economic slack bus in a cooperative game theoretic approach formulated to allocate transmission costs [120]. Results demonstrate a fair characteristic of the approach, leading to an equilibrium price vector. The authors in [121] introduce a novel method that uses the equivalent bilateral exchanges principle. This guarantees the slack bus independence, counter-flows recognition, and transmission usage charges always positive. A new technique to allocate transmission costs in markets with interconnected regions is presented in [122]. The technique aims for improving the nodal methodology in certain aspects, and it permits the definition of zonal tariffs. Reference [123] develops a novel methodology to TCA at systems with high participation of RES. It is based on the equivalent bilateral exchange and intends to provide tariffs with low dispersion. The TCA problem is dealt with a procedure based on circuit theory in [124], which considers the physical network usage. It allows a simple implementation and behaves similarly to the underlying electrical laws employed to derive it. A new technique is presented in [125] to allocate transmission fixed costs taking into account the power flow equation and modifying the network impedance matrix. It is an independent slack bus technique, which provides a fair estimation based on the current contribution of each agent. An approach based on circuit theory that employs equal sharing principle and orthogonal projections is developed in [126]. Power flow contributions are identified through employing game theory solutions, and a concept of effective contribution is defined. The approach provides results little influenceable by counter-flows and with locational characteristics. A method to allocate transmission cost based on a set of line utility factors is presented in [127]. The method evaluates the power transaction impact on each network element through the relation between line flows with generators and loads. Transaction costs are allocated to the participants proportionally to the power flow ratio occasioned by each participant.

4.3.1.5 Newfound Approaches

In [128], a new method, is presented, that decomposes transmission costs into two components: one associated with the used network capacity and the other related to the unused capacity available. Flexibility is its main feature because it allows the use of economic and technical criteria at single or interconnected systems. A methodology based on real power tracing is developed to obtain ex-ante point-of-connection (POC) rates for decentralized market agents [129]. POC charges consider the network usage, and they recover sunk costs, providing price signals. An approach to allocate transmission costs by controlling the nodal prices of electricity is introduced in [130], through generation injections and nodal penalties into the classic economic dispatch. As a result, the approach charges more from intensive system users and provides credits to agents that contribute to alleviate the network usage. Distinct schemes are suggested in [131] to deal with the TCA problem in a deregulated environment based on different TSO visions for generation and load. The schemes encourage the efficient contribution of users to recover the network costs. A comprehensive design of transmission charges is developed in [132] to retrieve the regulated network costs. The charges aim for encouraging the users to internalize transmission costs in local decisions, interfering the minimum as possible in short-term behaviours, because this must be made for regulatory mechanisms in the operational time range. In [133], a method based on the magnetic field caused by currents is proposed. The electro-magnetic field extension is identified and employed to determine the network usage. A drawback is the need to calculate the magnetic induction intensity, what is a complex task. A new technique based on load following and correlation factors is presented in [134]. The technique estimates the transmission cost of each agent before its entrance in the market, and it takes in account the reactive loads in its framework. The slack bus dependence is its major drawback.

4.3.2 Literature: Broad Findings

In the literature, there is no consensus about the best TCA method. The evaluation criteria are highly dependent on specific local characteristics and particular interests. Concerned to the number of recent and relevant publications in the area of TCA, it may be noted a low quantity. It happens because currently the publications are focused on energy price advances in modern electricity markets. Losses and congestion are issues already considered by the nodal pricing, the most sophisticated expression to form nodal prices of energy, also called spot prices and LMPs. Therefore, most of the TCA methods do not consider these two issues. As the transmission portion does not represent a very high percentage of the electricity bill in many countries, regulatory agencies do not prioritize the modernization of TCA methods. However, this area is becoming more important nowadays with DG, since the cost of transmission charges may be decisive for the location of new facilities. Furthermore, many publications emphasize economic aspects and neglect operational aspects of electric power systems. A strategy of maximizing the use of an asset to extract the maximum benefits from facilities is not adequate because it decreases their lifetime and hazards the whole system reliability. The physical particularities on power systems must always be respected.

4.3.3 Publications with the Most Suitable Features

Different publications containing methods to specify the transmission network usage by stakeholders have been developed and applied around the world. Their strategies vary considerably due to distinct purposes by the regulators, and due to particular features on power systems in terms of political, economic, geographic, and electrical issues. However, the final TCA purpose is to specify the agent responsibility in grid investment. To achieve this purpose, the most suitable TCA publications are indicated in the following. They were judged as such because they meet the greatest number of principles and requirements previously exposed.

The TCA method presented in [132] provides locational tariffs while making temporal considerations. Thus, tariffs of new network users are computed before they take part in the system, and that provides the possibility of deciding their expansion investments. Another feature is that the transmission tariffs are not able to recover the total cost of the grid. The not-recovered fraction should be socialized among the load, according to the authors. Power flow quantities are attributed to each generator and load responsible for them. For this, the employment of an algorithm based on network electric utilization, such as average participations, which is used on the article, is recommended. This algorithm has the assumption that power inflows into a bus contribute to the outflows from the bus proportionally to the volume (of the later). The need of transmission grid reinforcement is evaluated by the method associating incremental changes in the flows with generator and load activities. The attribution of responsibility about incremental flows created by new generators and loads is a tough task that is introduced by the method. The principle adopted to deal with this task is intuitive and with a simple mathematical framework. The first step in the method is to specify which fraction of the cost concerned to each line is allocated according to cost-causality principles. After, the cost of each fraction that is used in the lines is divided into a portion to be paid by generators and the remaining portion to be paid by loads. These portions may be defined based on aggregated incremental flows that are expected to be produced by each type of agent. The discussion of several distinct issues is dealt with thoroughly and didactically in [132] and is a useful guide. In terms of practical considerations, broad ideas and strategies are provided by this reference. The great merit of the method is to consider an expressive set of requirements in its framework.

In [109], a TCA approach comprised by two components is proposed to keep the grid discipline in terms of unscheduled power flows. The first component, scheduled incremental cost (SIC), is allocated to transmission network users according to their injections or withdrawals. This component is associated with the planned load growth rate. Whereas the second one, unscheduled interchange cost (UIC), deals with the unscheduled interchange in load or generation that may emerge with high uncertainty. Incremental active and reactive power flow is employed to assess and to allocate these components across the agents. An innovative element employed by the article is the generator participation factor concept. It is used to determine the scheduled incremental generation to supply the load growth rate. The SIC component is formulated to capture the future investment in transmission grid reinforcement. It corresponds to a long-term signal which is strongly coupled with the fundamental principle of allocation, wherein the responsibility of network users in grid investment must be the conceptual basis. This component may be added to the nominal transmission tariff as an incremental cost component. On the other hand, the UIC component aims to impose a heavy penalty on users since unscheduled power flows constitute the main problem in some countries, according to the authors. So, this component intends to discipline the users, increasing the grid predictability. The UIC corresponds to the variation between the present value and the future value that must be invested on a component, constituting an additional charge to users. Positive UIC values represent a penalty to be imposed on undisciplined users, whereas negative values constitute a discount to disciplined users. The equations presented in the article are simple and direct without complex steps. A relevant aspect of the method is the chosen operational scenario and the reference values to be used in the analysis, as they will strongly influence the results. The approach considers both active and reactive power flows that become the model rather representative in terms of operational issues.

A TCA method that distinguishes generation and load technologies based on LRIC pricing is proposed by [107]. The method provides positive transmission tariffs for agents who contribute to the congestion (loads) and negative tariffs for those who decrease the congestion (generators) in critical areas. The proposal distinguishes generation technologies and the tariffs are updated based on changes in generation mix. So, the proposed method intends to induce a proper generation behaviour, reducing congestion and postponing transmission expansion cost. The growth of generation cost due to transmission limits is defined as the congestion cost, which is used as a trigger to transmission network investment. Congestion management may be considered a better alternative until its cost exceeds the annualized network investment cost. The employed equations are straight and the mathematical framework is perfectly understandable and based on present values of annualized investment cost. The main contribution of this article is the development of distinct tariffs to reduce network congestion. Although this mechanism considers short-term conditions, which is not the usual framework that takes into account long-term variables, it may be useful for an electrical energy system. Moreover than reducing the congestion, it may increase system discipline and generation flexibility through tariff distinction. In a context of deep insertion of RES, this kind of mechanism must be deployed and evaluated because discipline and flexibility are scarce resources in an environment with high presence of generation uncertainty. Although the network model used is linear, the insights generated by this article are very expressive. The charging philosophy used to define costs and benefits among the agents is adherent with the main principle that considers the responsibility of each user in grid investment.

In [124], a physically based network usage method is proposed to allocate the transmission costs. The TCA method consists of three direct steps. First, the active power flow of any transmission line is mathematically split and associated with nodal currents. After, the line cost is allocated to all loads and generators. Finally, this procedure is repeated for all transmission lines. The formulated proposal is based on the impedance matrix of the system with a detailed model that includes the actual shunt admittances. An outstanding feature highlighted by the authors is the proximity effect provided by the method. This means that generators and loads use more intensively the lines electrically close to them, what is supported by the circuit theory. Two other features are the slack bus independence and the fact that is not necessary to split costs between generators and loads before running the process. To process the method firstly it is required to run a full power flow to determine the flow directions and the current injections in all buses. The equations are based on electrical features of the system with complete grid representation. The math development involves simple manipulations that allow direct understanding. The model may represent generation and load values on the same bus. It is a contributory feature because it gives flexibility and representativeness to the method. Thus, independent rankings (to generators and loads) may be created to be used in distinct applications, such as locations to new agents, capacity evaluation of assets, and network expansion planning.

In [125], it is proposed a TCA method based on power flow equation in a math framework very similar to [124]. Active power flows are expressed in terms of generator currents and load currents through power invariant matrixes that represent modifications of the impedance matrix. The theoretical basis is quite complete and provides a guide with the necessary stages to generate the equations employed by the method. As in [124], the method published in [125] uses a π equivalent model to represent transmission lines. Through manipulations, a coefficient that measures the electrical distance between a specific line and each bus is obtained. Firstly, the power into the lines is calculated in terms of generator currents, defining the total load current as the sum of all single load currents. After, generator currents are expressed in terms of load currents, using information from a power flow solution. As result, two conversion matrixes are obtained with invariant properties, which are used to calculate the real power flows. Specifically, on the calculation process, the following variables are used: bus voltage, electrical distance coefficient, conversion matrixes, and injection currents. In short, transmission costs are calculated considering power flows into the lines and the contribution of each load and generator in the line usage. A flexible feature of the proposal is the possibility of assigning real power flows only for load buses or generator buses. The flows may be still allocated considering simultaneously a 50–50% proportion between generators and loads. Finally, another interesting feature is the independence of a slack bus, which avoids the controversial issues on this topic.

A TCA approach based on the principle of equivalent bilateral exchanges (EBE) is presented in [121]. According to this principle each load is attributed to a fraction of each generation, and analogously each generator is attributed to a fraction of each generation uniformly. The EBE method constitutes in a flow-based network cost allocation with two major features. Firstly, a slack bus does not need to be defined. Secondly, counter-flows are allowed, which limits the volatility of charge outcomes. This method uses a direct current load flow model to emphasize clarity and concision on the work. However, the framework may employ more representative load flow models. With vectors containing numerical results obtained from an optimal power flow, the EBE principle may be applied. The equivalent bilateral power exchange (GD) is defined as the portion of a generation that supplies a specific load bus. Thus, each individual generator and load may be decomposed into linear combinations of EBE. With this, the effect of GD on power flows is determined by an operation that is independent of the slack bus and satisfies the Kirchhoff laws. The equations employed are simple and direct, characterizing the method as a suitable proposal to diverse applications. Three network usage properties that may be derived from the EBE principle: generator and loads at the same bus does not use the network; nonzero GDs involving different buses affect all line flows (except in some radial subnetworks); and even if the power flow in a line is zero, the usage of this line is nonzero. The last property is particularly important to provide stability to the method. The method produces tariffs with low volatility comparatively with other approaches, but the locational characteristic is kept. Finally, the results indicate great adherence to the fundamental principle of allocating costs proportionally to agent benefits.

4.4 Conclusions and Future Directions

It is increasingly difficult to quantify the economic benefits that each transmission facility provides to each user of the network. Therefore, actual TCA methods need to be improved and new approaches developed based on some good insights, which may be drawn from the discussed literature, and also based on the fundaments indicated in this chapter. New approaches must be evaluated considering the particularities of each application obviously. But they must capture the current dynamic processes that the power sector is going through, and they should also induce the agents to the optimal system usage as a way to fully exploit its capabilities, retarding the expansions. We may say that suitable methods to allocate the transmission costs are straight and without a mathematical framework that depends on arbitrary and subjective decisions.

With regard to decision-making processes, it is clear that decisions are becoming increasingly scattered, but coordinated. Therefore, TCA methods must have clear and fair strategies with robust approaches that prevent the contamination of results caused by specific factors, such as the slack bus choice and the simulation scenario. All the modifications, imposed by the energy transition process, force the TCA methods to be inserted into a more integrated and shorter-term planning. Integrated, as they must be part of a systemic planning, where distribution networks consider the TCA charges in their load management, through switching feeders, and where RES consider the locations with cheaper TCA charges to install new units. About the shorter-term planning, faster dynamics that arise from the energy transition force shorter-planning horizons and more frequent long-term planning updates.

Finally, the electricity consumption may not be represented anymore by a static amount. The consumption behaviour needs to be incorporated into TCA models. Its evolution along the time (historical data) must be considered and the response action (demand response) must be evaluated. This kind of modelling will allow that TCA methods form fair tariffs for individual agents, according to the costs and benefits that they cause to the system. With this, individual behaviours will be induced in a way that leads the power system to a better use of its global resources. As the result, transmission system gaps will be reduced, avoiding unnecessary expansions of facilities and optimizing the existing resources.

References

1. C. Defeuilley, Energy transition and the future(s) of the electricity sector. Util. Policy **57**, 97–105 (2019)
2. R. York, S.E. Bell, Energy transitions or additions? Why a transition from fossil fuels requires more than the growth of renewable energy. Energy Res. Soc. Sci. **51**, 40–43 (2019)
3. P. Guo, J. Kong, Y. Guo, X. Liu, Identifying the influencing factors of the sustainable energy transitions in China. J. Clean. Prod. **215**, 757–766 (2019)
4. B. Chen, R. Xiong, H. Li, Q. Sun, J. Yang, Pathways for sustainable energy transition. J. Clean. Prod. **228**, 1564–1571 (2019)
5. B. Batinge, J.K. Musango, A.C. Brent, Sustainable energy transition framework for unmet electricity markets. Energy Policy **129**, 1090–1099 (2019)
6. P.M. De Quevedo, G. Munoz-Delgado, J. Contreras, Impact of electric vehicles on the expansion planning of distribution systems considering renewable energy, storage, and charging stations. IEEE Trans. Smart Grid **10**, 794–804 (2019)
7. S.R. Sinsel, R.L. Riemke, V.H. Hoffmann, Challenges and solution technologies for the integration of variable renewable energy sources—a review. Renew. Energy **145**, 2271–2285 (2020)
8. H. Mehrjerdi, R. Hemmati, Coordination of vehicle-to-home and renewable capacity resources for energy management in resilience and self-healing building. Renew. Energy **146**, 568–579 (2020)
9. L. Han, T. Morstyn, M. McCulloch, Incentivizing prosumer coalitions with energy management using cooperative game theory. IEEE Trans. Power Syst. **34**, 303–313 (2019)
10. T. Li, M. Dong, Residential energy storage management with bidirectional energy control. IEEE Trans. Smart Grid **10**, 3596–3611 (2019)
11. A.G. Olabi, Circular economy and renewable energy. Energy **181**, 450–454 (2019)
12. E. Hooshmand, A. Rabiee, Robust model for optimal allocation of renewable energy sources, energy storage systems and demand response in distribution systems via information gap decision theory. IET Gener. Transm. Distrib. **13**, 511–520 (2019)
13. I.C. Gonzalez-Romero, S. Wogrin, T. Gomez, Proactive transmission expansion planning with storage considerations. Energy Strateg. Rev. **24**, 154–165 (2019)

14. M. Shahidehpour, H. Yamin, Z. Li, *Market Operations in Electric Power Systems: Forecasting, Scheduling, and Risk Management* (Wiley, New York, 2002)
15. D.R. Biggar, M.R. Hesamzadeh, *The Economics of Electricity Markets* (Wiley, Chichester, 2014)
16. P. Mäntysaari, *EU Electricity Trade Law* (Springer, Heidelberg, 2015)
17. IEA, *Re-powering Markets* (International Energy Agency, Paris, 2016)
18. S. Xiong, X. Ma, J. Ji, The impact of industrial structure efficiency on provincial industrial energy efficiency in China. J. Clean. Prod. **215**, 952–962 (2019)
19. M. Lynch, S. Nolan, M.T. Devine, M. O'Malley, The impacts of demand response participation in capacity markets. Appl. Energy **250**, 444–451 (2019)
20. A.M. Elshurafa, H.M. Farag, D.A. Hobbs, Blind spots in energy transition policy: case studies from Germany and USA. Energy Rep. **5**, 20–28 (2019)
21. B. Mohandes, M.S. El Moursi, N. Hatziargyriou, S. El Khatib, A review of power system flexibility with high penetration of renewables. IEEE Trans. Power Syst. **34**, 3140–3155 (2019)
22. T. Ahmed, A.A. Bhatti, Do power sector reforms affect electricity prices in selected Asian countries? Energy Policy **129**, 1253–1260 (2019)
23. K. Saxena, R. Bhakar, Impact of LRIC pricing and demand response on generation and transmission expansion planning. IET Gener. Transm. Distrib. **13**, 679–685 (2019)
24. O. Borne, K. Korte, Y. Perez, M. Petit, A. Purkus, Barriers to entry in frequency-regulation services markets: review of the status quo and options for improvements. Renew. Sust. Energ. Rev. **81**, 605–614 (2018)
25. R. Pérez Odeh, D. Watts, M. Negrete-Pincetic, Portfolio applications in electricity markets review: private investor and manager perspective trends. Renew. Sustain. Energy Rev. **81**, 192–204 (2018)
26. A. Shimbar, S.B. Ebrahimi, Political risk and valuation of renewable energy investments in developing countries. Renew. Energy **145**, 1325–1333 (2020)
27. J. Zhang, J. Jorgenson, T. Markel, K. Walkowicz, Value to the grid from managed charging based on California's high renewables study. IEEE Trans. Power Syst. **34**, 831–840 (2019)
28. E. Mengelkamp, J. Gärttner, K. Rock, S. Kessler, L. Orsini, C. Weinhardt, Designing microgrid energy markets: A case study: The Brooklyn Microgrid. Appl. Energy **210**, 870–880 (2018)
29. S. Zhang, P. Andrews-Speed, S. Li, To what extent will China's ongoing electricity market reforms assist the integration of renewable energy? Energy Policy **114**, 165–172 (2018)
30. W. Geffert, K. Strunk, Beyond net metering: a model for pricing services provided by and to distributed generation owners. Electr. J. **30**, 36–43 (2017)
31. Ö. Okur, N. Voulis, P. Heijnen, Z. Lukszo, Aggregator-mediated demand response: minimizing imbalances caused by uncertainty of solar generation. Appl. Energy **247**, 426–437 (2019)
32. J. Valdes, A.B. Poque González, L. Ramirez Camargo, M. Valin Fenández, Y. Masip Macia, W. Dorner, Industry, flexibility, and demand response: applying German energy transition lessons in Chile. Energy Res. Soc. Sci. **54**, 12–25 (2019)
33. M. Gaspari, A. Lorenzoni, The governance for distributed energy resources in the Italian electricity market: a driver for innovation? Renew. Sust. Energ. Rev. **82**, 3623–3632 (2018)
34. C. Wadström, E. Wittberg, G.S. Uddin, R. Jayasekera, Role of renewable energy on industrial output in Canada. Energy Econ. **81**, 626–638 (2019)
35. M. Keay, J. Rhys, D. Robinson, *Evolution of Global Electricity Markets* (Elsevier, Waltham, 2013)
36. A. Nikoobakht, J. Aghaei, M. Shafie-Khah, J.P.S. Catalão, Assessing increased flexibility of energy storage and demand response to accommodate a high penetration of renewable energy sources. IEEE Trans. Sustain. Energy **10**, 659–669 (2019)
37. M. Basu, Optimal generation scheduling of hydrothermal system with demand side management considering uncertainty and outage of renewable energy sources. Renew. Energy **146**, 530–542 (2020)

38. D.P. Brown, A. Eckert, H. Eckert, Electricity markets in transition: market distortions associated with retail price controls. Electr. J. **30**, 32–37 (2017)
39. M. Pesaran H.A, P.D. Huy, V.K. Ramachandaramurthy, A review of the optimal allocation of distributed generation: objectives, constraints, methods, and algorithms. Renew. Sust. Energ. Rev. **75**, 293–312 (2017)
40. A.K. Singh, S.K. Parida, A review on distributed generation allocation and planning in deregulated electricity market. Renew. Sust. Energ. Rev. **82**, 4132–4141 (2018)
41. M. Pourbehzadi, T. Niknam, J. Aghaei, G. Mokryani, M. Shafie-khah, J.P.S. Catalão, Optimal operation of hybrid AC/DC microgrids under uncertainty of renewable energy resources: a comprehensive review. Int. J. Electr. Power Energy Syst. **109**, 139–159 (2019)
42. H. Chahkandi Nejad, S. Tavakoli, N. Ghadimi, S. Korjani, S. Nojavan, H. Pashaei-Didani, Reliability based optimal allocation of distributed generations in transmission systems under demand response program. Electr. Power Syst. Res. **176**, 105952 (2019)
43. H. Wang, Z. Yan, X. Xu, K. He, Probabilistic power flow analysis of microgrid with renewable energy. Int. J. Electr. Power Energy Syst. **114**, 105393 (2020)
44. M. Emmanuel, R. Rayudu, Evolution of dispatchable photovoltaic system integration with the electric power network for smart grid applications: a review. Renew. Sust. Energ. Rev. **67**, 207–224 (2017)
45. X. Yang, C. Gu, F. Li, Network Pricing for Smart Grids Considering Customers' Diversified Contribution to System, in *International Conference on European Energy Market*, Dresden (2017)
46. P. Zhuang, H. Liang, M. Pomphrey, Stochastic multi-timescale energy management of greenhouses with renewable energy sources. IEEE Trans. Sustain. Energy **10**, 905–917 (2019)
47. I. Abdelmotteleb, T. Gómez, J.P. Chaves Ávila, J. Reneses, Designing efficient distribution network charges in the context of active customers. Appl. Energy **210**, 815–826 (2018)
48. R. Passey, N. Haghdadi, A. Bruce, I. MacGill, Designing more cost reflective electricity network tariffs with demand charges. Energy Policy **109**, 642–649 (2017)
49. O. Lutz, V. Olavarria, R. Hollinger, C. Wittwer, B. Koch, Dynamic Tariff Design for a Robust Smart Grid Concept: An Analysis of Global vs Local Incentives, in *Innovative Smart Grid Technologies Conference Europe*, IEEE, Torino (2017)
50. M.M. Santos, A.R. Abaide, M. Sperandio, Distribution networks expansion planning under the perspective of the locational transmission network use of system tariffs. Electr. Power Syst. Res. **128**, 123–133 (2015)
51. A. Pazderin, N. Mukhlynin, Technical and Economic Model of Energy Transmission and Distribution Based on the Smart Metering Technologies, in *International Symposium on Industrial Electronics*, IEEE, Edinburgh (2017)
52. S. Nojavan, R. Nourollahi, H. Pashaei-Didani, K. Zare, Uncertainty-based electricity procurement by retailer using robust optimization approach in the presence of demand response exchange. Int. J. Electr. Power Energy Syst. **105**, 237–248 (2019)
53. X. Fang, B.M. Hodge, E. Du, C. Kang, F. Li, Introducing uncertainty components in locational marginal prices for pricing wind power and load uncertainties. IEEE Trans. Power Syst. **34**, 2013–2024 (2019)
54. J. Zhu, Y. Yuan, W. Wang, Multi-stage active management of renewable-rich power distribution network to promote the renewable energy consumption and mitigate the system uncertainty. Int. J. Electr. Power Energy Syst. **111**, 436–446 (2019)
55. A. Ghasemi, S.S. Mortazavi, E. Mashhour, Integration of nodal hourly pricing in day-ahead SDC (smart distribution company) optimization framework to effectively activate demand response. Energy **86**, 649–660 (2015)
56. O. Zinaman, M. Miller, A. Adil, D. Arent, J. Cochran, R. Vora, S. Aggarwal, M. Bipath, C. Linvill, A. David, M. Futch, R. Kaufman, E.V. Arcos, J.M. Valenzuela, E. Martinot, D. Noll, M. Bazilian, R.K. Pillai, Power systems of the future. Electr. J. **28**, 113–126 (2015)
57. J. Wu, B. Zhang, Y. Jiang, P. Bie, H. Li, Chance-constrained stochastic congestion management of power systems considering uncertainty of wind power and demand side response. Int. J. Electr. Power Energy Syst. **107**, 703–714 (2019)

58. W. Huang, N. Zhang, J. Yang, Y. Wang, C. Kang, Optimal configuration planning of multi-energy systems considering distributed renewable energy. IEEE Trans. Smart Grid **10**, 1452–1464 (2019)
59. F. Roques, D. Finon, Adapting electricity markets to decarbonisation and security of supply objectives: toward a hybrid regime? Energy Policy **105**, 584–596 (2017)
60. N.C. Figueiredo, P.P. da Silva, D. Bunn, Weather and market specificities in the regional transmission of renewable energy price effects. Energy **114**, 188–200 (2016)
61. P. Codani, Y. Perez, M. Petit, Financial shortfall for electric vehicles: Economic impacts of transmission system operators market designs. Energy **113**, 422–431 (2016)
62. G. Fridgen, M. Kahlen, W. Ketter, A. Rieger, M. Thimmel, One rate does not fit all: an empirical analysis of electricity tariffs for residential microgrids. Appl. Energy **210**, 800–814 (2018)
63. F. Hinz, M. Schmidt, D. Möst, Regional distribution effects of different electricity network tariff designs with a distributed generation structure: the case of Germany. Energy Policy **113**, 97–111 (2018)
64. Q. Wang, C. Zhang, Y. Ding, G. Xydis, J. Wang, J. Østergaard, Review of real-time electricity markets for integrating distributed energy resources and demand response. Appl. Energy **138**, 695–706 (2015)
65. R.C. Duthu, T.H. Bradley, An evaluation of customer-optimized distributed generation in New England utility and real-time markets. Electr. J. **28**, 70–85 (2015)
66. G. Barbieri, J.M. Coulondre, Day-Ahead and Real-Time Operational Signals for the European Electricity Market, in *International Conferences on European Energy Market*, Florence (2012)
67. J. Villar, R. Bessa, M. Matos, Flexibility products and markets: literature review. Electr. Power Syst. Res. **154**, 329–340 (2018)
68. A. Roos, Designing a joint market for procurement of transmission and distribution system services from demand flexibility. Renew. Energy Focus **21**, 16–24 (2017)
69. T. Liu, B. Sun, X. Tan, D.H.K. Tsang, Market for Multi-Dimensional Flexibility with Parametric Demand Response Bidding, in *North American Power Symposium*, Morgantown (2017)
70. A.F. Meyabadi, M.H. Deihimi, A review of demand-side management: reconsidering theoretical framework. Renew. Sust. Energ. Rev. **80**, 367–379 (2017)
71. S. Burger, J.P. Chaves-Ávila, C. Batlle, I.J. Pérez-Arriaga, A review of the value of aggregators in electricity systems. Renew. Sust. Energ. Rev. **77**, 395–405 (2017)
72. Y. Zhang, W. Chen, W. Gao, A survey on the development status and challenges of smart grids in main driver countries. Renew. Sust. Energ. Rev. **79**, 137–147 (2017)
73. N. IqtiyaniIlham, M. Hasanuzzaman, M. Hosenuzzaman, European smart grid prospects, policies, and challenges. Renew. Sust. Energ. Rev. **67**, 776–790 (2017)
74. J. Soares, M.A.F. Ghazvini, N. Borges, Z. Vale, Dynamic electricity pricing for electric vehicles using stochastic programming. Energy **122**, 111–127 (2017)
75. B. Muruganantham, R. Gnanadass, N.P. Padhy, Challenges with renewable energy sources and storage in practical distribution systems. Renew. Sust. Energ. Rev. **73**, 125–134 (2017)
76. G. Castagneto Gissey, P.E. Dodds, J. Radcliffe, Market and regulatory barriers to electrical energy storage innovation. Renew. Sust. Energ. Rev. **82**, 781–790 (2018)
77. R.A. Verzijlbergh, L.J. De Vries, G.P.J. Dijkema, P.M. Herder, Institutional challenges caused by the integration of renewable energy sources in the European electricity sector. Renew. Sust. Energ. Rev. **75**, 660–667 (2017)
78. H. de Faria, F.B.M. Trigoso, J.A.M. Cavalcanti, Review of distributed generation with photovoltaic grid connected systems in Brazil: challenges and prospects. Renew. Sust. Energ. Rev. **75**, 469–475 (2017)
79. M.N. Naz, M.I. Mushtaq, M. Naeem, M. Iqbal, M.W. Altaf, M. Haneef, Multicriteria decision making for resource management in renewable energy assisted microgrids. Renew. Sust. Energ. Rev. **71**, 323–341 (2017)

80. J. Hu, R. Harmsen, W. Crijns-Graus, E. Worrell, M. van den Broek, Identifying barriers to large-scale integration of variable renewable electricity into the electricity market: a literature review of market design. Renew. Sust. Energ. Rev. **81**, 2181–2195 (2018)
81. S. Jain, S. Kalambe, G. Agnihotri, A. Mishra, Distributed generation deployment: State-of-the-art of distribution system planning in sustainable era. Renew. Sust. Energ. Rev. **77**, 363–385 (2017)
82. B. Singh, J. Sharma, A review on distributed generation planning. Renew. Sust. Energ. Rev. **76**, 529–544 (2017)
83. E. Mengelkamp, P. Staudt, J. Garttner, C. Weinhardt, Trading on Local Energy Markets: A Comparison of Market Designs and Bidding Strategies, in *International Conferences on European Energy Market*, Dresden (2017)
84. Y. He, B. Wang, D. Li, M. Du, K. Huang, T. Xia, China's electricity transmission and distribution tariff mechanism based on sustainable development. Int. J. Electr. Power Energy Syst. **64**, 902–910 (2015)
85. Y.X. He, J. Jiao, R.J. Chen, H. Shu, The optimization of Chinese power grid investment based on transmission and distribution tariff policy: a system dynamics approach. Energy Policy **113**, 112–122 (2018)
86. Z. Li, Q. Guo, H. Sun, J. Wang, A new LMP-sensitivity-based heterogeneous decomposition for transmission and distribution coordinated economic dispatch. IEEE Trans. Smart Grid **9**, 931–941 (2018)
87. N.H. Radzi, K. Iskandar, M.N. Abdullah, M.S. Kamaruddin, S.A. Jumaat, R. Aziz, Investigation on Cost Reflective Network Pricing and Modified Cost Reflective Network Pricing Methods for Transmission Service Charges, in *International Conferences on Sustainable and Renewable Energy Engineering*, Hiroshima (2017)
88. M. Junqueira, L.C. da Costa, L.A. Barroso, G.C. Oliveira, L.M. Thomé, M.V. Pereira, An Aumann-Shapley approach to allocate transmission service cost among network users in electricity markets. IEEE Trans. Power Syst. **22**, 1532–1546 (2007)
89. F. Porrua, S. Member, R. Chabar, L.M. Thomé, L.A. Barroso, M. Pereira, Incorporating Large-Scale Renewable to the Transmission Grid: Technical and Regulatory Issues, in *Power & Energy Society General Meeting*, IEEE, Calgary (2009)
90. K. Janda, J. Málek, L. Recka, Influence of renewable energy sources on transmission networks in Central Europe. Energy Policy **108**, 524–537 (2017)
91. B. Lin, W. Wu, Cost of long distance electricity transmission in China. Energy Policy **109**, 132–140 (2017)
92. Z. Ming, P. Lilin, F. Qiannan, Z. Yingjie, Trans-regional electricity transmission in China: Status, issues and strategies. Renew. Sust. Energ. Rev. **66**, 572–583 (2016)
93. C.G. Beckman, India's new cost allocation method for inter-state transmission: implications for the future of the Indian. Electr. J. **26**, 40–50 (2013)
94. J. Bushnell, A. Ibarra-Yúnez, N. Pappas, Electricity transmission cost allocation and network efficiency: implications for Mexico's liberalized power market. Util. Policy **59**, 100932 (2019)
95. P. Mastropietro, L.A. Barroso, C. Batlle, Power transmission regulation in a liberalised context: an analysis of innovative solutions in south American markets. Util. Policy **33**, 1–9 (2015)
96. R.L. Fares, C.W. King, Trends in transmission, distribution, and administration costs for U.S. investor-owned electric utilities. Energy Policy **105**, 354–362 (2017)
97. I.J. Pérez-Arriaga, *Regulation of the Power Sector* (Springer, Heidelberg, 2013)
98. J. Casazza, F. Delea, *Understanding Electric Power Systems* (Wiley, Hoboken, 2010)
99. F. Lévêque, *Transport Pricing of Electricity Networks* (Kluwer Academic, Boston, 2003)
100. D.A. Lima, A. Padilha-Feltrin, J. Contreras, An overview on network cost allocation methods. Electr. Power Syst. Res. **79**, 750–758 (2009)
101. Z.J.Z. Jing, X.D.X. Duan, F.W.F. Wen, Y.N.Y. Ni, F.F. Wu, Review of Transmission Fixed Costs Allocation Methods, in *Power Engineering Society General Meeting*, IEEE, Toronto (2003)

102. J. Pan, Y. Teklu, S. Rahman, K. Jun, Review of usage-based transmission cost allocation methods under open access. IEEE Trans. Power Syst. **15**, 1218–1224 (2000)
103. F. Rubio-Odériz, Marginal pricing of transmission services: a comparative analysis of network cost allocation methods. IEEE Trans. Power Syst. **15**, 448–454 (2000)
104. X. Gao, P. You, Allocation of Cross-Regional and Cross-Provincial Transmission Cost to Trade on Power Market Directly, in *Conference on Energy Internet and Energy System Integration*, IEEE, Beijing (2017)
105. J. Nikoukar, M.R. Haghifam, Transmission cost allocation based on the use of system and considering the congestion cost. Int. J. Electr. Power Energy Syst. **43**, 961–968 (2012)
106. J. Cervantes, F.F. Choobineh, A Methodology for Allocating the Annual Revenue Requirement of a Transmission Network, in *Power & Energy Society General Meeting*, IEEE, Portland (2018)
107. J. Li, Z. Zhang, C. Gu, F. Li, Long-Run Incremental Pricing Based Transmission Charging Method Distinguishing Demand and Generation Technologies, in *General Meeting |Conference & Exposition*, IEEE, National Harbor (2014)
108. É. Telles, D.A. Lima, A. Street, J. Contreras, Min-max long run marginal cost to allocate transmission tariffs for transmission users. Electr. Power Syst. Res. **101**, 25–35 (2013)
109. B. Kharbas, H. Tiwari, M. Fozdar, Scheduled incremental and unscheduled interchange cost components of transmission tariff allocation: a novel approach for maintaining the grid discipline. IET Gener. Transm. Distrib. **8**, 1754–1766 (2014)
110. E. Telles, D.A. Lima, J. Contreras, N. Alguacil, A new transmission tariff allocation model based on Bilevel programming. IEEE Trans. Power Syst. **32**, 2204–2213 (2017)
111. M.E. Cebeci, O.B. Tor, C. Gencoglu, A.N. Guven, A Geographical Zoning Methodology for Nodal Transmission Use-of-System Tariffs, in *Power and Energy Society General Meeting*, IEEE, San Diego (2012)
112. R. Mehta, S.A. Soman, Efficient Cost Allocation for Transmission System, in *International Conference on Power System Technology*, IEEE, Wollongong (2016)
113. O.B. Tör, Separation of power systems into a unique set of zones based on transmission usage of network tariffs and transmission loss tariffs. Int. J. Electr. Power Energy Syst. **69**, 367–379 (2015)
114. A. Sedaghati, Cost of transmission system usage based on an economic measure. IEEE Trans. Power Syst. **21**, 466–473 (2006)
115. M.S.S. Rao, S.A. Soman, P. Chitkara, R.K. Gajbhiye, N. Hemachandra, B.L. Menezes, Min-max fair power flow tracing for transmission system usage cost allocation: A large system perspective. IEEE Trans. Power Syst. **25**, 1457–1468 (2010)
116. H.A. Gil, F.D. Galiana, A.J. Conejo, Multiarea transmission network cost allocation. IEEE Trans. Power Syst. **20**, 1293–1301 (2005)
117. M. Roustaei, M.K. Sheikh-El-Eslami, H. Seifi, Transmission cost allocation based on the users' benefits. Int. J. Electr. Power Energy Syst. **61**, 547–552 (2014)
118. J.M. Zolezzi, H. Rudnick, Transmission cost allocation by cooperative games and coalition formation. IEEE Trans. Power Syst. **17**, 1008–1015 (2002)
119. R. Bhakar, V.S. Sriram, N.P. Padhy, H.O. Gupta, Transmission embedded cost allocation in restructured environment: a game-theoretic approach. Electr. Power Components Syst. **37**, 970–981 (2009)
120. M.S.S. Rao, S.A. Soman, Marginal pricing of transmission services using min-max fairness policy. IEEE Trans. Power Syst. **30**, 573–584 (2015)
121. F.D. Galiana, A.J. Conejo, H.A. Gil, Transmission network cost allocation based on equivalent bilateral exchanges. IEEE Trans. Power Syst. **18**, 1425–1431 (2003)
122. A.M. Leite Da Silva, J.G.D.C. Costa, L.H.L. Lima, C.R.R. Dornellas, Z.S. Machado, J.C.O. Mello, *Transmission Network Cost Allocation Via Nodal Methodology Considering Different Dispatching Scenarios and Tariff Zones* (Manchester PowerTech, IEEE, Manchester, 2017)
123. E.T. Carlos, D.A. Lima, Transmission Tariff Allocation under High Renewable Source Penetration and Congestion Conditions, in *Power Systems Computation Conference*, Wroclaw (2014)

124. A.J. Conejo, J. Contreras, D.A. Lima, A. Padilha-Feltrin, Zbus transmission network cost allocation. IEEE Trans. Power Syst. **22**, 342–349 (2007)
125. J. Nikoukar, M.R. Haghifam, A. Parastar, Electrical power and energy systems transmission cost allocation based on the modified Z-bus. Int. J. Electr. Power Energy Syst. **42**, 31–37 (2012)
126. S. Pouyafar, M.T. Hagh, K. Zare, Development of a circuit-theory based transmission cost allocation method by orthogonal projection and equal-sharing principles. CSEE J. Power Energy Syst. **5**, 159–170 (2019)
127. K. Visakha, D. Thukaram, L. Jenkins, An approach for evaluation of transmission costs of real power contracts in deregulated systems. Electr. Power Syst. Res. **70**, 141–151 (2004)
128. A.M.L. Da Silva, J.G. De Carvalho Costa, L.H.L. Lima, A new methodology for cost allocation of transmission systems in interconnected energy markets. IEEE Trans. Power Syst. **28**, 740–748 (2013)
129. A.R. Abhyankar, S.A. Khaparde, Electricity transmission pricing: tracing based point-of-connection tariff. Int. J. Electr. Power Energy Syst. **31**, 59–66 (2009)
130. H.A. Gil, F.D. Galiana, E.L. Silva, Nodal price control: a mechanism for transmission network cost allocation. IEEE Trans. Power Syst. **21**, 3–10 (2006)
131. A.A. Abou El Ela, R.A. El-Sehiemy, Transmission usage cost allocation schemes. Electr. Power Syst. Res. **79**, 926–936 (2009)
132. L. Olmos, I.J. Pérez-Arriaga, A comprehensive approach for computation and implementation of efficient electricity transmission network charges. Energy Policy **37**, 5285–5295 (2009)
133. X. Gao, P. You, M. Wen, Fixed Cost Allocation Based on Current Electromagnetic Fields on Power Market, in *Conference on Energy Internet and Energy System Integration*, IEEE, Beijing (2018)
134. M. Shivaie, M. Kiani-Moghaddam, M. Ansari, Transmission-service pricing by incorporating load following and correlation factors within a restructured environment. Electr. Power Syst. Res. **163**, 538–546 (2018)

Chapter 5
Quantifying the Effect of Autonomous Demand Response Program on Self-Scheduling of Multi-carrier Residential Energy Hub

Amin Namvar, Farhad Samadi Gazijahani, and Javad Salehi

Nomenclature

Sets and Indices

c	Index of controllable loads
ce	Index of carbon emission
ch	Index of charging
dch	Index of discharging
e	Index of power
es	Index of energy storage
g	Index of natural gas
GB	Index of gas boiler
h	Index of heat
l	Index of load
Net	Index of network
t	Index of time (h)
uc	Index of uncontrollable loads

Parameters

α_t	Natural gas distribution coefficient between CHP and Boiler
$\eta_{CHP,\,e}$	Efficiency of CHP power generation

A. Namvar · F. S. Gazijahani (✉) · J. Salehi
Department of Electrical Engineering, Azarbaijan Shahid Madani University, Tabriz, Iran
e-mail: a.namvar@azaruniv.ac.ir; f.samadi@azaruniv.ac.ir; j.salehi@azaruniv.ac.ir

S. Nojavan, K. Zare (eds.), *Electricity Markets*,
https://doi.org/10.1007/978-3-030-36979-8_5

$\eta_{\text{CHP, h}}$	Efficiency of CHP heat generation
$\eta_{\text{e, L}}$	Efficiency of electric loads
$\eta_{\text{es, ch}}$	Energy storage charge efficiency
$\eta_{\text{es, dch}}$	Energy storage discharge efficiency
η_{GB}	Efficiency of heat generation by boiler
$\eta_{\text{h, L}}$	Efficiency of thermal loads
μ	Mean value of sunlight
π_{ce}	Price of carbon emission (cent/kW)
π_{e}^{t}	Price of purchasing power (cent/kW)
π_{g}^{t}	Price of purchasing natural gas (cent/kW)
σ	Standard deviation
$C_{\text{CHP}}^{\max}$	Maximum natural gas imported into CHP (kW)
DSM^{t}	Participation of load in the proposed DSM at t
$E_{\text{c, e}}$	Energy consumed by controllable electric loads in 24 h (kWh)
$E_{\text{c, h}}$	Energy consumed by controllable thermal loads in 24 h (kWh)
E_{es}^{0}	Initial value of energy storage SOC (kWh)
$E_{\text{es}}^{\max}$	Upper bound of energy storage SOC (kWh)
$E_{\text{es}}^{\min}$	Lower bound of energy storage SOC (kWh)
FF	Fill factor
$H_{\max}^{t}$	Upper bound of controllable thermal loads at t (kW)
$H_{\min}^{t}$	Lower bound of controllable thermal loads at t (kW)
H_{uc}^{t}	Uncontrollable thermal loads at t (kW)
I_{Mpp}	Current at maximum power point (A)
I_{sc}	Short circuit current (A)
K_{i}	Current temperature coefficient (A °C)
K_{MPPT}	Maximum power temperature coefficient
K_{v}	Voltage temperature coefficient (V °C)
lDSM^{t}	Load curtailed by DSM program at period t
Load^{t}	Load after DSM running at t (kW)
Load_{0}^{t}	Load before DSM running at t (kW)
N_{OT}	Nominal operating temperature of cell (°C)
$P_{\text{es,ch}}^{\max}$	Upper bound of energy storage charging (kW)
$P_{\text{es,dch}}^{\max}$	Upper bound of energy storage discharging (kW)
$P_{\max}^{t}$	Upper bound of controllable electric loads at t (kW)
$P_{\min}^{t}$	Lower bound of controllable electric loads at t (kW)
P_{PV}	Output power of solar panel (kW)
P_{STC}	Output power of solar panel at standard test conditions (kW)
P_{sy}	Output power of the PV module during state y (kW)
P_{uc}^{t}	Uncontrollable electric loads at t (kW)
s	Solar radiation
s_{ay}	Average solar irradiance of state y
S_{STC}	Solar irradiance at standard test conditions
T_{A}	Temperature of ambient (°C)
T_{cy}	Cell temperature over state y (°C)

V_{MPP}	Voltage at maximum power point (V)
V_{oc}	Open-circuit voltage (V)

Variables

E_{es}^{24}	SOC of battery at hour 24 (kWh)
E_{es}^{t}	SOC of battery at t (kWh)
G_{Net}^{t}	Imported natural gas at t (kW)
H_{c}^{t}	Controllable thermal loads at t (kW)
H_{CHP}^{t}	Heat generated by CHP at t (kW)
H_{GB}^{t}	Heat generated by boiler at t (kW)
$H_{hs,ch}^{t}$	Charged heat of heat storage at t (kW)
$H_{hs,dch}^{t}$	Discharged heat of heat storage at t (kW)
H_{L}^{t}	Total thermal loads at t (kW)
$l_{es,ch}^{t}/l_{es,dch}^{t}$	Binary values preventing battery from charging/discharging simultaneously
P_{c}^{t}	Controllable electric loads at t (kW)
P_{CHP}^{t}	Power generated by CHP at t (kW)
$P_{es,ch}^{t}$	Charged power of energy storage at t (kW)
$P_{es,dch}^{t}$	Discharged power of energy storage at t (kW)
P_{L}^{t}	Total electric loads at t (kW)
P_{Net}^{t}	Power imported from grid at t (kW)

Functions

C	Total objective function (cent)
C_{ce}	Cost of carbon emission (cent)
C_{e}	Cost of purchasing power (cent)
C_{g}	Cost of purchasing natural gas (cent)

5.1 Introduction

The gradual decrease in fossil fuels as one of the most important sources of energy production as well as the environmental pollution problem has created many concerns in the world so that many international treaties (i.e., Paris Treaty) have signed [1]. These issues have caused to increase the attention to the subject of energy management and environmental protection at the international level [2]. Currently, energy resources such as electricity and natural gas networks are independently managed and operated that these matters cause to reduce energy efficiency and as a result, reduce the reliability in energy supply, increase the operating costs and excessive pollution. Energy consumers need different energy carriers to meet their living needs, but the most important energy carriers used by the consumers are natural gas and electricity because they are easy to operate and also many appliances depend on electricity and natural gas.

In conventional systems these needs are provided independently, that is, natural gas and electricity are supplied to consumers individually. However, with the advancement of the technology of combined heat and power (CHP) generation systems as an effective factor in the supply of natural gas and electricity, in addition to the gas needs of the consumers, they can simultaneously meet their thermal and electrical requirements. CHP systems can have an efficiency of between 60 and 80% which will increase the efficiency of energy supply [3]. On the other hand, combined cycle power plants with more than 60% efficiency, which are economically viable, have been able to conquer the power market. With the growth of global energy consumption and the environmental impacts of fossil fuels used in conventional power plants, the tendency to use renewable sources has increased. In addition, natural gas plays an important role in the global energy market by producing electricity in large scale namely in gas-fired power plants and in small scale in CHP systems. As a result, combining different sources of energy from renewable sources to natural gas in one set can facilitate the achievement of a sustainable energy network [4].

Since natural gas and electricity are interconnected, the operation of integrated energy systems increases the efficiency of the supply of energy to customers who need natural gas and electricity and heat. But since one of the most important issues regarding energy supply is economic subject, in order to improve consumer comfort and reduce government spending, we will try to minimize the cost of operation, namely to purchase natural gas and electricity. Another important issue in supplying energy is the matter of air pollution, which must be carefully investigated. To this end, we need to determine how much and when to use the source of energy to minimize operating costs, for example, it should be determined when the storage will be charged or discharged, or how much renewable resource production per hour is, and the rest of the resources and equipment are similarly. In the direction of optimizing utilization costs, we are faced with the uncertainty of renewable resources. The production capacity of wind turbines and solar panels is uncertain, since power generation by wind turbines and solar panels depends on the speed of wind and the amount of sunlight, respectively, and this makes it impossible for us to accurately describe the generation of electricity by these sources. Although the wind and sunlight are predictable, it cannot be commented on precisely and definitively, and this affects the process of modeling and simulation of the problem. On the other hand, there are several methods, including scenario production, for modeling uncertainties that we use to model the uncertainty of the output of renewable resources and consider these uncertainties in the optimization problem. On the other hand, there are constraints and limitations that will add to the optimization problem. Among these constraints are the limited capacity of gas pipelines and power lines, the limitation of wind turbine and solar cells production, and the storage capacity limitations.

The need for energy is one of the most fundamental human needs, so that without energy resources, such as electricity and natural gas, human life will be impossible. Regarding the gradual completion of fossil fuels, including natural gas, and the dependence of electricity generation on natural gas in gas-fired units and the problem of air pollution, energy management becomes increasingly important. Utilization of energy resources (gas and electricity), in addition to reducing energy efficiency, will result in excessive consumption of these resources, which will have economic consequences. The simultaneous operation of energy resources can prevent the above problems, because if we know what kind of energy source and when and how much must be used, we can manage energy resources and solve economic problems caused by excess energy consumption. On the other hand, greenhouse gas emissions and increased energy needs will lead us to utilize new sources and technologies. New technologies, such as CHP systems, have made possible the operation of energy resources with high profits. Reducing greenhouse gas emissions, improving reliability and efficient operation, have been considered as the advantages of combining different energy networks [5].

In this section, we will investigate the research on energy hubs. With the advancement of technology, the issue of energy management has been given particular attention. Considerable research has been done on energy management in the form of energy hub in various aspects. Some of the research related to energy hub done are the reduction of operating costs [5, 6], reduction of air pollution [7, 8] and increase in profits due to the sale of energy in the market [9, 10]. Among the technologies used in the energy hub, renewable sources are one of the most important equipment, because they cause to reduce the dependence of the energy hub on the grid and increase the reliability of energy supply and also decrease air pollution [11]. In some studies, the output of renewable resources has been deterministically modeled [12, 13], and in some other research outputs of these resources have been modeled uncertainly [13, 14]. Electric vehicle [15, 16] is one of the other technologies that has attracted particular attention to the fact that, in addition to reducing air pollution, it delivers its stored energy to energy hub at peak time. One of the most important equipment used in energy hub is energy storage [17, 18], which plays a key role in energy management by storing energy at off-peak times and delivering energy stored at peak hours. Another important equipment is CHP systems [19, 20], which produces electricity and heat from natural gas to meet part of the needs of consumers in the energy hub.

In short, the contributions of this chapter are as follows:

- A novel cost-emission based modeling for energy management in residential sectors
- Utilizing various equipment such as renewable resources and co-generation devices to reduce the cost of operation and air pollution
- Considering responsible loads to investigate the effect of the demand response program
- Formulating the model as a mixed-integer linear programming

The remainder of this chapter is organized as follows: In Sect. 5.2, the concept of the energy hub and its general structure are expressed. Mathematical modeling and problem formulation of the proposed scheme is shown in Sect. 5.3. In Sect. 5.4, simulation results are presented and discussed. Finally, in Sect. 5.5, the conclusions of this chapter are given.

5.2 Energy Hub

Today energy plays an indelible role in the development of human societies. Especially the electrical energy that can easily change into different forms of energy and eliminate the needs of consumers. Recently, the concept of energy hub [21, 22] has been proposed for the use of integrated energy systems. The energy hub is a super node that receives different energy carriers at its input, and then determines which technology and energy carrier to meet the needs of the subscriber according to planning [5]. In a typical energy hub, its entrances are natural gas and electricity, and its outputs are electricity and heat. Structure of the energy hub is composed of different equipment such as a CHP system for generating electricity and heat from natural gas, a transformer for converting electrical voltage levels, an electric heater for generating heat from electricity and energy storages for storing electricity and heat. Subscribers who feed on energy hub can be residential, industrial, and commercial consumers. The energy hub supplies its consumers and sells its energy surplus to the grid. Figure 5.1 shows the general model of an energy hub. As you can see, the electrical energy is converted to an acceptable voltage level by the transformer after entering the energy hub, and then a part of it has been given to consumers and the other part is stored and its surplus is sold to the network. Natural gas after converting to the heat and electricity is given to consumers.

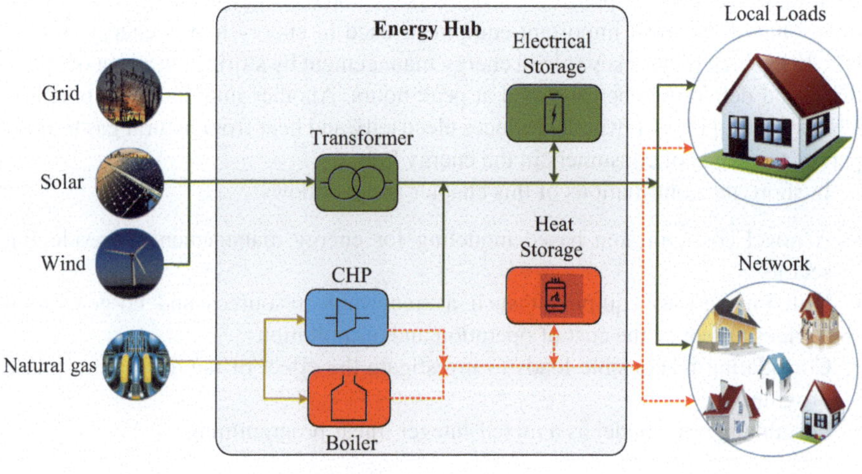

Fig. 5.1 Overview of energy hub

5.3 Problem Formulation

In this section, a residential building is considered as an energy hub whose general structure is shown in Fig. 5.2. Various and sometimes conflicting issues, such as system reliability, environmental protection, profitability, comfortable life, and economic should be considered in optimal utilization of residential energy hub [23]. The proposed energy hub for a smart home receives two energy carriers, including natural gas and electricity, at its entrances. The energy hub is also composed of various equipment and tools including solar panels, power, and heat storage units and a CHP system to meet the needs of its consumers, which require heat and power. The consumers connected to the smart energy hub divide into two categories. The first type is uncontrollable loads that have an invariable profile, and the second category is controllable loads that have specific energy consumption and the operating time of these loads is controllable. The electricity loads are supplied by the electricity purchased from the grid, electricity generated by the CHP unit, solar panels, and battery. The heat loads are fed by the boiler, the heat storage, and thermal energy generated by the CHP unit.

Air pollution today is one of the most serious global concerns that should to be addressed seriously and the most important way to prevent its release is to reduce greenhouse gas emissions from fossil fuels. One of the important tasks that can be done to increase the efficiency of the energy hub and thereby reduce air pollution

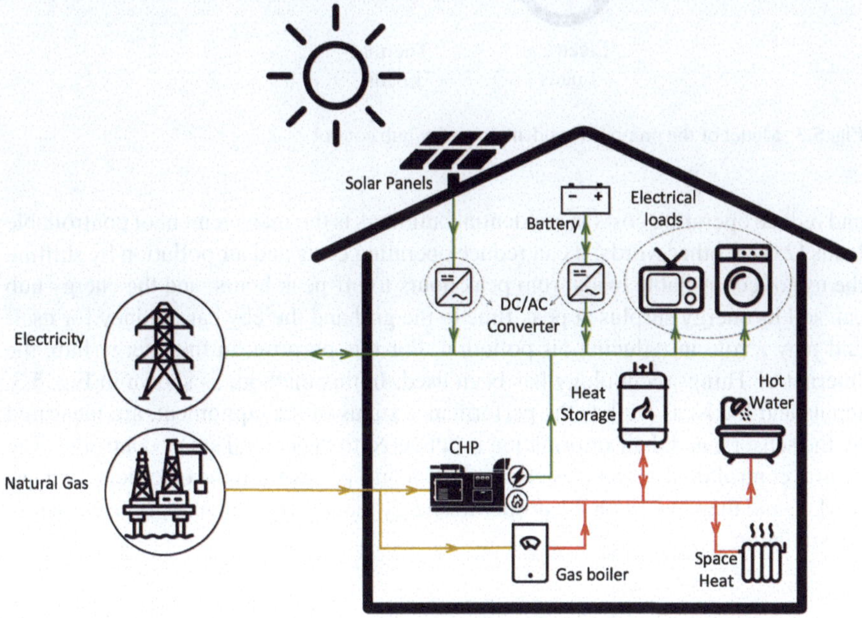

Fig. 5.2 The proposed residential energy hub

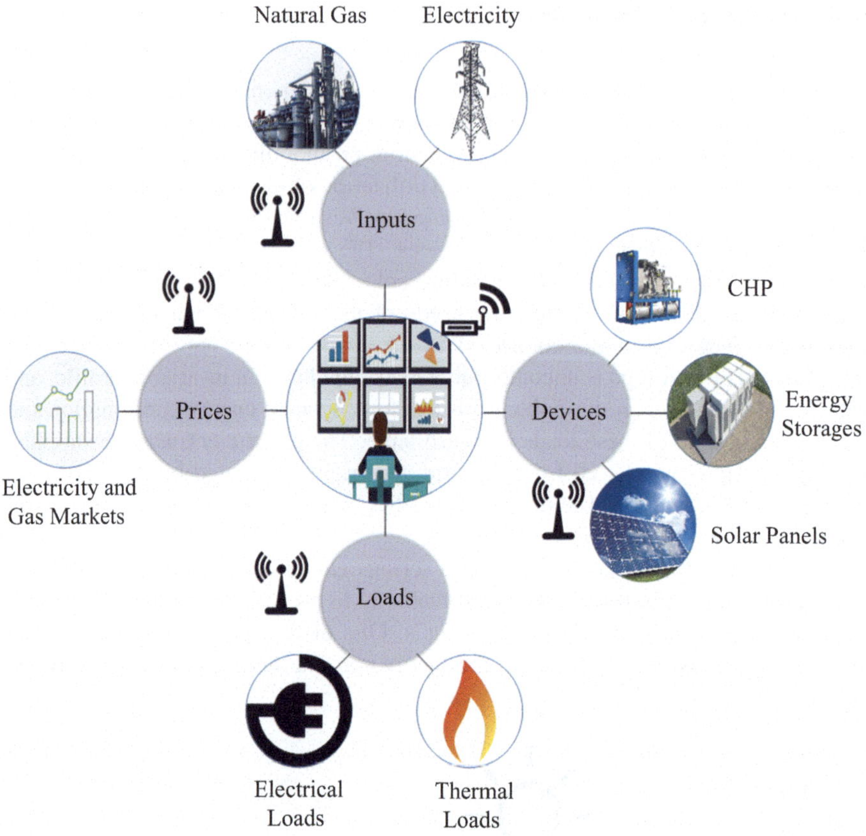

Fig. 5.3 Model of the proposed residential energy hub control

and reduce operating costs in residential buildings is the management of controllable loads [24]. In other words, it can reduce operating costs and air pollution by shifting the use of controllable loads from peak hours to off-peak hours, and the energy hub can sell its energy surplus at peak time to the grid and thereby earn money for itself and play a role in reducing air pollution. For this purpose, in the energy hub, the Internet of Things technology has been used. In this method, as shown in Fig. 5.3, inputs and loads, as well as the performance status of the equipment, are measured by the sensors, and their information is delivered to the central smart controller. The central controller also receives electricity and gas prices from the electricity and gas market, and then, based on the proposed energy management plan, optimizes energy consumption.

5.3.1 Component Modelling

In this section, introducing and modeling the components of the energy hub and the objective function in the form of mathematical formulas are discussed to be used in the optimization problem.

5.3.1.1 Energy Storage

Nowadays, due to the rise in the price of energy carriers and the sharp fluctuation of energy prices in the spot market, the use of storage in the energy sector has increased significantly. On the other hand, electrical storages also increase the use of renewable resources, as fluctuations in the production of these resources are controlled, which increases the quality of the system, reduces energy costs, and increases system profits [25]. Therefore, energy storage units are one of the most important and most profitable parts of the energy hub [26, 27]. Energy storage has been considered for economic benefit and reliability. Using energy storage units in energy hubs helps greatly reduce the cost of purchasing energy because they can be charged at low energy costs and supply consumers when energy is high. In the proposed model of this chapter, electric and thermal storage devices are used, and their mathematical equations are as follows.

Battery Energy Storage

As stated, the storage device is charged at low energy cost hours and discharged at high energy cost hours to feed local loads [28]. In this section, the mathematical equations of the electric storage are shown, which shows the state of charge of the battery in Eq. (5.1). Equation (5.2) shows the minimum and maximum storage capacity. Equation (5.3) emphasizes that the energy stored in the battery at hour 24 is equal to its initial energy value. Equations (5.4) and (5.5) show the maximum charge and discharge power of the storage device. Equation (5.6) also prevents charging and discharging battery simultaneously.

$$E_{\mathrm{es}}^{t}=E_{\mathrm{es}}^{0}+\sum_{h=1}^{t}\left(\eta_{\mathrm{es,ch}}P_{\mathrm{es,ch}}^{h}-\frac{P_{\mathrm{es,dch}}^{h}}{\eta_{\mathrm{es,dch}}}\right) \tag{5.1}$$

$$E_{\mathrm{es}}^{\min}\leq E_{\mathrm{es}}^{t}\leq E_{\mathrm{es}}^{\max} \tag{5.2}$$

$$E_{\mathrm{es}}^{24}=E_{\mathrm{es}}^{0} \tag{5.3}$$

$$0 \leq P_{\text{es,ch}}^{t} \leq P_{\text{es,ch}}^{\max} I_{\text{es,ch}}^{t} \tag{5.4}$$

$$0 \leq P_{\text{es,dch}}^{t} \leq P_{\text{es,dch}}^{\max} I_{\text{es,dch}}^{t} \tag{5.5}$$

$$0 \leq I_{\text{es,ch}}^{t} + I_{\text{es,dch}}^{t} \leq 1 \tag{5.6}$$

Heat Storage

Heat storage is another storage device used in this modeling that stores thermal energy and, if necessary, provides it to consumers. In this section, the mathematical modeling of heat storage is shown. Equation (5.7) shows the state of charge. Equation (5.8) indicates the minimum and maximum storage capacity. Equation (5.9) emphasizes that the energy stored in the heat storage at hour 24 is equal to its initial energy value. Equations (5.10) and (5.11) show the maximum charge and discharge power of the storage device. Equation (5.12) also prevents charging and discharging the storage simultaneously.

$$E_{\text{es}}^{t} = E_{\text{es}}^{0} + \sum_{h=1}^{t} \left(\eta_{\text{es,ch}} P_{\text{es,ch}}^{h} - \frac{P_{\text{es,dch}}^{h}}{\eta_{\text{es,dch}}} \right) \tag{5.7}$$

$$E_{\text{es}}^{\min} \leq E_{\text{es}}^{t} \leq E_{\text{es}}^{\max} \tag{5.8}$$

$$E_{\text{es}}^{24} = E_{\text{es}}^{0} \tag{5.9}$$

$$0 \leq P_{\text{es,ch}}^{t} \leq P_{\text{es,ch}}^{\max} I_{\text{es,ch}}^{t} \tag{5.10}$$

$$0 \leq P_{\text{es,dch}}^{t} \leq P_{\text{es,dch}}^{\max} I_{\text{es,dch}}^{t} \tag{5.11}$$

$$0 \leq I_{\text{es,ch}}^{t} + I_{\text{es,dch}}^{t} \leq 1 \tag{5.12}$$

5.3.1.2 CHP Unit

CHP unit is one of the most important technologies used in energy hub. This unit receives natural gas at its input and generates electricity and heat. This device, which is the most important factor in the connection between natural gas and electricity,

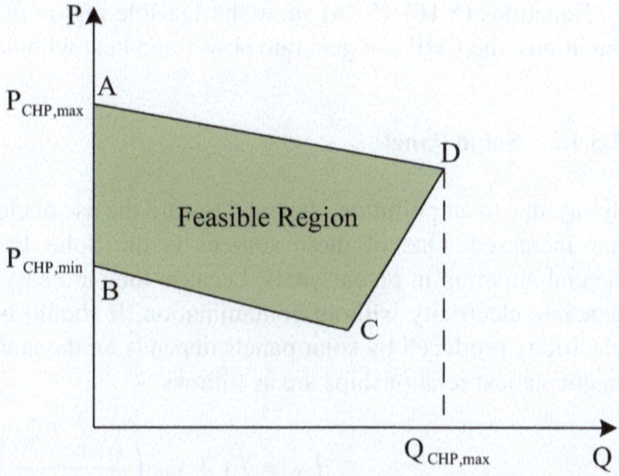

Fig. 5.4 Feasible region of CHP unit

has been considered for high efficiency. The CHP unit has a feasible region for operating as shown in Fig. 5.4. Electricity and heat generation by CHP depend on each other, which means that it generates a certain amount of electricity for a certain amount of heat production. The mathematical equations for CHP unit are shown in Eqs. (5.13)–(5.15).

$$P_{\mathrm{CHP}}^{t} = \eta_{\mathrm{CHP,e}} \alpha_t G_{\mathrm{Net}}^{t} \tag{5.13}$$

$$H_{\mathrm{CHP}}^{t} = \eta_{\mathrm{CHP,h}} \alpha_t G_{\mathrm{Net}}^{t} \tag{5.14}$$

$$\alpha_t G_{\mathrm{Net}}^{t} \le C_{\mathrm{CHP}}^{\max} \tag{5.15}$$

Equations (5.13) and (5.14), respectively, represent the electrical and thermal power generated by the CHP unit, and Eq. (5.15) denotes the input of natural gas to the CHP unit for the production of electricity and heat.

$$P_{\mathrm{CHP}}^{t} - P_A - \frac{P_A - P_B}{H_A - H_B}\left(H_{\mathrm{CHP}}^{t} - H_A\right) \le 0 \tag{5.16}$$

$$P_{\mathrm{CHP}}^{t} - P_B - \frac{P_B - P_C}{H_B - H_C}\left(H_{\mathrm{CHP}}^{t} - H_B\right) \ge 0 \tag{5.17}$$

$$P_{\mathrm{CHP}}^{t} - P_C - \frac{P_C - P_D}{H_C - H_D}\left(H_{\mathrm{CHP}}^{t} - H_C\right) \ge 0 \tag{5.18}$$

Equations (5.16)–(5.18) show the feasible region of the CHP. According to the equations, the CHP can generate power and heat within the feasible region.

5.3.1.3 Solar Panel

Today, due to air pollution, the trend toward the use of clean and renewable resources has increased. One of these sources is the Solar Panels, which have attracted special attention in recent years, because they are easy to install and use and also generate electricity without contamination. It should be noted that the amount of electricity produced by solar panels depends on the amount of sunlight [29] whose mathematical relationships are as follows.

$$T_{cy} = T_A + s_{ay}\left(\frac{N_{\mathrm{OT}} - 20}{0.8}\right) \tag{5.19}$$

$$I_y = s_{ay}\left(I_{\mathrm{sc}} + K_i\left(T_c - 25\right)\right) \tag{5.20}$$

$$V_y = V_{\mathrm{oc}} - K_v T_{cy} \tag{5.21}$$

$$P_{sy}\left(s_{ay}\right) = N.FF.V_y.I_y \tag{5.22}$$

$$FF = \frac{V_{\mathrm{MPP}} I_{\mathrm{MPP}}}{V_{\mathrm{oc}} I_{\mathrm{sc}}} \tag{5.23}$$

As it was said, the output of renewable resources is uncertain. In this modelling, the normal distribution function is used to generate scenarios for the output of solar panels, the equation of which is given from Eq. (5.24) [23].

$$f(s) = \frac{1}{\sqrt{2\pi}\sigma} \mathrm{e}^{-\frac{(s-\mu)^2}{2\sigma^2}} \tag{5.24}$$

5.3.1.4 Load Modelling

Recently, consumer demand management systems have attracted much attention, especially in smart cities, as an effective tool for optimizing demand management at peak time [30]. In fact, demand management programs aim at changing consumption pattern to reduce costs and increase system reliability, which encourages consumers to use their programs and activities to optimize their consumption [31]. This can be a great benefit for consumers in terms of lower costs and better energy consumption control. In this modeling, loads are divided into two categories of electrical loads and thermal loads, each of which consists of two groups of controllable loads and uncontrollable loads. In other words, some of the electrical and thermal loads have a certain amount and time of consumption that cannot be controlled. On the other hand, some other loads have a certain amount of consumption that cannot be controlled, but their operating time can be controlled, and they can be shifted from peak time to off-peak time. The modeling of controllable loads is as follows.

Electrical Loads

Using demand-side management programs, consumers shift their consumption from peak hours that the cost of electricity is high to off-peak hours to reduce operating costs. Modeling of electric loads is as follows.

$$P_{\mathrm{e}}^{t} = P_{\mathrm{uc,e}}^{t} + P_{\mathrm{c,e}}^{t} \tag{5.25}$$

$$P_{\mathrm{min,e}}^{t} \leq P_{\mathrm{c,e}}^{t} \leq P_{\mathrm{max,e}}^{t} \tag{5.26}$$

$$\sum_{t=1}^{24} P_{\mathrm{c,e}}^{t} = E_{\mathrm{c,e}} \tag{5.27}$$

Equation (5.25) denotes that the electric loads of the system consist of controllable and uncontrollable loads. Equation (5.26) shows the minimum and maximum amount of consumable power for the electric loads. Equation (5.27) also states that the total controllable load power in 24 h is equal to the total energy consumed by these devices.

Heat Storage

In this modeling, demand-side management program is also used for controllable heat loads, which is modeled as follows.

$$P_{\mathrm{h}}^{t} = P_{\mathrm{uc,h}}^{t} + P_{\mathrm{c,h}}^{t} \tag{5.28}$$

$$P_{\mathrm{min,h}}^{t} \leq P_{\mathrm{c,h}}^{t} \leq P_{\mathrm{max,h}}^{t} \tag{5.29}$$

$$\sum_{t=1}^{24} P_{\mathrm{c,h}}^{t} = E_{\mathrm{c,h}} \tag{5.30}$$

Equation (5.28) denotes that the thermal loads of the system consist of controllable and uncontrollable loads. Equation (5.29) shows the minimum and maximum amount of consumable power for the thermal loads. Equation (5.30) also states that the total controllable loads power in 24 h is equal to the total energy consumed by these devices.

5.3.1.5 Uncertainty Modeling

Engineering modeling always is accompanied by uncertainties. There are a lot of methods such as scenario generation, robust optimization to model uncertainties. In this chapter, the scenario generation method has been applied to meet the renewable resource uncertainty. In this method at the first, numerous scenarios have been generated from probability density functions, and then the scenario reduction method has been applied to decrease the number of the scenarios. In the end, one of the scenarios that has the greatest probability value has been chosen as the expected scenario. The mathematical formulas for scenario generation and reduction are as follows:

$$\phi_{WS} = \left\{ \left(WS^1, \psi_{WS}^1 \right), \left(WS^2, \psi_{WS}^2 \right), \ldots, \left(WS^n, \psi_{WS}^n \right) \right\} \tag{5.31}$$

$$S = \prod_{WS} \phi_{WS} \tag{5.32}$$

$$\sum_{s \in S} \psi_{WS} = 1 \tag{5.33}$$

In this chapter, Normal PDF has been applied to the scenario generation for the PV output at each hour. Equation (5.31) indicates the number of the scenarios and their probability. Equation (5.32) shows the set of the scenarios. Equation (5.33) expresses that the sum of the probabilities must equal to 1.

After generating the scenarios, the scenario reduction method is utilized in order to decrease the burden of calculations. The mathematical formulas of the scenario reduction method are as follows:

$$S_1 = \arg \left[\underset{s' \in S}{\text{Min}} \sum_{s \in S} \psi^S W\left(S, S'\right) \right] S = \{S_1\} \tag{5.34}$$

$$S_n = \arg \left[\underset{s' \in S}{\text{Min}} \sum_{s \in S} \psi^S \underset{s'' \in S}{\text{Min}} W\left(S, S''\right) \right] \tag{5.35}$$

5.3.1.6 Heat and Power Balance

The balance of power and heat is the most important constraint in this model, which indicates that the amount of generated power and heat must be equal to the demands. According to Fig. 5.2, the equations of power and heat are written as follows:

$$P_{\mathrm{net}}^{t}+P_{PV}^{t}+\alpha_{t}P_{\mathrm{CHP}}^{t}+P_{\mathrm{Bat},d}^{t}=P_{\mathrm{e}}^{t}+P_{\mathrm{Bat},c}^{t} \tag{5.36}$$

$$(1-\alpha_{t})\,H_{\mathrm{GB}}^{t}+\alpha_{t}H_{\mathrm{CHP}}^{t}+H_{\mathrm{HS},d}^{t}=P_{\mathrm{h}}^{t}+H_{\mathrm{HS},c}^{t} \tag{5.37}$$

$$H_{\mathrm{GB}}^{t}=\ \eta_{\mathrm{GB}}G_{\mathrm{net}}^{t} \tag{5.38}$$

Equation (5.36) represents the balance of electrical power. Equation (5.37) also expresses the heat equilibrium equation, and Eq. (5.38) is the amount of thermal power generated by the boiler.

5.3.1.7 Objective Function

The purpose of this modeling is to reduce operating costs, including electricity and natural gas purchase costs, as well as to reduce the carbon emissions in air using various equipment and planning that are modeled as follows:

$$C=\ \min\ (C_{\mathrm{E}}+C_{\mathrm{G}}+C_{\mathrm{C}}) \tag{5.39}$$

$$C_{\mathrm{E}}=\ \sum_{t=1}^{24}\ \pi_{\mathrm{E}}^{t}P_{\mathrm{Grid}}^{t} \tag{5.40}$$

$$C_{\mathrm{G}}=\ \sum_{t=1}^{24}\ \pi_{\mathrm{G}}^{t}G_{\mathrm{net}}^{t} \tag{5.41}$$

$$\begin{gathered}C_{\mathrm{C}}=\ \pi_{\mathrm{C}}\sum_{t=1}^{24}\ \left(\beta_{\mathrm{e}}P_{\mathrm{Grid}}^{t}+\beta_{\mathrm{g}}G_{\mathrm{net}}^{t}\right)\\ \text{s.t.}:(3-1)\ \text{to}\ (3-30)\end{gathered} \tag{5.42}$$

Equation (5.39) represents the objective function, consisting of three functions that are the purchase cost of electricity, the cost of purchasing natural gas, and the cost of carbon emissions in the air, and the equation of each of them is obtained from Eqs. (5.40) to (5.42).

5.4 Numerical Results

5.4.1 Data

In this section, the technical information of the system, including the price of energy carriers, consumer information, and equipment information, will be shown for use in the simulation process. Table 5.1 shows the price of electricity and natural gas in the time of use tariff. The amount of the natural gas distribution coefficient between the CHP and the boiler units is presented in Table 5.2. Also, information about electricity and heat storage devices is given in Table 5.3. The amount of daily energy consumption, the minimum and maximum allowable power consumption, and the operating time for the controllable electric and thermal equipment are shown in Table 5.4. In addition, the pattern of uncontrollable electric and thermal loads for different times are shown in Fig. 5.5. The CHP unit has an electrical and thermal

Table 5.1 Price of energy carriers in TOU tariff

	Electricity			Natural gas	
	Off-peak	Mid-peak	On-peak	Off-peak	On-peak
Hour	1–8	13–17	9–12	1–9	10–14
		22–24	18–21	15–18	19–21
				22–24	
Price (cent/kWh)	7	10	14	2	6

Table 5.2 Dispatch factor at different hours

Hour	1	2	3	4	5	6	7	8
α_t	0.813	0.83	0.776	0.741	0.852	0.717	0.738	0.741
Hour	9	10	11	12	13	14	15	16
α_t	0.801	0.834	0.769	1	0.858	1	1	1
Hour	17	18	19	20	21	22	23	24
α_t	0.834	0.745	0.714	0.666	0.741	0.77	0.675	0.705

Table 5.3 Information of energy storages

	$P_{es,ch}^{max}$ (kW)	$P_{es,ch}^{max}$ (kW)	E_{es}^{max} (kWh)	E_{es}^{min} (kWh)	E_{es}^{0} (kWh)	$\eta_{es,\,ch}$	$\eta_{es,\,dch}$
Battery	0.7	0.9	5	1	2	0.88	0.88
Heat storage	0.5	0.5	3	0.5	2	0.6	0.8

Table 5.4 Controllable loads data

Electric load				Thermal load			
P_{max} (kW)	P_{min}^{t} (kW)	$E_{c,\,e}$ (kWh)	Operating time (h)	H_{max}^{t} (kW)	H_{min}^{t} (kW)	$E_{c,\,h}$ (kWh)	Operating time (Hour)
0.55	0.3	5.5	6–17	0.4	0.25	3.5	8–18

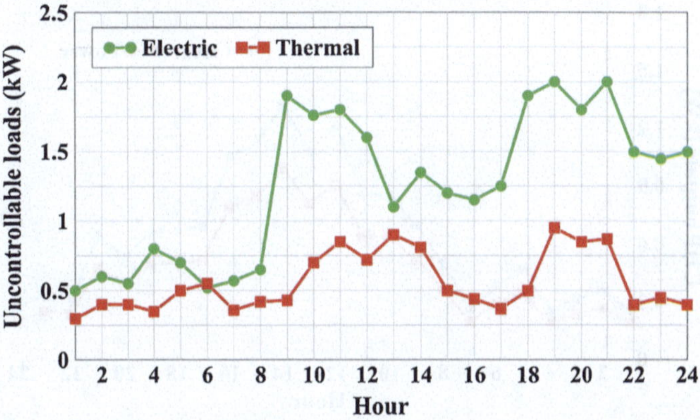

Fig. 5.5 Profile of uncontrollable loads

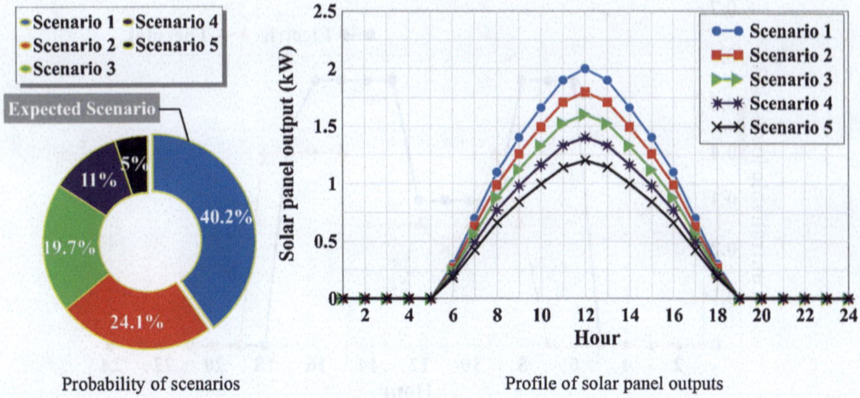

Fig. 5.6 Solar panel outputs in different scenarios

efficiency of 45% and 35%, respectively, and the maximum input of natural gas to it is 2.5 kW. Five scenarios are generated for modelling of solar panel output uncertainty that are shown in Fig. 5.6.

5.4.2 *Results*

The CHP unit, by producing electricity from natural gas, can partly reduce operating costs because in this modeling the price of natural gas is less than electricity. Figure 5.7 shows the amount of electricity and heat generated by the CHP unit. As can be seen, at different times, the ratio of generated heat to the generated electricity is equal, and this is due to the approximate modeling of the CHP unit. Figure 5.8 also shows the controllable electrical and thermal load profile. It can be seen that

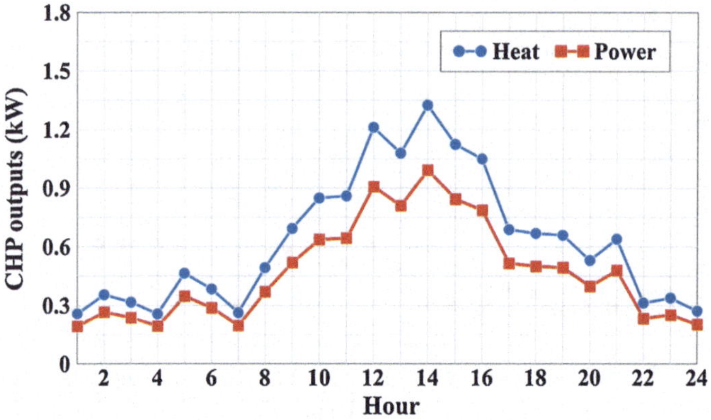

Fig. 5.7 CHP outputs

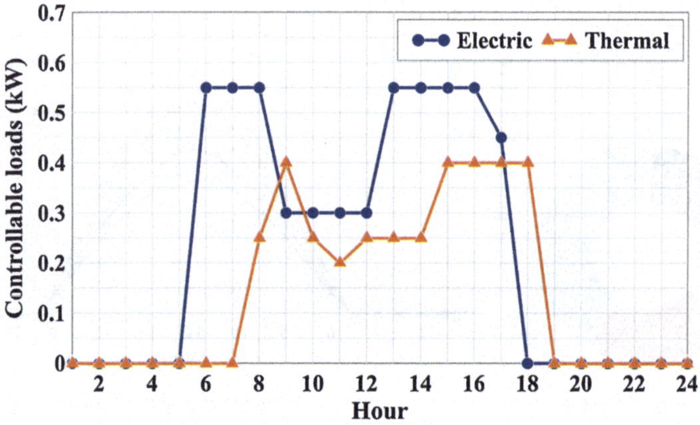

Fig. 5.8 Profile of controllable loads

when the price of energy carriers is low, these loads work at their maximum power, and when the price of energy carriers is high, they work at their minimum power, that is, they are transferred to off-peak hours to reduce the costs of operation.

Energy storages store energy when the price of energy carriers is low, and give stored energy to consumers when their cost is high. Figure 5.9 represents the state of charge of electrical and thermal storage. For example, the battery is charged between hour 13 and 17, with low electricity price, and discharged from hour 18 to 21, when the price of electricity is high. Similarly, the heat storage is charged between hour 15 and 18, with low natural gas prices, and discharged when the price of natural gas is high.

Solar panels, as clean energy sources, have a significant impact on reducing carbon emissions, as well as reducing the cost of purchasing natural gas and

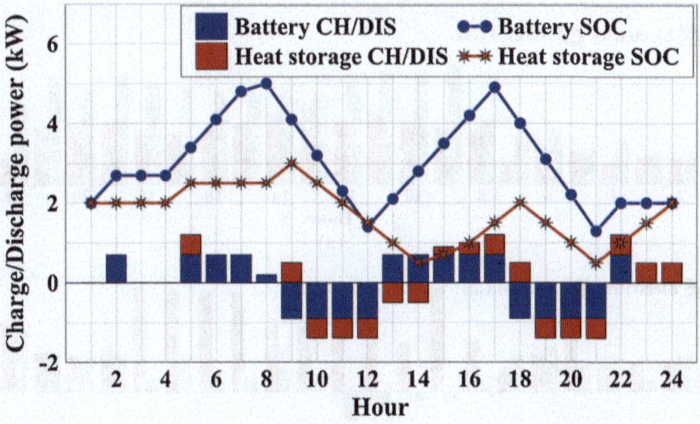

Fig. 5.9 SOC of energy storages

Table 5.5 Impact of solar panels output on cost of operation (Cent)

Scenario	Electricity	Natural gas	Carbon emission
Scenario 1	166.05	158.55	6.04
Scenario 2	184.44	158.55	6.32
Scenario 3	202.84	158.55	6.59
Scenario 4	221.23	158.55	6.87
Scenario 5	239.62	158.55	7.14

electricity from the networks. On the other hand, the output of the solar panels is uncertain, and for this reason, using the Normal probability distribution function, five scenarios have been generated to examine the effect of solar panels on the energy hub. Table 5.5 shows the impact of each scenario on operating costs. As can be seen, with the increase in the amount of electricity produced by solar panels, the costs of air pollution, electricity, and natural gas have been reduced.

After completing simulations, the operating in the form of energy hub is compared with the base case. Figure 5.10 shows the electricity and natural gas purchases in this simulation. In the base case, the energy carriers were individually operated and the amount of energy input to the energy hub per hour was dependent on the amount of load consumed at that hour. With adding different equipment to proposed residential building, the energy management has been optimized. The CHP unit by generating power from natural gas, energy storages by storing energy at off-peak times and delivering at peak times, and solar panels by generating power from sunlight, help to reduce the operating costs and carbon emission. According to Fig. 5.10, in the base case, the energy consumption at the peak times is high, but in the operating in the form of the energy hub, the energy consumption shifts to off-peak times, and also the energy hub can sell energy to the grid. Table 5.6 shows the overall simulation results, including electricity, natural gas and carbon emissions, as well as total operating costs, which are the total cost of each step of the simulation.

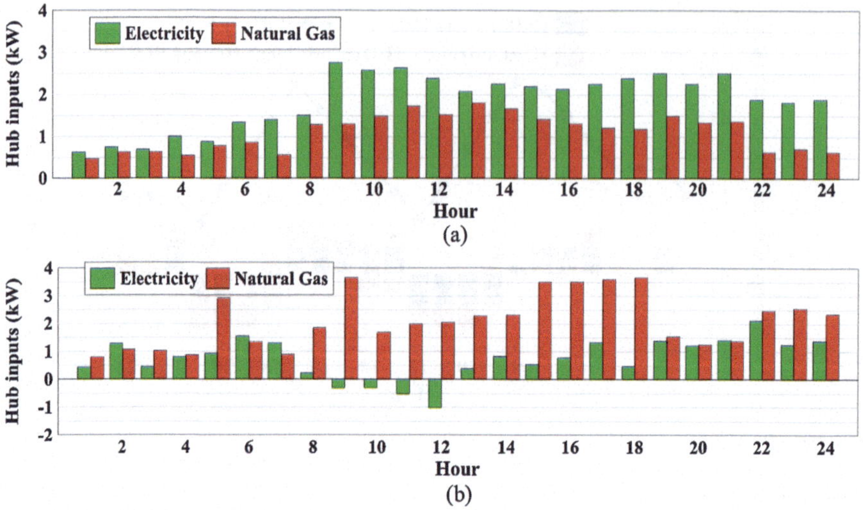

Fig. 5.10 Energy hub inputs. (**a**) Base case, (**b**) energy hub

Table 5.6 Cost of operation (Cent)

Case	Electricity	Natural gas	Carbon emission	Total
Base case	500.9	102.25	9.162	612.31
Energy hub	166.05	158.65	6.04	330.65

5.5 Conclusion

So doing simulations and reviewing the results will be discussed in this chapter by the conclusion of the modeling. In this chapter, optimal utilization of integrated energy systems in the form of energy hubs was used to reduce operating costs and reduce air pollution. In modelling, a smart residential building was considered as an energy hub controlled by the Internet of Things technology. Simulations were carried out in the presence of various equipment such as the CHP unit, power and heat storage equipment and renewable resources, and the role of each of them was investigated. Electricity and natural gas pricing was based on the time of use tariff in order to see the impact of the price of energy carriers on optimal utilization. It was observed that the existence of a CHP unit by generating electricity and heat from natural gas and supplying part of the needs of consumers reduced the cost of purchasing electricity in peak hours. In addition, the energy storage equipment had an important role in reducing costs by storing energy at peak hours when the energy costs are low and delivering the stored energy to subscribers at off-peak times when energy costs are high. The presence of renewable resources as clean energy sources contributed greatly to reducing operating costs and reducing air pollution.

In the future works, various studies can be investigated as follow:

- Adding other energy carriers such as water and district cooling into the model
- Employing more efficient uncertainty modeling methods such as Information Gap Decision theory or robust optimization
- Modelling the interactions between different energy hubs with private owners through multi-agent based approaches
- Considering both energy and gas markets with detailed specifications
- Deploying cutting-edge technologies within the energy hubs

References

1. M. Shamshirband et al., Look-ahead risk-averse power scheduling of heterogeneous electric vehicles aggregations enabling V2G and G2V systems based on information gap decision theory. Electr. Pow. Syst. Res. **173**, 56–70 (2019)
2. J. Salehi et al., Scenario-based co-optimization of neighboring multi carrier smart buildings under demand response exchange. J. Clean. Prod. **235**, 1483–1498 (2019)
3. S. Nojavan, M. Majidi, K. Zare, Optimal scheduling of heating and power hubs under economic and environment issues in the presence of peak load management. Energ. Conver. Manage. **156**, 34–44 (2018)
4. F.S. Gazijahani, J. Salehi, Reliability constrained two-stage optimization of multiple renewable-based microgrids incorporating critical energy peak pricing demand response program using robust optimization approach. Energy **161**, 999–1015 (2018)
5. M. Vahid-Pakdel et al., Stochastic optimization of energy hub operation with consideration of thermal energy market and demand response. Energ. Conver. Manage. **145**, 117–128 (2017)
6. F.S. Gazijahani, J. Salehi, Game theory based profit maximization model for microgrid aggregators with presence of EDRP using information gap decision theory. IEEE Syst. J. **13**(2), 1767–1775 (2019)
7. M. Majidi, S. Nojavan, K. Zare, A cost-emission framework for hub energy system under demand response program. Energy **134**, 157–166 (2017)
8. A. Maroufmashat, A. Elkamel, M. Fowler, S. Sattari, R. Roshandel, A. Hajimiragha, et al., Modeling and optimization of a network of energy hubs to improve economic and emission considerations. Energy **93**, 2546–2558 (2015)
9. G. Moghaddam, M. Saniei, E. Mashhour, A comprehensive model for self-scheduling an energy hub to supply cooling, heating and electrical demands of a building. Energy **94**, 157–170 (2016)
10. A. Sheikhi, S. Bahrami, A.M. Ranjbar, An autonomous demand response program for electricity and natural gas networks in smart energy hubs. Energy **89**, 490–499 (2015)
11. A. Abdolahi et al., Probabilistic multi-objective arbitrage of dispersed energy storage systems for optimal congestion management of active distribution networks including solar/wind/CHP hybrid energy system. J. Renew. Sustain. Energy **10**(4), 045502 (2018)
12. E. Fabrizio, V. Corrado, M. Filippi, A model to design and optimize multi-energy systems in buildings at the design concept stage. Renew. Energy **35**, 644–655 (2010)
13. A. Najafi-Ghalelou et al., Robust scheduling of thermal, cooling and electrical hub energy system under market price uncertainty. Appl. Therm. Eng. **149**, 862–880 (2019)
14. M.-W. Tian et al., Risk-based stochastic scheduling of energy hub system in the presence of heating network and thermal energy management. Appl. Therm. Eng. **159**, 113825 (2019)

15. M.-R. Haghifam, S. Pazouki, S. Pazouki, Renewables and plug in electric vehicles modeling on electricity and gas infrastructures scheduling in presence of responsive demand, in *2013 3rd International Conference on Electric Power and Energy Conversion Systems (EPECS)*, 2013, pp. 1–6
16. M. Rastegar, M. Fotuhi-Firuzabad, M. Lehtonen, Home load management in a residential energy hub. Electr. Pow. Syst. Res. **119**, 322–328 (2015)
17. F.S. Gazijahani, J. Salehi, IGDT based complementarity approach for dealing with strategic decision making of price maker VPP considering demand flexibility. IEEE Trans. Ind. Inf., **16**, 1–9 (2019)
18. S. Pazouki, M.-R. Haghifam, A. Moser, Uncertainty modeling in optimal operation of energy hub in presence of wind, storage and demand response. Int. J. Electri. Power Energy Syst. **61**, 335–345 (2014)
19. M. Shamshirband et al., Decentralized trading of plug-in electric vehicle aggregation agents for optimal energy management of smart renewable penetrated microgrids with the aim of CO_2 emission reduction. J. Clean. Prod. **200**, 622–640 (2018)
20. M. Smith, P. Few, J. Twidell, Technical and operational performance of a small-scale comined heat-and-power (CHP) plant. Energy **20**, 1205–1214 (1995)
21. M.C. Bozchalui, S.A. Hashmi, H. Hassen, C.A. Cañizares, K. Bhattacharya, Optimal operation of residential energy hubs in smart grids. IEEE Trans. Smart Grid **3**, 1755–1766 (2012)
22. A. Najafi-Ghalelou, S. Nojavan, K. Zare, Heating and power hub models for robust performance of smart building using information gap decision theory. Int. J. Electr. Pow. Energy Syst. **98**, 23–35 (2018)
23. F.S. Gazijahani, S.N. Ravadanegh, J. Salehi, Stochastic multi-objective model for optimal energy exchange optimization of networked microgrids with presence of renewable generation under risk-based strategies. ISA Trans. **73**, 100–111 (2018)
24. F.S. Gazijahani, J. Salehi, Integrated DR and reconfiguration scheduling for optimal operation of microgrids using Hong's point estimate method. Int. J. Electr. Pow. Energy Syst. **99**, 481–492 (2018)
25. F.S. Gazijahani, J. Salehi, Robust design of microgrids with reconfigurable topology under severe uncertainty. IEEE Trans. Sustain. Energy **9**(2), 559–569 (2018)
26. J. Rajasekharan, V. Koivunen, Optimal energy consumption model for smart grid households with energy storage. IEEE J. Select. Top. Signal Process. **8**, 1154–1166 (2014)
27. P. Salyani et al., Chance constrained simultaneous optimization of substations, feeders, renewable and non-renewable distributed generations in distribution network. Electr. Pow. Syst. Res. **158**, 56–69 (2018)
28. S. Bahramara, H. Golpîra, Robust optimization of micro-grids operation problem in the presence of electric vehicles. Sustain. Cities Soc. **37**, 388–395 (2018)
29. F.S. Gazijahani, J. Salehi, Optimal bi-level model for stochastic risk-based planning of microgrids under uncertainty. IEEE Trans. Ind. Inf. **14**(7), 3054–3064 (2018)
30. L. Gelazanskas, K.A. Gamage, Demand side management in smart grid: a review and proposals for future direction. Sustain. Cities Soc. **11**, 22–30 (2014)
31. F.S. Gazijahani, J. Salehi, Stochastic multi-objective framework for optimal dynamic planning of interconnected microgrids. IET Renew. Power Gener. **11**(14), 1749–1759 (2017)

Chapter 6
Offering Strategy of Thermal-Photovoltaic-Storage Based Generation Company in Day-Ahead Market

Hooman Khaloie, Amir Abdollahi, Sayyad Nojavan, Miadreza Shafie-Khah, Amjad Anvari-Moghaddam, Pierluigi Siano, and João P. S. Catalão

6.1 Introduction

Sustainability and environmentally friendly as well as diminishing fossil fuel consumption are among the main benefits of turning to clean energy sources. However, these sources of energy are not free from defects, high investment costs, the intermittent output power of some of these resources (e.g., wind and solar units), and dependence on climate can be enumerated as the disadvantages of renewable energy sources. Nevertheless, the advantages of renewable energy sources preponderate over its disadvantages. In 2016, 52.4% of the electricity

H. Khaloie (✉) · A. Abdollahi
Department of Electrical Engineering, Shahid Bahonar University of Kerman, Kerman, Iran
e-mail: hoomankhaloiee@gmail.com; a.abdollahi@uk.ac.ir

S. Nojavan
Department of Electrical Engineering, University of Bonab, Bonab, Iran
e-mail: sayyad.nojavan@bonabu.ac.ir

M. Shafie-Khah
School of Technology and Innovations, University of Vaasa, Vaasa, Finland
e-mail: mshafiek@univaasa.fi

A. Anvari-Moghaddam
Department of Energy Technology, Aalborg University, Aalborg East, Denmark
e-mail: aam@et.aau.dk

P. Siano
Department of Management & Innovation Systems, University of Salerno, Fisciano, Italy
e-mail: psiano@unisa.it

J. P. S. Catalão
Faculty of Engineering of the University of Porto, Porto, Portugal
INESC TEC, Porto, Portugal
e-mail: catalao@ubi.pt

S. Nojavan, K. Zare (eds.), *Electricity Markets*,
https://doi.org/10.1007/978-3-030-36979-8_6

consumed by Danish consumers was supplied by renewable energy sources, which 37.6 and 2% were the shares of wind and photovoltaic units, respectively [1]. It should be mentioned that in 2017 43.7 % of Denmark's electricity demand was supplied through wind power share, which until now was the highest percentage of wind power contribution in Denmark's electricity industry [2].

Renewable energy sources with a large capacity or a group of renewable energy sources owned by a generation company (GenCo) must design appropriate offering strategies to achieve the maximum profit by participating in various electricity markets. Different attitudes and approaches in this problem, along with the representation of various mathematical models in accordance with the real technical specifications of generation units, are among the unique aspects of these studies in the literature of offering strategy problem. The offering strategy of a pumped-storage power plant in energy and ancillary services market is studied in [3, 4]. Contrary to [3], authors in [4] considered the risk associated with price forecasting errors of target markets using the covariance matrix in the process of maximizing profits. In [5], a risk-based offering strategy for a sample GenCo is proposed. Modeling the uncertainty associated with rival's behavior with the Monte Carlo technique and optimizing the whole proposed problem via SPSO-TVAC (self-organizing particle swarm optimization time-varying acceleration coefficients) is the main contribution of this work. In [6], an optimal offering strategy model for an emission-constrained GenCo is proposed. The authors modeled the electricity market price uncertainty through a set of scenarios while several emission allowances are considered to evaluate the impact of this parameter on GenCo's expected profit.

Another challenge faced by researchers in the optimal offering strategy problem is how to deal with the unspecified nature of parameters playing key roles in the optimization process. To this end, various approaches have been proposed by researchers of this field to deal with the uncertainties of the bidding strategy problem. Uncertainty management through a set of scenarios in the form of stochastic programming has been considered in [7]. That paper focuses on the offering strategy of a wind-hydro-pumped storage system, while water inflows for the reservoirs, markets prices, and wind power output are the considered uncertainties in this work. A stochastic bi-level self-scheduling framework for a GenCo in coordination with an electric vehicle load aggregator is suggested in [8], while the uncertainties related to wind power production and driving pattern of electric vehicle owners are modeled using appropriate scenario generation techniques. Also, authors in [9] have proposed a coordinated offering strategy for combined heat and power (CHP) units and renewable energy sources through the concept of the virtual power plant while the uncertain sources are taken into account with numerous scenarios. Robust optimization is another common approach in engineering and economic studies that assists the decision-maker in designing a suitable strategy for the worst realization of the uncertain parameters [10]. Kabiri Renani et al. [10] developed the SS problem for a transmission-constrained GenCo with incomplete market information while the robust optimization is used to deal with locational market prices (LMPs) and wind power production. In [11], the authors have developed a novel method for optimal participating of the wind power producers (WPP) in the day-ahead (DA) electricity market while the uncertainty associated with wind power and electricity prices are considered via stochastic scenarios. The authors benefit from kernel density estimation for modeling wind power uncertainty. In [12], short-term offering strategy for a price maker wind power producer has been introduced. The considered WPP in that paper is treated as a price-taker agent in the day-ahead market while its treatment in the balancing market is like a price-maker agent. Information gap decision theory (IGDT) [13], interval optimization [14], and hybrid probabilistic–possibilistic techniques [15, 16] are other approaches that have been repeatedly investigated by diverse researchers to cope with uncertainties in electricity market issues.

This chapter provides a risk-constrained offering strategy for a thermal-photovoltaic-battery storage (TPVBS) GenCo in the DA market. The uncertainty that stems from the DA and imbalance prices as well as photovoltaic (PV) production are taken into consideration via a set of scenarios. The offering strategy problem is formulated as a multi-stage stochastic programming problem while the emission limitations concerning the thermal units are incorporated in the offering process and the associated risk is modeled through conditional value at risk (CVaR) technique. The optimal offering strategy of the TPVBS system is examined in various risk levels, especially in both emission-constrained and emission-free conditions, and finally, appropriate offering curves will be obtained.

In the next section, the uncertainty modeling of input parameters, including electricity market prices and output power of the PV system, are described. Then the precise formulation of the proposed problem is presented. In the next section, numerical studies are conducted, and the simulation results are discussed. Eventually, the research findings are explained.

6.2 Uncertainty Modeling

In this chapter, uncertain sources are split into two categories: electricity prices and renewable production. The price of electricity in various markets is the most substantial factor affecting the offering strategy problem, which is entirely faced with many uncertainties. On the other hand, the output of the PV site is proportional to the solar irradiance, which is an uncertain parameter. Despite the almost zero irradiance during night-long, it is not even possible to consider a specified value for this parameter throughout the daylight. A variety of factors, including season and climatic conditions have the potential to affect the solar irradiance. For example, during certain hours of the daylight, solar radiation may be at the highest level, but due to specific weather conditions, such as cloudy weather, this potential can be significantly reduced. In the present chapter, normal and beta distributions are utilized to characterize the market prices and solar irradiance, respectively [17].

After modeling the probabilistic behavior of uncertain parameters with proper distribution functions, the roulette wheel technique (RWT) will be applied for scenario generation [18]. To this end, first, the continuous probability density

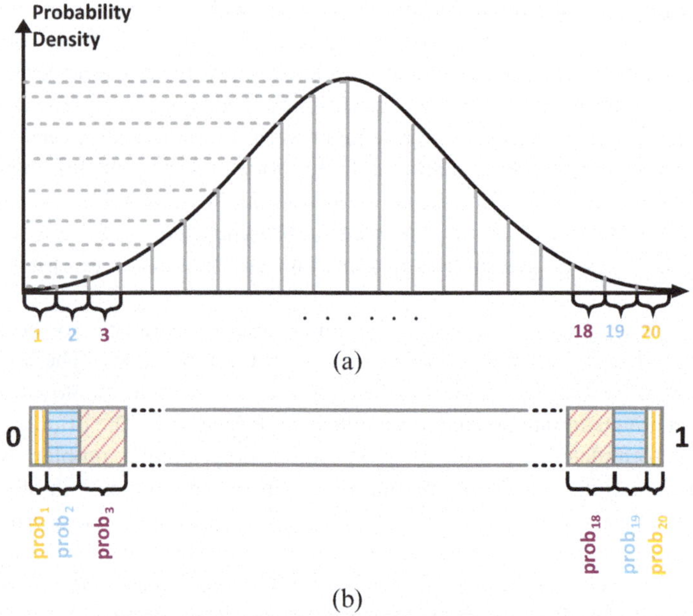

Fig. 6.1 A typical PDF and its relevant roulette wheel technique. (**a**) PDF of electricity prices. (**b**) Roulette wheel technique

functions (PDF) of each parameter are divided into 20 levels with their relevant normalized probabilities as depicted in Fig. 6.1a for the normal PDF. It is noteworthy to say that the number of levels for each parameter is selected in such a way that it does not reduce the precision of the proposed method and not raise the intricacy of the problem [18]. Next, as shown in Fig. 6.1, the interval [0, 1] is occupied by the different levels of discretized probability density function concerning their normalized probabilities. Then, a random number in the range of [0, 1] pertaining to each uncertain parameter is generated. This random number will be allocated to a specified level of the roulette wheel, which will represent the corresponding realization of the uncertain parameter in each scenario. This procedure will be reiterated till the required number of scenarios is attained. It is undeniable that considering a large number of scenarios will lead to an intractable problem. To this end, fast forward reduction technique is employed to reduce the initially generated scenarios [19]. Consequently, by applying this method, the initial scenarios pertaining to the electricity market prices (DA and imbalance prices) and solar irradiance are reduced to ten scenarios for each separate parameter. Eventually, the final set of scenarios for the proposed offering strategy problem will contain 1000 scenarios (10^3). It is worth highlighting that the current chapter does not cope with the correlation between electricity prices and renewable power production. A survey on the correlation between all uncertain parameters entails a new topic which is outside the scope of this chapter.

6.3 Problem Formulation

The offering strategy problem from the perspective of GenCos is an issue to maximize total profit in the intended scheduling horizon. In this problem, a suitable strategy for the participation of TPVBS system in the DA market is provided. The scheduling period is 24 h, and the uncertainty that originates from market prices (DA and imbalance prices) and production power of the PV site are characterized via appropriate scenarios. The proposed decision framework in the offering strategy problem is divided into three stages, which the classification of these decisions is presented in Table 6.1.

In the following subsections, at first, the objective function of the coordinated operation of all three sources, i.e., thermal units, PV site, and BSS, is presented, and then, the relevant constraints of the offering strategy problem will be entirely described.

6.3.1 *Objective Function*

The CVaR-based objective function of the suggested offering strategy for a sample TPVBS system shown in Fig. 6.2 with the aim of profit maximization is developed as a mixed integer programming (MIP) problem as follows:

$$\text{Max}\ \ PF^{TPVBS} = \sum_{s=1}^{S} prob_s \times \left[\sum_{t=1}^{T} \left\{ \left(\vartheta_{t,s}^{DA} \chi_{t,s}^{DA,th} \right) + \left(\vartheta_{t,s}^{DA} \chi_{t,s}^{DA,PV} \right) \right. \right.$$
$$+ \left(\vartheta_{t,s}^{DA} \chi_{t,s}^{DA,BS,dis} \right) - \left(\vartheta_{t,s}^{DA} \chi_{t}^{DA,BS,ch} \right)$$
$$+ \left(\vartheta_{t,s}^{DA} \rho_{t,s}^{+} \delta_{t,s}^{+} \right) - \left(\vartheta_{t,s}^{DA} \rho_{t,s}^{-} \delta_{t,s}^{-} \right)$$

Table 6.1 Classification of decisions in the proposed three-stage stochastic programming framework

First stage decisions	Charging power of BSS and operation status of BSS and thermal units
Second stage decisions	Offering curves of the TPVBS system in the DA market
Third stage decisions	Imbalance costs/incomes in the balancing market due to energy deviations in this market

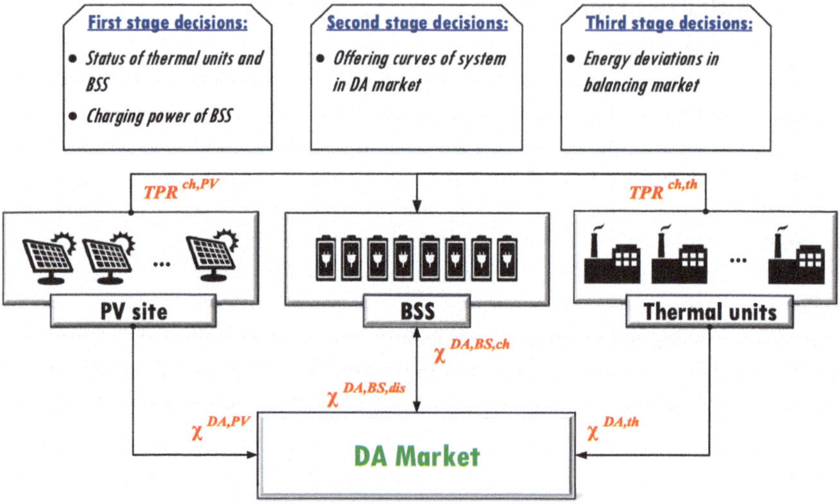

Fig. 6.2 Schematic of the proposed GenCo

$$-\sum_{g=1}^{G} CF_{g,t,s}\left(PR_{g,t,s}^{DA,th}+PR_{g,t}^{ch}\right)\Big\}\Big] - \sum_{t=1}^{T}\sum_{g=1}^{G}\left(U_{g,t}+D_{g,t}\right)$$

$$+\beta\left(\gamma-\frac{1}{1-\alpha}\sum_{s=1}^{S} prob_s \eta_s\right) \tag{6.1}$$

where the first two parentheses are related to the participation of thermal units and PV site in the DA market, respectively. The next two parentheses represent the income and expense terms of BSS for selling/buying energy in/from the DA market. The third row refers to income and expense of TPVBS system in the balancing market, while the fourth row calculates the costs arising from thermal units for the energy production as well as their start-up and shut-down. Finally, the last row represents the risk modeling term, namely CVaR.

6.3.2 Emission Constraint

In this chapter, it assumed that our TPVBS system is an emission-constrained power producer, which in certain circumstances, it cannot exceed the specified level of emission during the scheduling period. Equation (6.2) calculates the total expected emission of thermal units while the emission limitation of TPVBS system is imposed by (6.3).

$$EM^{TPVBS} = \sum_{s=1}^{S} prob_s \times \left[\sum_{\kappa}^{SO_2, NO_X} \sum_{g=1}^{G} E_{\kappa,g} \times \left(PR_{g,t,s}^{DA,th} + PR_{g,t}^{ch} \right) \right] \tag{6.2}$$

$$EM^{TPVBS} \le E_{\max} \tag{6.3}$$

6.3.3 CVaR Constraints

The constraints related to the applied risk index, i.e., CVaR, are expressed by the following equations:

$$\begin{aligned} &\left[\sum_{t=1}^{T} \left\{ -\left(\vartheta_{t,s}^{DA} \chi_{t,s}^{DA,th} \right) - \left(\vartheta_{t,s}^{DA} \chi_{t,s}^{DA,PV} \right) - \left(\vartheta_{t,s}^{DA} \chi_{t,s}^{DA,BS,dis} \right) \right. \right. \\ &+ \left(\vartheta_{t,s}^{DA} \chi_{t}^{DA,BS,ch} \right) - \left(\vartheta_{t,s}^{DA} \rho_{t,s}^{+} \delta_{t,s}^{+} \right) \\ &\left. \left. + \left(\vartheta_{t,s}^{DA} \rho_{t,s}^{-} \delta_{t,s}^{-} \right) + \sum_{g=1}^{G} CF_{g,t,s} \left(PR_{g,t,s}^{DA,th} + PR_{g,t}^{ch} \right) \right\} \right] \\ &+ \sum_{t=1}^{T} \sum_{g=1}^{N_G} \left(U_{g,t} + D_{g,t} \right) + \gamma - \eta_s \le 0, \quad \forall s \end{aligned} \tag{6.4}$$

$$\eta_s \ge 0, \quad \forall s \tag{6.5}$$

6.3.4 Imbalance Constraints

Constraints (6.6)–(6.8) are utilized to address the imbalances in the offering strategy of TPVBS system. Constraints (6.6) and (6.7) are fulfilled to, respectively, limit the negative and positive energy deviations of TPVBS system in the balancing market while Eq. (6.8) calculates the total energy deviations in the aforementioned market.

$$0 \le \delta_{t,s}^{-} \le CAP^{PV} + \sum_{g=1}^{G} CAP_g^{th} u_{g,t} + CAP^{dis} v_t^{dis}, \quad \forall t, \forall s \tag{6.6}$$

$$0 \le \delta_{t,s}^{+} \le \chi_{t,s}^{DA,th} + \chi_{t,s}^{DA,BS,dis} + RP_{t,s}^{PV} - TPR_t^{ch,PV}, \quad \forall t, \forall s \tag{6.7}$$

$$\delta^{+}_{t,s} - \delta^{-}_{t,s} = \left(\chi^{DA,th}_{t,s} + \chi^{DA,BS,dis}_{t,s} + RP^{PV}_{t,s} - TPR^{ch,PV}_{t}\right) - \left(\chi^{DA,th}_{t,s} + \chi^{DA,BS,dis}_{t,s} + \chi^{DA,PV}_{t,s}\right), \quad \forall t, \forall s \tag{6.8}$$

6.3.5 BSS Constraints

The operational constraints of the BSS are introduced in this subsection. The total provided energy by all thermal units for charging the BSS is represented in (6.9). Constraints (6.10) and (6.11) enforce the limitations pertaining to the maximum charging and discharging capacities of BSS. Constraint (6.12) prevents concurrent discharging and charging of BSS. The energy level of BSS will be updated according to (6.13) while the boundaries of this energy level are imposed in (6.14).

$$\sum_{g=1}^{G} PR^{ch}_{g,t} = TPR^{ch,th}_{t}, \quad \forall t \tag{6.9}$$

$$0 \leq \chi^{DA,BS,dis}_{t,s} \leq CAP^{dis} \upsilon^{dis}_{t}, \quad \forall t, \forall s \tag{6.10}$$

$$0 \leq \chi^{DA,BS,ch}_{t} + TPR^{ch,th}_{t} + TPR^{ch,PV}_{t} \leq CAP^{ch} \upsilon^{ch}_{t}, \quad \forall t, \forall s \tag{6.11}$$

$$\upsilon^{dis}_{t} + \upsilon^{ch}_{t} \leq 1, \quad \forall t \tag{6.12}$$

$$EL^{BS}_{t,s} = EL^{BS}_{t-1,s} - \left(\frac{1}{\Upsilon^{BS,dis}}\right)\left(\chi^{DA,BS,dis}_{t,s}\right) + \Upsilon^{BS,ch}\left(\chi^{DA,BS,ch}_{t} + TPR^{ch,th}_{t} + TPR^{ch,PV}_{t}\right), \quad \forall t, \forall s \tag{6.13}$$

$$0 \leq EL^{BS}_{t,s} \leq EL^{BS,\mathrm{Max}}, \quad \forall t, \forall s \tag{6.14}$$

6.3.6 Thermal Units Constraints

Thermal units are subject to several technical limitations which each of them will be individually presented in the following. Equation (6.15) computes the aggregate amount of units' offer in the DA market, while constraint (6.16) ensures that the offered energy by each thermal unit should be bound within its allowable production limit. Constraint (6.17) limits the provided power by each thermal unit for charging the BSS. Constraints (6.18) and (6.19) are utilized to model the start-up and shut-down costs of thermal units. Finally, the technical limitations pertaining to minimum up/down times as well as ramp-up/-down rates of each thermal unit are imposed by (6.20)–(6.25).

$$\sum_{g=1}^{G} PR_{g,t,s}^{DA,th} = \chi_{t,s}^{DA,th}, \quad \forall t, \forall s \tag{6.15}$$

$$\mathrm{MIN}_g^{th} u_{g,t} \le PR_{g,t,s}^{DA,th} + PR_{g,t}^{ch} \le CAP_g^{th} u_{g,t}, \quad \forall g, \forall t, \forall s \tag{6.16}$$

$$0 \le PR_{g,t}^{ch} \le CAP^{ch} u_{g,t}, \quad \forall g, \forall t \tag{6.17}$$

$$0 \le U_{g,t} \ge UC_g x_{g,t}, \quad \forall g, \forall t \tag{6.18}$$

$$0 \le D_{g,t} \ge DC_g y_{g,t}, \quad \forall g, \forall t \tag{6.19}$$

$$\sum_{n=t-UT_g+1}^{t} x_{g,t} \le u_{g,t}, \quad \forall g, \forall t \tag{6.20}$$

$$\left(\sum_{n=t-DT_g+1}^{t} y_{g,t} \right) + u_{g,t} \le 1, \quad \forall g, \forall t \tag{6.21}$$

$$y_{g,t-1} - u_{g,t} + x_{g,t} - y_{g,t} = 0, \quad \forall g, \forall t \tag{6.22}$$

$$PR_{g,t,s}^{DA,th} + PR_{g,t}^{ch} = PR_{g,t,s}^{tot,th}, \quad \forall g, \forall t, \forall s \tag{6.23}$$

$$PR_{g,t,s}^{tot,th} \le PR_{g,t-1,s}^{tot,th} + RU_g u_{g,t-1} + SRU_g x_{g,t}, \quad \forall g, \forall t, \forall s \tag{6.24}$$

$$PR_{g,t-1,s}^{tot,th} \le PR_{g,t,s}^{tot,th} + RD_g u_{g,t} + SRD_g y_{g,t}, \quad \forall g, \forall t, \forall s \tag{6.25}$$

6.3.7 PV System Constraints

Equations (6.26)–(6.29) are applied to bound the DA offer of the PV system, the charging power provided by the PV system for BSS, and the aggregate amount of DA power and the charging power within the maximum capacity of PV site.

$$0 \leq \chi_{t,s}^{DA,PV} \leq CAP^{PV}, \quad \forall t, \forall s \tag{6.26}$$

$$0 \leq TPR_t^{ch,PV} \leq CAP^{PV}, \quad \forall t \tag{6.27}$$

$$0 \leq TPR_t^{ch,PV} \leq CAP^{ch}, \quad \forall t \tag{6.28}$$

$$0 \leq TPR_t^{ch,PV} + \chi_{t,s}^{DA,PV} \leq CAP^{PV}, \quad \forall t, \forall s \tag{6.29}$$

6.3.8 Offering Curves Constraints

In many electricity markets, the power producer will be asked to submit non-decreasing energy offers in the electricity markets. Consider two different scenarios s and $\widetilde{s}$ that $\vartheta_{t,s}^{DA}$ is greater than $\vartheta_{t,\widetilde{s}}^{DA}$. The non-decreasing constraints will enforce that the offering quantity for a specific hour t in scenario s should be greater than or equal to the bidding quantity in the scenario $\widetilde{s}$. In fact, these constraints prevent the submit of inconsequent offers by the power producer in the electricity markets. The non-decreasing energy offer in the DA market is modeled according to the following Eq. (6.30):

$$\chi_{t,s}^{DA,\Gamma} \leq \chi_{t,\widetilde{s}}^{DA,\Gamma}, \quad \forall s, \widetilde{s} : \left[\vartheta_{t,s}^{DA} \leq \vartheta_{t,\widetilde{s}}^{DA}\right], \quad \forall t \quad \& \quad \Gamma = th/PV/BS, dis \tag{6.30}$$

$$\chi_{t,s}^{DA,\Gamma} = \chi_{t,\widetilde{s}}^{DA,\Gamma}, \quad \forall s, \widetilde{s} : \left[\vartheta_{t,s}^{DA} = \vartheta_{t,\widetilde{s}}^{DA}\right], \quad \forall t \quad \& \quad \Gamma = th/PV/BS, dis \tag{6.31}$$

where Eq. (6.31) is used to ensure that energy offers in two distinct scenarios with the same realization of electricity prices must be identical. This limitation is called non-anticipativity constraint.

6.4 Numerical Results

6.4.1 Input Data

In this section, the simulation results related to the offering strategy of a TPVBS system are presented. The considered GenCo in this chapter comprises a PV site, a BSS, and a thermal power plant with the nominal capacities of 150 MW, 50 MW, and 794 MW, respectively. The technical specifications of the BSS have been shown in Table 6.2. The nominal capacity of BSS has been assumed 50 MW while its discharging and charging efficiencies are equal to 0.95 and 0.8, respectively. Data on the characteristics of every thermal unit has been provided in Tables 6.3 and 6.4. As can be seen from this table, the considered power plant includes fourteen units, in which their quadratic cost function has been linearized with four blocks [20]. The historical data of the first half of 2018 has been utilized for the uncertainty modeling of electricity prices [21], and solar irradiance [22] has been given in Fig. 6.3. The value of α is set to 0.95. The intended problem has been formulated as a MIP problem which CPLEX under general algebraic modeling system (GAMS) has been employed to solve the suggested offering strategy problem.

6.4.2 Simulation Results

First, the simulation results of the offering strategy of TPVBS system in the DA market will be presented, and accordingly, the effect of imposing emission limitations on the offering strategy problem will be investigated. In other words, in the first study, the system maximizes its expected profit by ignoring constraint (6.3), whereas in the second study, the results of the offering strategy problem are examined under various emission limits.

Table 6.2 Information on BSS

Parameter	Value	Unit
$\Upsilon^{BS,dis}$	95	%
$\Upsilon^{BS,ch}$	80	%
CAP^{dis}	50	MW
CAP^{ch}	50	MW
EL^{BS}	250	MWh

Table 6.3 Data on the cost curve and the emission rate of each thermal unit

Generation units	Piece wise linearization parameters (MW)				Cost pertaining to each block (€/MW)				Emission ratios (lbs/MWh)	
	MIN	$P(1)$	$P(2)$	CAP	$C(1)$	$C(2)$	$C(3)$	$C(4)$	$E_{NO_X,g}$	$E_{SO_2,g}$
G1–G5	2.4	6	9.6	12	48.41	48.78	51.84	55.4	2.513	1.005
G6–G9	15.8	16	19.8	20	54.58	55.42	67.82	68.28	1.834	0.734
G10–G13	15.2	38	60.8	76	36.46	36.96	38.89	40.97	6.889	2.755
G14	140	227.5	280	350	35.08	35.66	36.09	36.72	18.371	7.348

Table 6.4 Technical data of each thermal unit

Generation units	RU_g and RD_g SRU_g and SRD_g (MW/hr)	UC_g (€)	DC_g (€)	UT_g (hr)	DT_g (hr)
G1–G5	12	87.4	8.74	4	2
G6–G9	20	15	1.5	1	1
G10–G13	35	715.2	71.52	8	4
G14	180	2298	229.8	4	4

The results of risk-based offering strategy for a TPVBS system have been reported in Table 6.5. According to this table, in the risk-neutral scheduling, i.e., $\beta = 0$, the expected profit, CVaR, and expected emission of TPVBS system are, respectively, equal to €244,454.898, €177,110.864, and 270,586.518 lbs. By changing the system's attitude towards a more conservative approach, i.e., increasing the value of β, the system's expected profit will lessen, and on the other side, the amount of CVaR will significantly grow. For example, by comparing two situations $\beta = 0$ and $\beta = 0.5$, it can be seen that the CVaR gain will be 3.8% while the expected profit will only reduce 0.07%.

Figure 6.4 illustrates the optimal participation of thermal units and PV site in the DA market for two separate scheduling approaches, i.e., $\beta = 0$ and $\beta = 4$. Overall, the participation level of these sources in the DA market by increasing parameter β will decrease. It stems from the fact that the system tends to lessen its participation in the market in the hope of diminishing its risk. The optimal behavior of BSS in the suggested offering model in two different modes of operation, namely risk-neutral and risk aversion, has been depicted in Fig. 6.5. By altering the operation mode of the system from a risk-neutral case to a risk aversion situation, it can be seen from these figures:

1. The charging period of BSS through thermal units will entirely change, except hour 1.
2. In the risk-neutral condition, the BSS does not benefit from the DA market for charging, while in the risk aversion situation, it purchases energy from the DA market at hours 7 and 16.
3. The stored energy level of the BSS system will considerably change. In the risk aversion case, it only includes one peak with the value of 155 MWh, while in the risk aversion state, it experiences two peaks of 110 MWh.

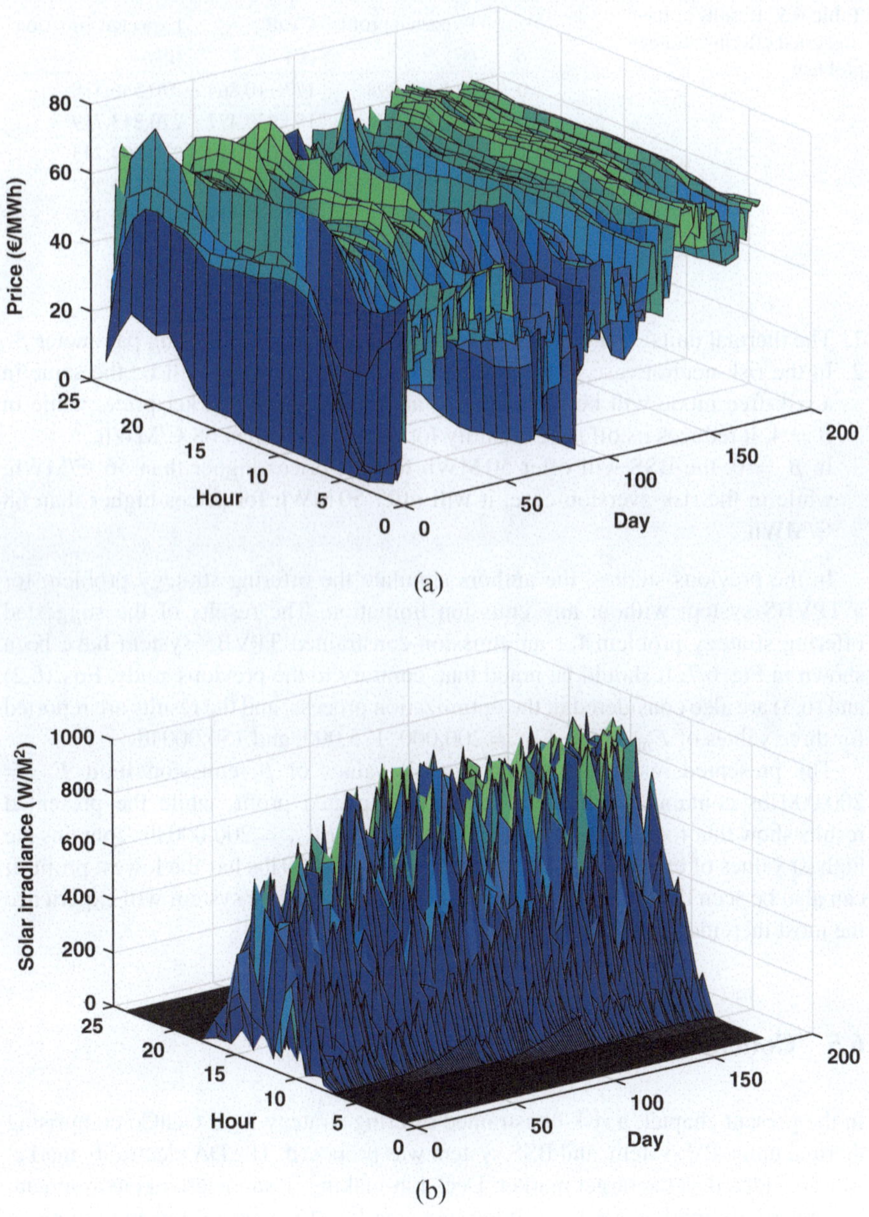

Fig. 6.3 Data on DA market price and solar irradiance. (**a**) DA market. (**b**) Solar irradiance

The offering curves of TPVBS system in the DA market for time interval $t = 14$ in two different values of β, i.e., $\beta = 0$ and $\beta = 4$ have been demonstrated in Fig. 6.6. It can be observed from these figures that:

Table 6.5 Results of the suggested offering strategy problem

β	Expected profit (€)	CVaR (€)	Expected emission (lbs)
0	244,454.898	177,110.864	270,586.518
0.5	244,278.972	183,839.477	270,843.719
1	243,485.075	185,033.449	270,909.745
2	242,642.108	185,668.329	270,556.397
4	242,421.594	185,737.515	270,556.397
6	242,313.791	185,783.973	270,556.397

1. The thermal units' strategy at this hour will not change by varying parameter β.
2. In the risk-neutral case, the participation of the PV system will be the same in a risk-free mode will be the same for all values of DA market price, while in $\beta = 4$, it reduces its offering quantity for prices lower than 68 €/MWh.
3. In $\beta = 0$, the BSS will offer 50 MWh for DA prices higher than 56 €/MWh, while in the risk aversion case, it will offer 50 MWh for prices higher than 68 €/MWh.

In the previous studies, the authors simulate the offering strategy problem for a TPVBS system without any emission limitation. The results of the suggested offering strategy problem for an emission-constrained TPVBS system have been shown in Fig. 6.7. It should be noted that, contrary to the previous study, Eqs. (6.2) and (6.3) are also considered in the optimization process, and the results are reported for three values of $E_{\max}$, i.e., $E_{\max}$= 200,000, 175,000, and 150,000 lbs.

The presented results show that for all values of β, emission limit $E_{\max}$= 200,000 lbs contains the highest values of expected profit, while the presented results show that for all values of β, emission limit $E_{\max} = 200{,}000$ lbs contains the highest values of expected profit, while $E_{\max} = 150{,}000$ lbs has the lowest profit. It can also be seen that by changing the $\beta = 0$ to $\beta = 0.5$, the system will experience the most increment in CVaR.

6.5 Conclusion

In the present chapter, a risk-constrained offering strategy for a GenCo comprising thermal units, PV system, and BSS system was proposed. The DA electricity market was considered as the target market. Decision-making in an uncertain environment, i.e., electricity market, requires addressing significant sources of uncertainty by an appropriate approach. To this end, all problem uncertainties, namely DA market price, imbalance price, and PV production, were characterized by a set of scenarios. Roulette wheel technique was employed to generate the desired number of scenarios, and finally, in order to prevent computational burden in the optimization stage, the fast forward reduction method was applied to reduce the initially generated scenarios. In the proposed methodology, an applicable risk measure, namely CVaR

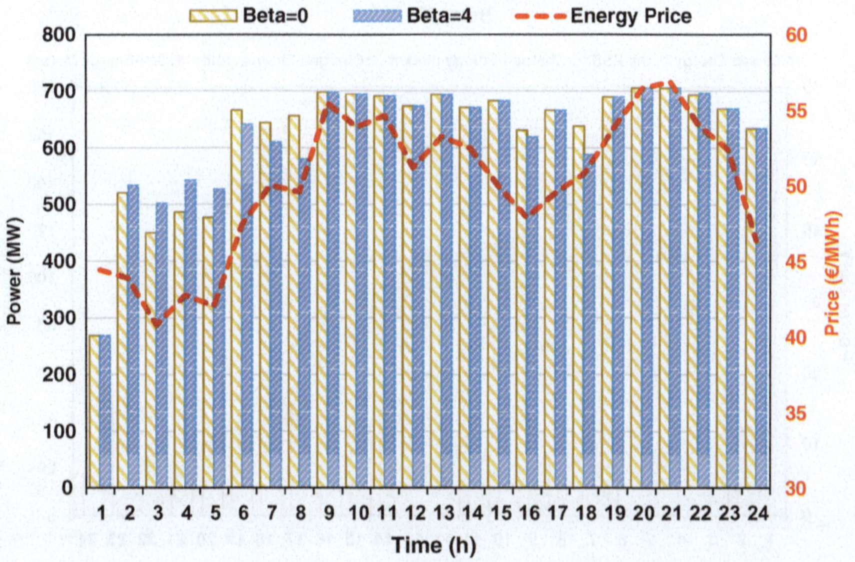

(a)

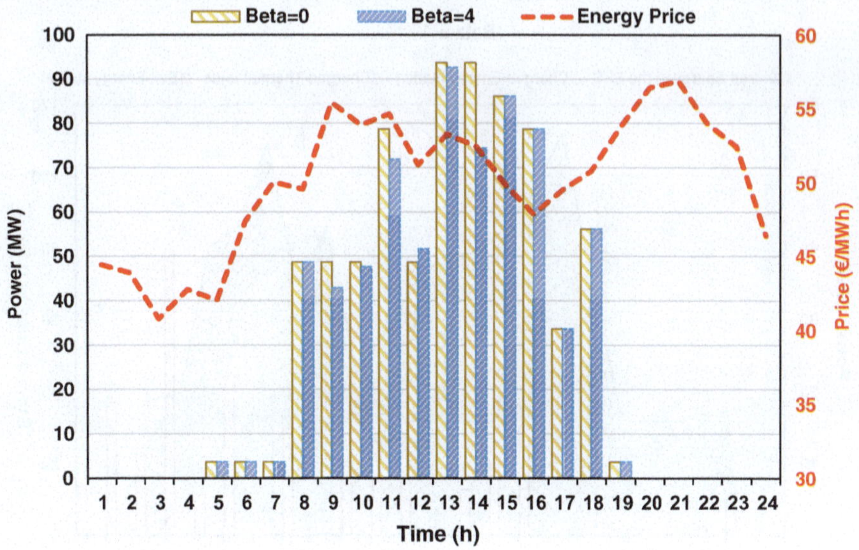

(b)

Fig. 6.4 Participation of thermal units and PV site in the DA market. (**a**) Thermal units. (**b**) PV site

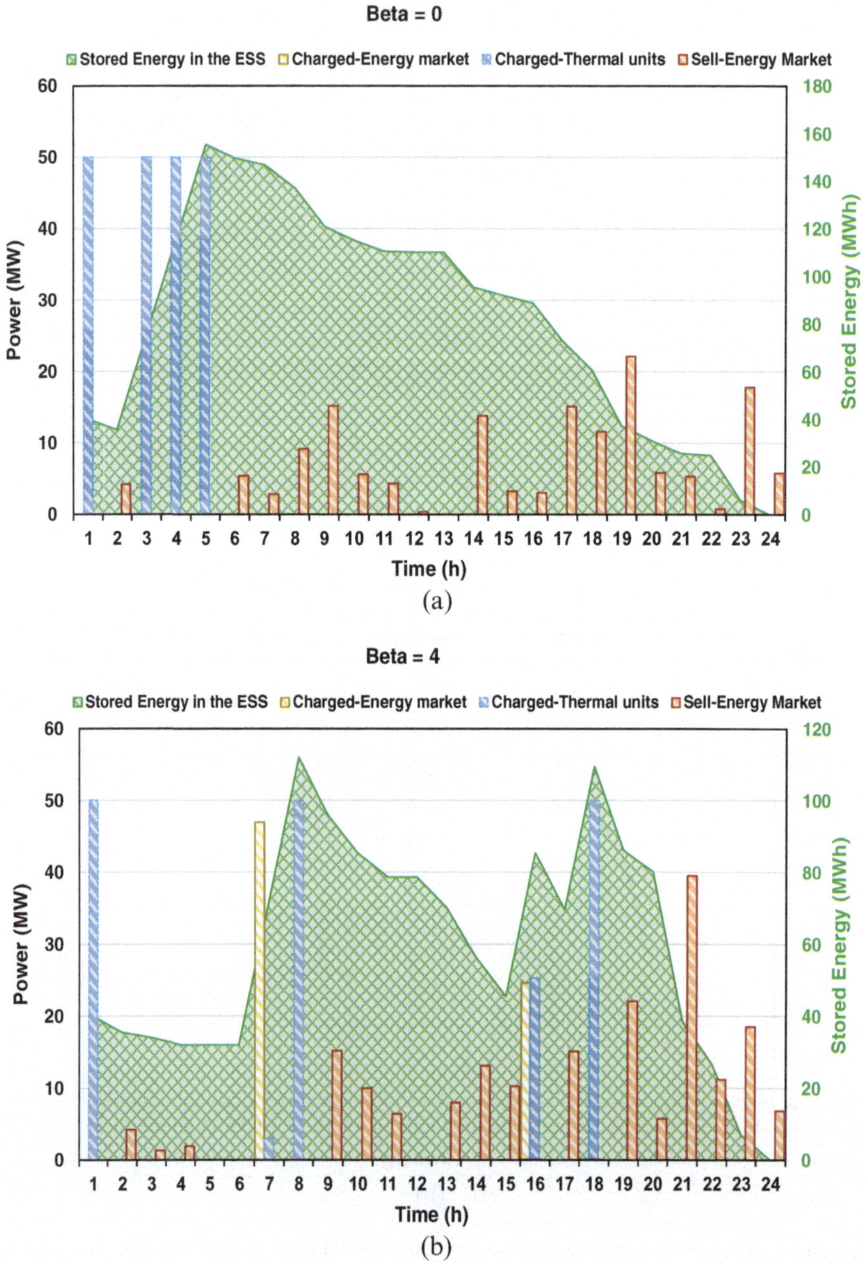

Fig. 6.5 Optimal behavior of BSS in the DA market. (**a**) Risk-neutral operation. (**b**) Risk aversion operation

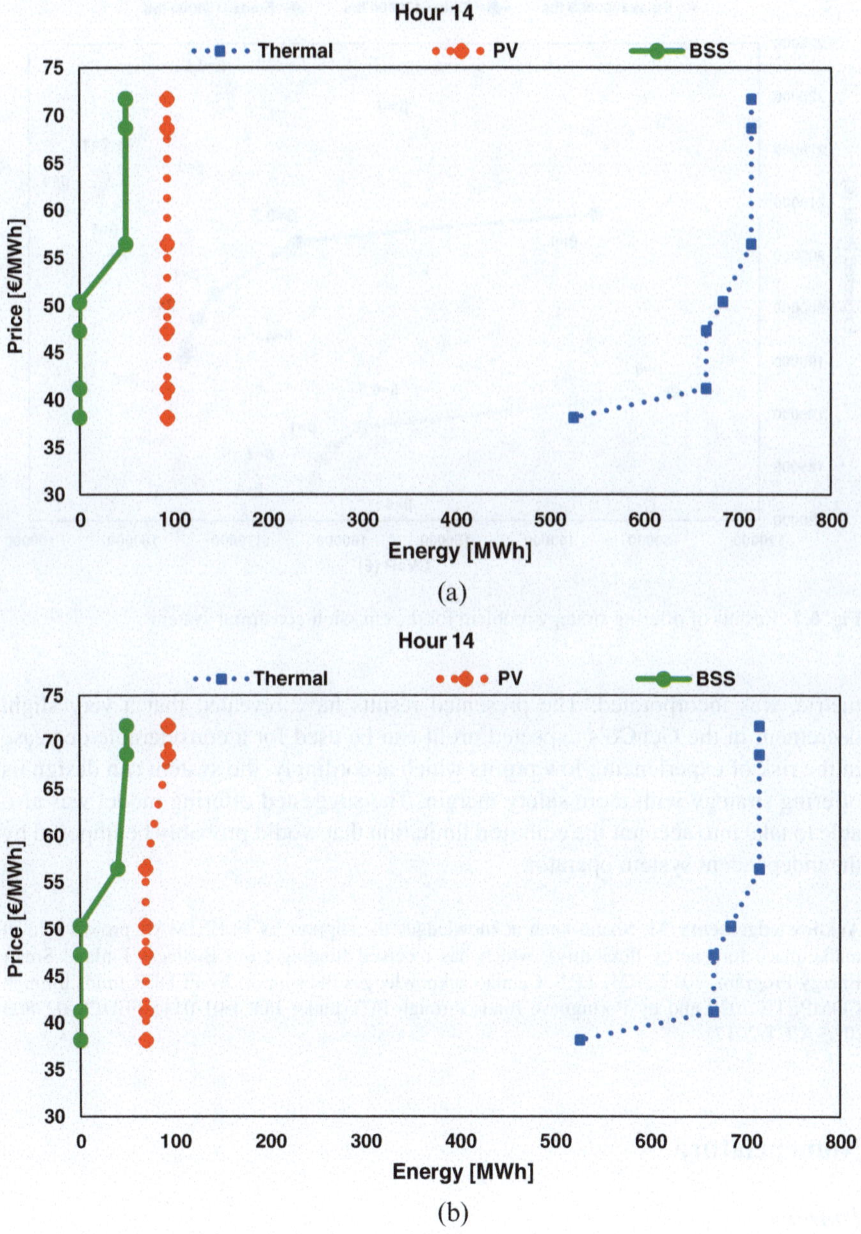

Fig. 6.6 Offering curves of TPVBS system in the DA market. (**a**) Risk-neutral case ($\beta = 0$). (**b**) Risk aversion case ($\beta = 4$)

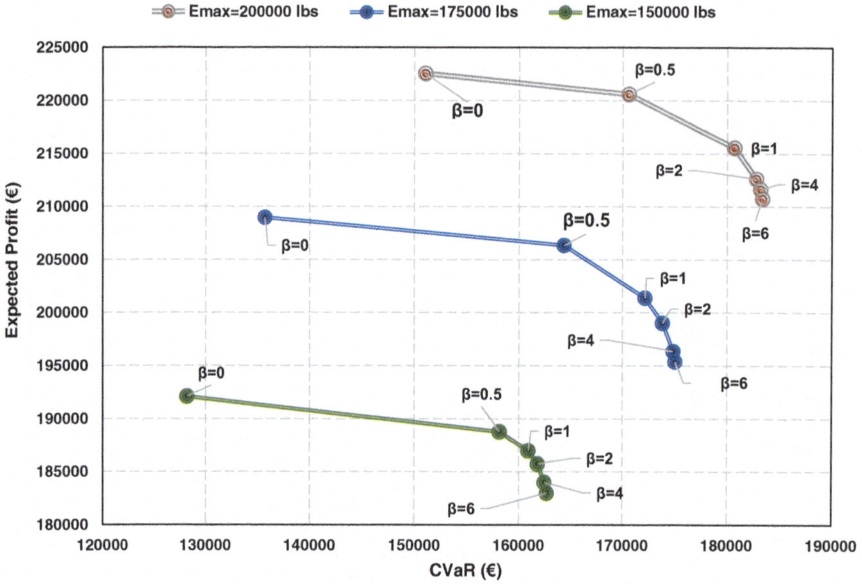

Fig. 6.7 Results of offering strategy problem for the emission-constraint system

metric, was incorporated. The presented results have revealed that a very slight decrement in the GenCo's expected profit can be used for a considerable decrease in the risk of experiencing low profits which accordingly, the system can design its offering strategy with more safety margin. The suggested offering model was also able to take into account the emission limitation that would probably be imposed by the independent system operator.

Acknowledgements M. Shafie-khah acknowledges the support by FLEXIMAR project (Novel marketplace for energy flexibility), which has received funding from Business Finland Smart Energy Program, 2017-2021. J.P.S. Catalão acknowledges the support by FEDER funds through COMPETE 2020 and by Portuguese funds through FCT under POCI-01-0145-FEDER-029803 (02/SAICT/2017).

Nomenclature

Indices

t Index indicating period
g Index indicating each thermal unit
s Index indicating scenario
k Index indicating emission type

Constants

$prob_s$	Probability of a scenario incidence
CAP^{PV}	Nominal capacity of the PV site, MW
UC_g/DC_g	Cost appertaining to start-up/shut-down of thermal units, €
DT_g/UT_g	Minimum down/up times of thermal units, hr
RU_g/RD_g	Rates appertaining to ramp up/down of thermal units, MW/hr
$E_{\max}$	Emission limitation of the system, lbs
$CAP_g^{th}/\mathrm{MIN}_g^{th}$	Upper/lower bound of permitted production of thermal units, MW
$P^{dis,\mathrm{Max}}/P^{ch,\mathrm{Max}}$	Maximum allowed charging/discharging power for ESS, MW
$PS_g^{th,S,\mathrm{Max}}$	Maximum allowable power of every thermal unit for taking part in spinning reserve market, MW.
$E_{k,g}$	Rate of emission appertaining to each emission type and each thermal unit, lbs/MWhr
SRU_g/SRD_g	Ramp limits appertaining to start-up/shut-down of thermal units, MW/hr
$C(L)$	Cost appertaining to segment of L in linearized cost curve of thermal units, €/MWh
$\Upsilon^{BS,dis}/\Upsilon^{BS,ch}$	BSS efficiencies appertaining to discharging/charging mode.
$EL^{BS,\mathrm{Max}}$	BSS maximum allowable stored energy, MWh

Variables

$\vartheta_{t,s}^{DA}$	Price appertaining to DA market, €/MW
$\chi_{t,s}^{DA,th}/\chi_{t,s}^{DA,PV}$	Offering quantity from thermal units/PV system in the DA market, MW markets, MW
$\chi_{t,s}^{DA,BS,dis}/\chi_{t,s}^{DA,BS,ch}$	Selling/purchasing quantity of BSS in the DA market, MW
$RP_{t,s}^{PV}$	Actual power of PV system, MW.
$PR_{g,t,s}^{tot,th}$	Final generated power of each thermal unit, MW
$\delta_{t,s}^{+}/\delta_{t,s}^{-}$	Upward/downward imbalance, MW
$U_{g,t}/D_{g,t}$	Cost appertaining to start-up/shut-down of thermal units, €
$CF_{g,t,s}()$	Cost function of thermal units
$PR_{g,t,s}^{DA,th}$	Offering quantity from each thermal unit in the DA market, MW
$PR_{g,t}^{ch}/TPR_t^{ch,th}/TPR_t^{ch,PV}$	Supplied charging power through each thermal unit/whole thermal units/ PV system for the BSS, MW

v_t^{dis}/v_t^{ch}	Binary variable appertaining to each operation mode of BSS, i.e., discharging/charging
$u_{g,t}/x_{g,t}/y_{g,t}$	Binary variable appertaining to online/start-up/shut-down status of thermal units
$EL_{t,s}^{BS}$	Stored energy in the BSS, MWh
$\rho_{t,s}^{+}/\rho_{t,s}^{+}$	Price ratios for upward/downward imbalance

References

1. Environmental Report 2017, Energinet. https://en.energinet.dk/About-our-reports/Reports/Environmental-Report-2017. Accessed 24 Jun 2019
2. Denmark Puts New Record in the Wind, Danish Energy. https://www.danskenergi.dk/nyheder/danmark-saetter-ny-rekord-vind. Accessed 14 Sep 2018
3. P. Kanakasabapathy, K. Shanti Swarup, Bidding strategy for pumped-storage plant in pool-based electricity market. Energ. Convers Manage. **51**, 572–579 (2010). https://doi.org/10.1016/j.enconman.2009.11.001
4. S.J. Kazempour, M.P. Moghaddam, M.R. Haghifam, G.R. Yousefi, Risk-constrained dynamic self-scheduling of a pumped-storage plant in the energy and ancillary service markets. Energ. Convers Manage. **50**, 1368–1375 (2009). https://doi.org/10.1016/j.enconman.2009.01.006
5. C. Boonchuay, W. Ongsakul, Optimal risky bidding strategy for a generating company by self-organising hierarchical particle swarm optimisation. Energ. Convers Manage. **52**, 1047–1053 (2011). https://doi.org/10.1016/j.enconman.2010.08.033
6. R. Laia, H.M.L. Pousinho, R. Melíco, V.M.F. Mendes, Self-scheduling and bidding strategies of thermal units with stochastic emission constraints. Energ. Convers Manage. **89**, 975–984 (2015). https://doi.org/10.1016/j.enconman.2014.10.063
7. A.A.S. De La Nieta, J. Contreras, J.P.S. Catalão, Optimal single wind hydro-pump storage bidding in day-ahead markets including bilateral contracts. IEEE Trans. Sustain Energ. **7**, 1284–1294 (2016). https://doi.org/10.1109/TSTE.2016.2544704
8. A. Tavakoli, M. Negnevistky, S. Saha, et al., Self-scheduling of a generating company with an EV load aggregator under an energy exchange strategy. IEEE Trans. Smart Grid **10**, 4253–4264 (2018)
9. J.Z. Riveros, K. Bruninx, K. Poncelet, W. D'haeseleer, Bidding strategies for virtual power plants considering CHPs and intermittent renewables. Energ. Conver. Manag. **103**, 408–418 (2015). https://doi.org/10.1016/j.enconman.2015.06.075
10. Y.K. Renani, M. Ehsan, M. Shahidehpour, Day-ahead self-scheduling of a transmission-constrained GenCo with variable generation units using the incomplete market information. IEEE Trans. Sustain Energ. **8**, 1260–1268 (2017). https://doi.org/10.1109/TSTE.2017.2672690
11. A. Botterud, Z. Zhou, J. Wang, et al., Wind power trading under uncertainty in LMP markets. IEEE Trans. Power Syst. **27**, 894–903 (2012). https://doi.org/10.1109/TPWRS.2011.2170442
12. M. Zugno, J.M. Morales, P. Pinson, H. Madsen. Pool strategy of a price-maker wind power producer. IEEE Trans Power Syst. **28**, 3440–3450 (2013). 10.1109/TPWRS.2013.2252633
13. S. Nojavan, K. Zare, B. Mohammadi-Ivatloo. Information gap decision theory-based risk-constrained bidding strategy of price-taker GenCo in joint energy and reserve markets. Electron. Power Compo. Syst. **45**, 49–62 (2017). https://doi.org/10.1080/15325008.2016.1231244
14. S. Nojavan, K. Zare, Interval optimization based performance of photo-voltaic/wind/FC/electrolyzer/electric vehicles in energy price determination for customers by electricity retailer. Sol. Energy. **171**, 580–592 (2018). https://doi.org/10.1016/j.solener.2018.07.004

15. H. Khaloie, A. Abdollahi, M. Rashidinejad, P. Siano, Risk-based probabilistic-possibilistic self-scheduling considering high-impact low-probability events uncertainty. Int. J. Electr. Power Energ. Syst. **110**, 598–612 (2019). https://doi.org/10.1016/j.ijepes.2019.03.021
16. H. Khaloie, A. Abdollahi, M. Rashidineiad, Risk-constrained self-scheduling and forward contracting under probabilistic-possibilistic uncertainties, in Iranian Conference on Electrical Engineering (ICEE). (IEEE, Piscataway, 2018), pp 1138–1143
17. H. Khaloie, A. Abdollahi, M. Shafie-Khah, et al., Co-optimized bidding strategy of an integrated wind-thermal-photovoltaic system in deregulated electricity market under uncertainties. J. Clean. Prod. **242**, 118434 (2020). https://doi.org/10.1016/j.jclepro.2019.118434
18. T. Niknam, M. Zare, J. Aghaei, Scenario-based multiobjective volt/var control in distribution networks including renewable energy sources. IEEE Trans. Power Deliv. **27**, 2004–2019 (2012)
19. N. Growe-Kuska, H. Heitsch, W. Romisch, Scenario reduction and scenario tree construction for power management problems, in 2003 IEEE Bologna Power Tech Conference Proceedings (IEEE, Piscataway, 2003), p. 7
20. H. Khaloie, A. Abdollahi, M. Shafie-Khah, et al., Coordinated wind-thermal-energy storage offering strategy in energy and spinning reserve markets using a multi-stage model. Appl. Energy **259**, 114168 (2020). https://doi.org/10.1016/j.apenergy.2019.114168
21. Bienvenido, ESIOS electricidad · datos · transparencia. https://www.esios.ree.es/es. Accessed 14 Mar 2019
22. Weather history+ - meteoblue. https://www.meteoblue.com/en/historyplus. Accessed 22 Apr 2019

Chapter 7
Risk-Based Purchasing Energy for Electricity Consumers by Retailer Using Information Gap Decision Theory Considering Demand Response Exchange

Ramin Nourollahi, Sayyad Nojavan, and Kazem Zare

Nomenclature

Parameters

C_{b}	Minimum expected cost of retailer
C_{o}	Critical cost for opportunity function
C_{r}	Critical cost for robustness function
$d(t)$	Time period
$f_{po}^{\mathrm{pen}}(t)$	Penalty of not running pool-order DR in time period t
$P_{f,b}^{\mathrm{DR,MAX}}(t)$	Highest demand in block b of forward DR f in time period t
$P_{f,b}^{\mathrm{MAX}}(t)$	Highest demand in block b of forward contract in time period t
$\overline{P}_{j}^{\mathrm{DR}}(t)$	Demand in jth step of reward-base DR in time period t
$P_{po}^{\mathrm{MAX}}(t)$	Highest demand in pool-order DR in time period t
$P^{\mathrm{req}}(t)$	Value of purchased power by retailer in period t
$\overline{R}_{j}^{\mathrm{DR}}(t)$	Highest value in jth step of reward-base DR in time period t
$\lambda_{po}(t)$	Price of pool-order DR in period t
$\lambda_{f,b}^{\mathrm{DR}}(t)$	Price of block b of forward DR f option in time period t
$\lambda_{f,b}^{\mathrm{F}}(t)$	Price of the block b of forward contract f in time period t

R. Nourollahi (✉) · K. Zare
Faculty of Electrical and Computer Engineering, University of Tabriz, Tabriz, Iran
e-mail: ramin.nourollahi96@ms.tabrizu.ac.ir; kazem.zare@tabrizu.ac.ir

S. Nojavan
Department of Electrical Engineering, University of Bonab, Bonab, Iran
e-mail: sayyad.nojavan@bonabu.ac.ir

S. Nojavan, K. Zare (eds.), *Electricity Markets*,
https://doi.org/10.1007/978-3-030-36979-8_7

$\tilde{\lambda}^{\mathrm{p}}(t)$	Forecasted pool market price
ω	Percentage increase in cost for retailer
Υ	Percentage decrease in cost for retailer

Numbers

N_{BDR}	Number of blocks in forward DR
N_{F}	Number of forward contracts
N_{FB}	Number of blocks in forward contracts
N_{FDR}	Number of contract in forward DR
N_J	Number of steps in reward-base DR
N_{po}	Number of pool-order options

Variables

$C(\mathrm{F})$	Total cost of forward contracts
$C(\mathrm{FDR})$	Total cost of forward DR program
$C(\mathrm{PO})$	Total cost of pool-order options
$EC(\mathrm{P})$	Total cost of power procurement from pool market
$EC(\mathrm{RDR})$	Total cost of reward-base DR
$P^{\mathrm{DR}}(t)$	Purchased power from reward-base DR in time period t
$P^{\mathrm{p}}(t)$	Purchased power from the pool market in time period t
$P_{po}(t)$	Purchased power from pool-order in time period t
$P_{f,b}^{\mathrm{DR}}(t)$	Purchased power from block b of forward DR f in time period t
$P_{f,b}^{\mathrm{F}}(t)$	Purchased power from block b of forward contract f in time period t
$R^{\mathrm{DR}}(t)$	Value of reward in time period t
$R_j^{\mathrm{DR}}(t)$	Value of reward of step j in time period t
$v_{\mathrm{DR},j}(t)$	Binary variable that shows which step is executed in time period t
$v_{po}(t)$	Binary variable which is 1 if pool-order is run in time period t
$\lambda^{\mathrm{p}}(t)$	Actual pool market price

Functions

$C(p, \lambda)$	Procurement cost function of retailer
$\hat{\alpha}\,(C_{\mathrm{r}})$	Robustness function
$\hat{\beta}\,(C_{\mathrm{o}})$	Opportunity function

7.1 Introduction

In the restructured electricity market retailer has an important role in procuring consumer energy which tends to reduce their energy procurement cost by using efficient DR programs. Also, in this markets retailer can sign contracts with aggregators and consumers to reduce their operation costs. Pool market and forward contracts are two main sources to supply consumer's demand by retailers, where retailers can buy their own power from these sources [1]. Also, DR programs are options that the retailer can use to improve the market efficiency [2]. Retailer can flatten the load curve using DR programs and reduce the cost of purchasing power at peak times [3]. Also, retailer should offer optimal bidding curves to the pool market in order to participate in the day-ahead market.

7.1.1 Literature Review

In [4], binary imperialist competitive algorithm and binary particle swarm optimization have been used to find optimal energy procurement for retailer using the combination of two algorithms. An appropriate optimal strategy is presented in [5] to procure energy of retailer from the pool market and bilateral contracts. In [6], a stochastic model helps retailers to maximizing their profits at an acceptable level of risk which is pool market price uncertainty in the short-term horizon and sets up appropriate contracts with suppliers and end users. Retailer problems in providing final selling prices to consumers and setting high prices for forward contracts are evaluated in [7], in which consumer's elasticity is considered against increasing or decreasing prices. Also, in order to analyze the retailer's risk in meeting with the uncertainties in the load and wholesale market price, a multistage stochastic optimization method has been introduced in [8]. One of the important issues for retailers is determining the optimum price for energy sales to consumers and procurement of its energy from the market which is studied in [9]. In [10], the financial risks caused by the uncertainty of the market price are addressed and solved in the mixed-integer stochastic optimization problem. Selling energy pricing to consumers with the time-of-use (TOU) rate is another problem for retailers which is addressed in [11]. In [12], in order to avoid the need to buy energy at the peak times from the pool market, shiftable load is proposed at the peak times according to TOU tariff. In [13], a linear programing method for mid-term contracts is provided, in which retailers offer the right consumer price considering the consumers' demand, market competition demand, and market prices. In [14], information gap decision theory (IGDT) approach is considered for handling and analyzing the uncertainty of pool price for mid-term scheduling. Comprehensive information on DR and its benefits as well as its positive effects in the electricity market is descripted in [15]. On the other hand, to properly handle multiple uncertainties inherent in the micro-grids, probabilistic energy management techniques are deployed in [16]. In [17] an algorithm for electricity market participant to obtain optimal bid under price uncertainty. The model is based on the known probability density functions of forecast prices. Reference [18] introduced a new approach to simulate electricity prices with hourly resolution for several months up to 3 years. In [19], the market elasticity caused by market price volatility is addressed. In [20], it can be seen that new DR schemes include several options for customer's satisfaction and improvement of load profiles. Retailers implant the coupon incentive-based demand response (CIDR) scheme in [21]. Authors in [22] first examined the implications of RES concentration on electricity price uncertainty and then decomposed the variance in electricity price into permanent and transitory components and explained what drives these trends. Also [23] used the toolkit of interest rate theory in the day-ahead electricity market in Asia.

In [24], distributed demand-side energy management strategy is proposed in which each user applied its best strategy to the total load and tariffs in the power distribution system. [25] mainly focuses on demand side management and demand response, including drivers and benefits, shiftable load scheduling methods, and peak shaving techniques. Authors in [26] model a residential customer in a multi-energy system (MES). In addition, demand response schemes have been classified based on their potential for field deployment in [27]. In [28] a multi-objective mixed integer linear programming model has been developed to minimization of peak load and cost of smart grids. Also, [29] determined selling price and compared by the retailer in the smart grid in three cases containing fixed pricing, time-of-use (TOU) pricing, and real-time pricing (RTP). In [30], an optimization model is proposed in which the consumer's hourly response to hourly changes in electricity prices. The beneficial results of TOU pricing is accessed through [31], in which the welfare of the TOU pricing is compared with fixed pricing. Technical aspects of DR programs have been investigated in order to control load management such as water heater, air conditioners, space heating, and cooling systems in [27, 32–35]. In [36], a new concept of DR is introduced as DR expected (DRX), in which DR is exchanged directly as a public good and traded between buyer and seller of DR. Also, the modified model of this design has been improved in [37]. For help to consumers in choosing a type of DR, three types of DRs are introduced and evaluated in [38]. The result of this evaluation is applied in several articles and the result can be seen in [39, 40]. In [41], a stochastic programming approach is used to measure the load curtailment capability of industry consumers in a short time period. In [42], a new demand response, called consumer preference, based demand response model introduced in a game-theoretic framework. As a good work, authors in [43] in an integrated energy system (IES) propose the pricing and operation strategy considering DR for an MG retailer. Retailers usually use DR programs to reduce their cost and risk. Several researches have addressed this issue to contain the uncertainty issues of load-serving entity, interruptible loads which is used in [44]. In [45], two interruptible load contracts, pay-in-advance and pay-as-you-go, are used to prevent power outages and reduce retailer damage at a time of falling prices. Also, in [46], self-production has been used to reduce the risk of market price fluctuations. In [47], it has been shown that interruptible loads can be used as energy sources for distribution companies (DISCOs). In [48], DISCOs use interruptible loads as energy sources in the day-ahead market. Also, in [49], in addition to interruptible loads, time-of-use and real-time pricing are used in order to reduce consumption of consumers. In [50, 51], a robust optimization method is used to model pool price uncertainty in order to obtain an optimal bidding strategy which is offered to the day-ahead market by the retailer.

In order to maximize the profit for suppliers and minimize the payments of customers, the building of bidding strategies is a major concern in the restructured power market because their profits depend on their bids. Still, now wide research work has been done on developing bidding strategies for generation-side market participant only and little work is done on demand-side participants. The problem of developing optimal bidding strategies for competitive generation companies was first introduced by David [52] and then surveyed by several researchers. Most of the researchers have used a linear bid function or quadratic bid function to build the bidding strategy for the electricity market participant. In [53, 54], a linear bid function is assumed to build the bidding strategies for the participant and the system is dispatched to maximize the social welfare. In [55], market clearing price with and without wind power has been evaluated in double sided bidding for linear bid and block bid trading model. Using graphical analysis, the MCP and schedules are determined under different market conditions in which quadratic bid function from both generating side and consumer side is considered [56]. In [57], a conceptual study is carried out on optimal bidding strategies of power suppliers in the operating Zhejiang provincial electricity market in which the stepwise bidding protocol is used. Finally, as a similar work [58] proposed a new framework in which demand response (DR) is incorporated as an energy resource of electricity retailers in addition to the commonly used forward contracts and pool markets.

In Table 7.1 reviewed papers above are clarified in order to compare difference between reviewed papers.

The work done in researches about DR programs can be categorized as follow:

1. The works in the articles focus on the basic concepts, formulation, and technical aspects of DR programs. Also, in a few articles, DR is traded directly between the seller and buyer of DR.
2. Few articles have reviewed DR from financial aspects in which most of these articles have been from the consumer's point of view.
3. Few articles have focused on DR options from the retailer's point of view.

7.1.2 Novelty and Contributions

According to the above, the work done in this study differs in four directions as follow:

1. DR is a public benefit and also is directly traded between the buyer and the seller.
2. This work proposes several schemes of DR for the retailer to implement them in accordance with his needs (Sect. 7.2, Fig. 7.1). This scheme covers the long

Table 7.1 Reviewed paper comparison

Reference	Considered DRP	Uncertain parameter	Uncertainty modeling	Time period	
[1]	—	1. Market price	IGDT	Short term (24 h)	Retailer
[3]	1. TOU	1. Market price 2. Demand 3. Outage RERs	Scenario	Short term (24 h)	Retailer
[5]	–	1. Market price 2. Demand	Scenario	1. Medium term 2. Short term	Retailer
[6]	–	1. Demand	Scenario	Long term	Retailer
[7]	–	1. Market price	Scenario	Long term	Retailer
[8]	–	1. Market price	Scenario	Long term	Retailer
[9]	–	1. Market price 2. Demand	Scenario	Medium term	Retailer
[10]	–	1. Market price 2. Demand	Scenario	Medium term	Retailer
[11]	–	1. Market price 2. Demand	Scenario	Long term	Retailer
[12]	1. TOU	1. Market price 2. Demand	Scenario	Medium term	Retailer
[13]	–	1. Market price 2. Demand 3. Rival-retailer prices	Scenario	Medium term	Retailer
[14]	TOU	1. Market price	IGDT	Medium term	Retailer
[20]	1. RTP 2. TOU 3. CPP	–	–	Short term	Consumers
[21]	1. RTP 2. TOU 3. CPP 4. PLP	–	–	Short term	Consumers
[24]	1. RTP 2. TOU 3. CPP	1. Market price 2. Demand	Scenario	Short term	Utility company and its customers/users
[29]	1. TOU	1. Market price	Scenario	Short term (24 h)	Retailer
[30]	1. RTP	1. Market price (for consumer)	Robust (for consumer)	Short term	Consumers
[31]	1. TOU	–	–	Short term	Electricity markets
[36]	1. Pool-based demand response	–	–	Short term	DR buyers and sellers

(continued)

Table 7.1 (continued)

Reference	Considered DRP	Uncertain parameter	Uncertainty modeling	Time period	
[37]	1. Demand response exchange (DRX)	–	–	Short term	DR buyers and DR sellers
[39]	1. CPP	–	–	Short term	Energy service provider
[40]	1. CPP	–	–	Short term	Utilities
[41]	1. Other DRP	1. Demand	Scenario	Medium term	Consumers
[44]	–	1. Demand	Scenario	Short term	Consumers
[45]	–	1. Market price	Scenario	Medium term	Retailer
[46]	–	1. Market price	Scenario	Long term	Large consumer
[47]	–	1. Market price	Scenario	Short term	Retailer
[48]	1. RTP 2. TOU	–	–	Short term	Electricity markets
[52]	–	1. Market price 2. Demand	Probabilistic methods	Short term	Consumers Bidding curve
[53]	–	Suppliers profit	Mont Carlo Method	Short term	Consumers Bidding curve
[58]	TOU	Market price	Robust optimiza-tion approach	Short term	Consumers Bidding curve
[61]	1. Reward-base	1. Demand	Scenario	Short term	Retailer
[62]	Other types	–	–	Short term	Ancillary service market
[63]	Other types	–	–	Short term	Electricity markets
This paper	1. Pool-order DR 2. Forward DR 3. Reward-base DR	1. Pool market price	IGDT approach	Long term	Electricity retailer

term or short term, which retailers choose according to their circumstances. These schemes are presented in the form of contracts which include: forward DR contracts, which are agreements for future periods in which a certain amount of energy is traded at a given price. The second option is a pool-order contract that will be used by the retailer at a time when the pool market is fluctuating; this contract is derived from the concept of well-known financial options referred in [59, 60]. Also, finally, a reward-base contract [61] is introduced, which is used as a real-time resource in proposed DR scheme. By using this option, the retailer

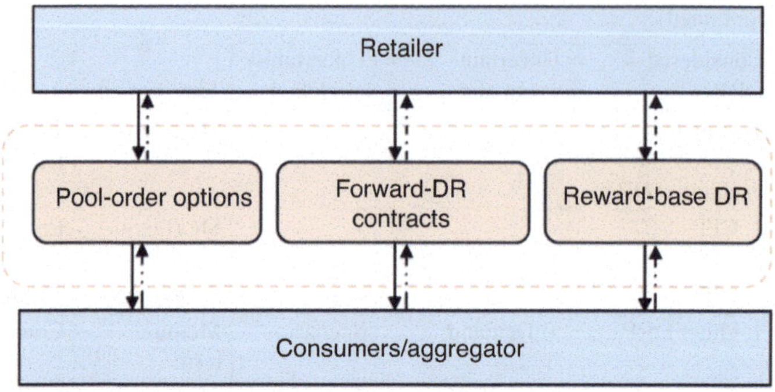

Fig. 7.1 Available DR programs for energy supply of retailers

can increase the amount of load reduction by increasing rewards. These schemes include long-term and short-term contracts that the retailer can use to procure their needed energy. From the viewpoint of difference between this work with previous works, proposed DR scheme allows a retailer to decide how to buy DR from aggregators and consumers.

3. Furthermore, IGDT technique is proposed to handling of market prices uncertainty. Using this technique, different strategies of increase and decrease in market prices are considered in which robustness and opportunity functions will provide the risk-averse and risk-taker strategies for retailer.
4. Finally, based on IGDT technique and using opportunity and robustness functions, optimal bidding strategy is obtained which the retailer can offer bidding strategy to day-ahead market to purchase its energy from the pool market. In general, IGDT analyzes effects of various amounts of deviation from optimal solution on the uncertain parameter.

According to mentioned contexts, novelty of this chapter can be summarized as follow:

1. Direct trade demand response between retailer and consumers.
2. Pool-order DR, forward DR, and reward-base DR are proposed as new DRP scheme.

7.1.3 Chapter Organization

The structure of this chapter is organized as follows: Sect. 7.2 introduces the mathematical formulation of the new DR scheme, market options, and objective function. The concept and formulation of the IGDT technique and the two important functions including opportunity and robustness will be presented in Sect. 7.3. The IGDT approach is applied to the base formulation to handling the uncertainty in the pool price in Sect. 7.4. The method for obtaining optimal bidding strategy is explained in Sect. 7.5. The input data and the detailed results are presented in Sect. 7.6. Finally, the conclusion of the chapter is presented in Sect. 7.7.

7.2 Problem Formulation

This chapter introduces a new design for direct DR exchange between retailer and consumer and aggregators; structure of this scheme is illustrated in Fig. 7.1. According to Fig. 7.1, retailer can use three options to buy DR from the consumers or aggregator in which each contract will be employed according to its price in different periods. Three DR options are set up in long-term and short-term contracts, and aggregated between consumers and retailers. In general consumers can sell their DR in different markets (for example, pool market or ancillary services [62, 63]), but this chapter considers only consumers who are able to sell their DR to retailers.

7.2.1 Objective Function and Power Balance Constraint

The objective function for retailer's cost is proposed in Eq. (7.1) in which total power procurement cost from the pool market, forward contracts, pool-order DR, forward DR, and reward-base DR are presented. Finally, power balance constraint is provided in Eq. (7.2) in which the total required power by the retailers is equal to the total demand reduction of the DRs and the power purchased from forward contract and pool market.

$$\text{Min } C(p, \lambda) = Ec(\text{P}) + C(\text{F}) + C(\text{PO}) + C(\text{FDR}) + C(\text{RDR}) \tag{7.1}$$

$$P^{\text{erq}}(t) = P^{\text{p}}(t) + P^{\text{F}}(t) + P_{po}^{\text{total}}(t) + P^{\text{FDR}}(t) + P^{\text{DR}}(t) \tag{7.2}$$

Equation (7.1) indicates that retailer total cost is equal to the cost of all used option, which includes the two market option such as pool market, forward contract, and three DR options such as pool-order DR, forward DR, and reward-base DR.

7.2.2 Wholesale Market Suppliers

Retailers can use two other market options other than DR in order to supply their consumer's energy. These two options include pool market and forward contract option.

7.2.2.1 Pool Market

The retailers can use the pool market as an option to buy or sell energy. But in this study, retailers use pool market only as a source to buy their energy. Also, in this study, pool price has been considered as an uncertain parameter in which IGDT technique is used to handle the pool price uncertainty in order to obtain optimal bidding strategy of retailers.

The total cost of purchasing energy from the pool market is calculated by Eq. (7.3).

$$EC\,(\mathrm{P}) = \sum_{t\in T} P^{\mathrm{P}}(t).\lambda^{\mathrm{P}}(t).d(t) \tag{7.3}$$

Equation (7.3) points out that the total cost of purchased power from pool market is equal to the amount of power purchased from pool market according to pool market price condition.

7.2.2.2 Forward Contract

Forward contract is an agreement to procure energy from market in which each contract has different blocks and each block has a specific price and volume. These blocks have a stepwise additive price.

Total forward contracts' cost is considered as follows:

$$C\,(\mathrm{F}) = \sum_{t\in T}\sum_{f=1}^{N_{\mathrm{F}}}\sum_{b=1}^{N_{\mathrm{FB}}} P_{f,b}^{\mathrm{F}}(t).\lambda_{f,b}^{\mathrm{F}}(t).d(t) \tag{7.4}$$

$$0 \le P_{f,b}^{\mathrm{F}}(t) \le P_{f,b}^{\mathrm{MAX}}(t) \tag{7.5}$$

$$P^{\mathrm{F}}(t) = \sum_{f=1}^{N_{\mathrm{F}}}\sum_{b=1}^{N_{\mathrm{FB}}} P_{f,b}^{\mathrm{F}}(t) \tag{7.6}$$

Equation (7.4) shows the total cost of forward contract for all blocks of all forward contracts in all periods. So, total cost of forward contract is equal to total used power from forward contracts which is equal to sum of the used power from each block of each forward contracts multiple price of the each block of the each forward DR contract in all periods.

Similar to other contracts, demand range of forward contract is shown in Eq. (7.5). The total demand of each block in forward contracts is shown in Eq. (7.6).

7.2.3 Pool-Order Option

Pool-order option is one of the options for retailers, which can be used to buy DR from the consumers or aggregator. In other words pool-order DR is a contract with few power volumes, which is proper when pool market price a low vitiations. Retailers usually use from this option when the market price increased. But, at the execution time use of this contract depends on profitability of this contract for the retailers. In other words, when the execution time of contract is coming, depending on the market price retailers will decide to run the contract or not. So, the execution of the contract depends on the market price at the time of execution of the contract and if the execution of the contract is not profitable, retailers with the payment of penalty to consumers or aggregator will be discontinued from the execution of the contract. In other words, if the cost of purchasing energy from the pool market is lower than the cost of pool-order options and penalties, the retailers will buy energy from the market. Figure 7.2 shows the structure of pool-order option.

Total cost of pool-order options is formulated as follows:

$$C\,(\mathrm{PO}) = \sum_{t \in T} \sum_{po=1}^{N_{po}} \left[P_{po}(t).\lambda_{po}(po).v_{po}(t).d(t) + \left(1 - v_{po}(t)\right).f_{po}^{\mathrm{pen}}(t) \right] \quad (7.7)$$

$$0 \le P_{po}(t) \le P_{po}^{\mathrm{Max}}(t) \quad \forall po = 1, 2, \ldots, N_{po} \quad (7.8)$$

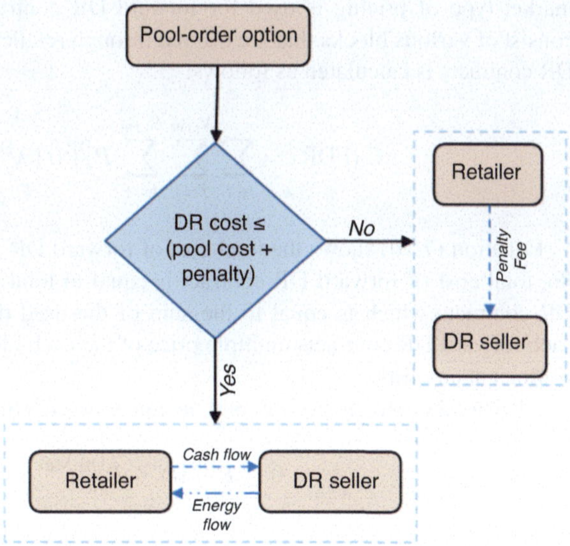

Fig. 7.2 Configuration of pool-order option

$$P_{po}^{\text{total}}(t) = \sum_{po=1}^{N_{po}} P_{po}(t).\nu_{po}(t) \tag{7.9}$$

Equation (7.7) shows the total cost of the pool-order option for all periods. This equation consists two parts, the first part is the cost of employed contract and second part of the fine is due to the not execute the contract as penalty cost. The power range of pool-order option is limited in Eq. (7.8). Also, total demand of all the executed pool-order options is shown in Eq. (7.9) which is equal to the sum of the total used contract *po*th, so that if pool-order *po*th is selected by consumers ν_{po} is one, otherwise it will be zero.

7.2.4 Forward DR

Forward DR contracts are agreements that are agreed for future periods in which a certain amount of energy is traded at a given price [55]. The pricing of forward DR contracts is usually done in two ways:

Over-the-Counter Market: In this type of pricing, both forward contract parties (seller and buyer) directly determine prices.

Exchange-Trade Market: This is a market in which standard contracts are traded at a specified volume and price. The benefits and prices in this type of pricing are determined through a centralized clearing house.

Because DR is directly traded by retailers and DR providers, over-the-counter market type of pricing is used for forward DR contracts. Forward DR contracts consist of various blocks that are offered through retailers. The total cost of forward DR contracts is calculated as follows:

$$C\,(\text{FDR}) = \sum_{t \in T} \sum_{f=1}^{N_{\text{FDR}}} \sum_{b=1}^{N_{\text{BDR}}} P_{f,b}^{\text{DR}}(t).\lambda_{f,b}^{\text{DR}}(t).d(t) \tag{7.10}$$

Equation (7.10) shows the total cost of forward DR for all blocks in all periods. So, total cost of forward DR contract is equal to total used demand from forward DR contracts which is equal to the sum of the used demand from each block of each forward DR contracts multiple price of the each block of the each forward DR contract at any time.

Also, total contracted power and the range of each block is introduced as follow:

$$0 \le P_{f,b}^{\text{DR}}(t) \le P_{f,b}^{\text{DR,Max}}(t) \tag{7.11}$$

$$P^{\mathrm{FDR}}(t) = \sum_{f=1}^{N_{\mathrm{FDR}}} \sum_{b=1}^{N_{BDR}} P_{f,b}^{\mathrm{DR}}(t) \tag{7.12}$$

Equation (7.11) shows the demand range of each block of forward DR contracts. Also, Eq. (7.12) shows total demand of all the executed forward DR contracts. In other words, total used demand by forward DR contracts is equal to sum of the used demand from each block of each forward DR contracts.

7.2.5 Reward-Base DR

The reward-base DR curve is shown in Fig. 7.3. According to this figure, the offered rewards by the retailers are increased while the volume of load reduction is increased in a stepwise curve. It should be mentioned that proposed reward function has been a linear function, which in this chapter, in order to simplify education to amateur consumers, is considered as a stepwise function. Therefore, the amount of traded DR in reward-base DR will depend on consumer's behavior.

Reward-base DR is modeled as follows:

$$P^{\mathrm{DR}}(t) = \sum_{j=1}^{N_J} \overline{P}_j^{\mathrm{DR}}(t).v_{\mathrm{DR},j}(t) \tag{7.13}$$

$$R^{\mathrm{DR}}(t) = \sum_{j=1}^{N_J} R_j^{\mathrm{DR}}(t) \tag{7.14}$$

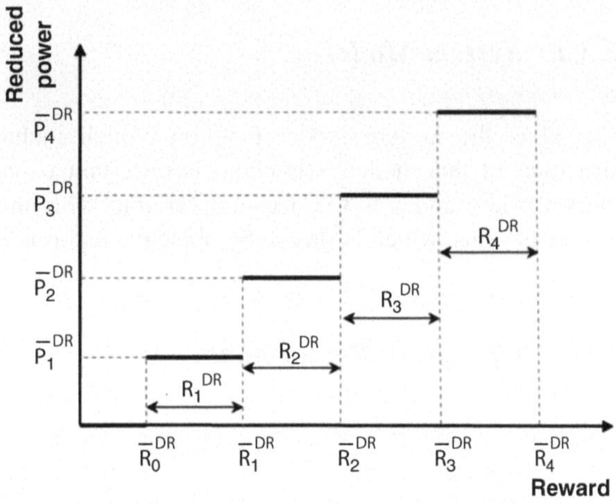

Fig. 7.3 The reward-base DR curve

$$\overline{R}_{j-1}^{\mathrm{DR}}(t).\nu_{\mathrm{DR},j}(t) \le R_j^{\mathrm{DR}}(t) \le \overline{R}_j^{\mathrm{DR}}(t).\nu_{\mathrm{DR},j}(t) \tag{7.15}$$

$$\sum_{j=1}^{N_J} \nu_{\mathrm{DR},j}(t) = 1 \tag{7.16}$$

Equation (7.13) shows the total demand reduced by consumers in reward-base DR which is equal to the sum of the selected steps demand by each consumer. Total reward paid by retailers because the reduced load is shown in Eq. (7.14), which is equal to total paid reward for reduced load. The range of each step is shown in Eq. (7.15). Equation (7.16) also shows that only one step can be selected from reward-base DR curve so that if step *j*th is selected by consumers v_j is one, otherwise it will be zero.

Total cost of reward-base DR is modeled as follows:

$$EC\,(\mathrm{RDR}) = \sum_{t \in T} \left[\sum_{j=1}^{N_J} \overline{P}_j^{\mathrm{DR}}(t).R_j^{\mathrm{DR}}(t).d(t) \right] \tag{7.17}$$

7.3 IGDT Technique

Uncertainty is usually onc of the main challenges in power system. This uncertainty may be detrimental or profitable for power system. One of the methods for handling these uncertainties is IGDT technique in which opportunity and robustness functions simulate cost and damage [64]. Each uncertainty is evaluated by this method in three sections: (a) system model, (b) operation requirements and (c) uncertainty model.

7.3.1 *System Model*

$C(p, \lambda)$ is the system model function, which include the system input/output structure. In this model, λ(lambda) is uncertain parameter (in this paper is pool market price) and p is the decision variable. This model can be expressed for a variety of aims, which in this study objective function is the retailers cost.

7.3.2 *Operation Requirements*

This section describes the operation requirements of the studied system in the form of different objective functions. The IGDT technique uses robustness and opportunity functions to evaluate uncertainty. The opportunity and robustness functions for retailers cost, which are objective functions, can be defined as follows:

$$\hat{\alpha}\left(C_{\mathrm{r}}\right)=\max_{\alpha}\left\{\alpha:\max\left(C\left(p,\lambda\right)\right)\leq C_{\mathrm{r}}\right\} \tag{7.18}$$

$$\hat{\beta}\left(C_{\mathrm{o}}\right)=\min_{\alpha}\left\{\alpha:\ \min\left(C\left(p,\lambda\right)\right)\leq C_{\mathrm{o}}\right\} \tag{7.19}$$

The immunity and degree of robustness to the uncertainty parameter versus high operation cost is measured in the robustness function. Furthermore, it illustrates the greatest level of uncertainty parameter at which the minimal requirements are always satisfied; therefore, a great $\hat{\alpha}$ value is favorable. In other words, the risk-aversion model of procurement strategy is provided in Eq. (7.18) which is robustness function. The decision will be robust versus uncertainty parameter for a great value of $\hat{\alpha}\left(C_{\mathrm{r}}\right)$.

Also, the favorable perspective of uncertainty parameter is expressed in the opportunity function. Furthermore, it provides the opportunity of obtaining benefits from low upstream grid prices. Here, $\hat{\beta}$ is defined as the minimum value of α which enables the possibility of low operation cost as a result of decisions. It should be noted that the opportunity function is the least value of α for which the operation cost of MG can be as small as a given value, C_{o}. Therefore, a small value of $\hat{\beta}$ is useful. A small value of $\hat{\beta}\left(C_{\mathrm{o}}\right)$ illustrates the condition in which the benefit is accessible versus low upstream grid prices. The related mathematical formulation of opportunity function in IGDT approach is expressed in Eq. (7.19), where C_{o} is generally smaller than C_{r} which are minimum and maximum variation in the uncertain parameter.

By analyzing and obtaining information from operation of the uncertain parameter λ, can be modeled this parameter (λ) using the IGDT technique. The modeling is described in the following section.

7.4 Proposed IGDT-Based Risk-Constraint Formulation

In this section, the IGDT technique is applied to the base formulation to handling the uncertain parameter at different risk levels.

7.4.1 Uncertainty Modeling

Equation (7.20) models a fractional info-gap uncertainty model. It should be noted that the uncertain parameter in this chapter is the pool price (λ).

$$U\left(\alpha,\tilde{\lambda}^{\mathrm{p}}(t)\right)=\left\{\lambda^{\mathrm{p}}(t)\ :\ \frac{\left|\lambda^{\mathrm{p}}(t)-\tilde{\lambda}^{\mathrm{p}}(t)\right|}{\tilde{\lambda}^{\mathrm{p}}(t)}\leq\alpha\right\},\qquad \alpha\geq 0 \tag{7.20}$$

Eq. (7.20) presents information gap uncertainty model, where the scale of gap depends on the forecasted parameter value $\tilde{\lambda}^{\mathrm{p}}(t)$.

7.4.2 Robustness Function (Risk-Averse Strategy)

In the robustness function, the maximum retailers' resistance (lowest loss against pool market price rises) against the pool price rise is shown by the parameter $\hat{\alpha}\,(C_{\mathrm{r}})$. This retailer's strategy against rising prices is risk-averse strategy. The robustness function is modeled as follows:

$$\hat{\alpha}\,(C_{\mathrm{r}}) = \max\left\{\alpha : \left(\max_{l \in U\left(\alpha, \tilde{\lambda}^{\mathrm{p}}(t)\right)} \cos t^{\mathrm{total}} \leq C_{\mathrm{r}} = (1+\omega)\,C_b\right)\right\} \tag{7.21}$$

The value of the opportunity function is obtained by minimizing α as follow:

$$\hat{\alpha}\,(C_{\mathrm{r}}) = \max\ \alpha \tag{7.22}$$

Subject to:

$$\mathrm{Max}\,\{Ec\,(\mathrm{P}) + C\,(\mathrm{F}) + C\,(\mathrm{PO}) + C\,(\mathrm{FDR}) + C\,(\mathrm{RDR})\} \leq C_{\mathrm{r}} \tag{7.23}$$

$$(1-\alpha)\,\tilde{\lambda}^{\mathrm{p}}(t) \leq \lambda^{\mathrm{p}}(t) \leq (1+\alpha)\,\tilde{\lambda}^{\mathrm{p}}(t) \tag{7.24}$$

$$\text{Equations (7.2)–(7.17)} \tag{7.25}$$

Since in the robustness function, maximum pool market price increase is obtained from$\lambda^{\mathrm{p}}(t) = (1+\alpha)\,\tilde{\lambda}^{\mathrm{p}}(t)$, the robustness function is reformulated as follows:

$$\hat{\alpha}\,(C_{\mathrm{r}}) = \max\ \alpha \tag{7.26}$$

Subject to:

$$\mathrm{Max}\,\{Ec\,(\mathrm{P}) + C\,(\mathrm{F}) + C\,(\mathrm{PO}) + C\,(\mathrm{FDR}) + C\,(\mathrm{RDR})\} \leq C_{\mathrm{r}} \tag{7.27}$$

$$\lambda^{\mathrm{p}}(t) = (1+\alpha)\,\tilde{\lambda}^{\mathrm{p}}(t) \tag{7.28}$$

$$\text{Equations (7.2)–(7.17)} \tag{7.29}$$

7.4.3 Opportunity Function (Risk-Taker Strategy)

Any reduction in the pool prices will be beneficial for retailers, which will be modeled by opportunity function. Using the opportunity function, the value of a reduction in pool price that guarantees a certain profit for the retailers is obtained. This retailer's performance is called risk-taker strategy. According to the above, the opportunity function derived from the IGDT technique is defined as follows:

$$\hat{\beta}\left(C_{\mathrm{o}}\right)=\min\left\{\alpha:\left(\min_{l\in U\left(\alpha,\tilde{\lambda}^{\mathrm{P}}(t)\right)}\cos t^{\mathrm{total}}\leq C_{\mathrm{o}}\right)=(1-\Upsilon)\,C_{b}\right\} \tag{7.30}$$

The value of the opportunity function is obtained by minimizing α as follow:

$$\hat{\beta}\left(C_{\mathrm{o}}\right)=\min\ \alpha \tag{7.31}$$

Subject to:

$$\mathrm{Min}\left\{Ec\left(\mathrm{P}\right)+C\left(\mathrm{F}\right)+C\left(\mathrm{PO}\right)+C\left(\mathrm{FDR}\right)+C\left(\mathrm{RDR}\right)\right\}\leq C_{\mathrm{o}} \tag{7.32}$$

$$(1-\alpha)\,\tilde{\lambda}^{\mathrm{P}}(t)\leq\lambda^{\mathrm{P}}(t)\leq(1+\alpha)\,\tilde{\lambda}^{\mathrm{P}}(t) \tag{7.33}$$

$$\text{Equations (7.2)–(7.17)} \tag{7.34}$$

Since in the opportunity function, the pool price reduction is considered than the real price is obtained from $\lambda^{\mathrm{P}}(t)=(1-\alpha)\,\tilde{\lambda}^{\mathrm{P}}(t)$. So the opportunity function is reformulated as follows:

$$\hat{\beta}\left(C_{\mathrm{o}}\right)=\min\ \alpha \tag{7.35}$$

Subject to:

$$\mathrm{Min}\left\{Ec\left(\mathrm{P}\right)+C\left(\mathrm{F}\right)+C\left(\mathrm{PO}\right)+C\left(\mathrm{FDR}\right)+C\left(\mathrm{RDR}\right)\right\}\leq C_{\mathrm{o}} \tag{7.36}$$

$$\lambda^{\mathrm{P}}(t)=(1-\alpha)\,\tilde{\lambda}^{\mathrm{P}}(t) \tag{7.37}$$

$$\text{Equations (7.2)–(7.17)} \tag{7.38}$$

7.4.4 Base Function (Risk-Neutral Strategy)

In risk-neutral strategy (without IGDT), retailers cost is calculated with the estimated value for pool market price.

The formulation of this strategy for the retailer is as follows:

$$\text{Min } C(p, \lambda) = Ec(\text{P}) + C(\text{F}) + C(\text{PO}) + C(\text{FDR}) + C(\text{RDR}) \tag{7.39}$$

Subject to

$$\text{Equations (7.2)–(7.17)} \tag{7.40}$$

7.5 Proposed Algorithm for Obtaining Optimal Bidding Strategy

To buy energy from the day-ahead market, retailers will need an appropriate buying strategy due to pool market price uncertainty. The appropriate buying strategy can be derived from the IGDT technique which is done using robustness and opportunity functions. Moreover, the proposed equations of IGDT technique are used to create the bidding curves for purchasing energy from day-ahead market. Day-ahead market is a future days market in which the pool price is uncertain. An instruction is presented to create an appropriate optimal bidding curve using opportunity and robustness functions. Figure 7.4 shows the performance of this curve in detail.

Different levels of cost are selected which are lower and higher than the expected cost (C_{ex}), i.e., $C_{\text{O2}} < C_{\text{O1}} < C_{\text{ex}} < C_{\text{R1}} < C_{\text{R2}}$. C_{R} and C_{O} is retailer's costs in the

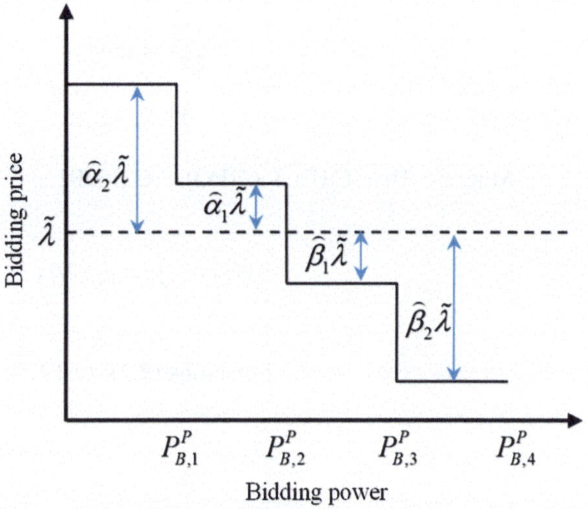

Fig. 7.4 The creation of a four-step bidding curve

robustness and opportunity functions, respectively. For each amount of costs, the optimum purchased price and the purchased power are obtained. Finally, using of the result based on IGDT technique, optimal bidding curve is obtained to bid to the day-ahead market. The proposed method for obtaining the optimal bidding curve is presented as follows:

1. The proposed cost function (1) is minimized subject to constraints (7.2)–(7.17) in which the obtained cost is set as without IGDT (risk-neutral) cost (C_{ex}).
2. After obtaining the result of the minimized cost from step 1, critical costs for robustness function as C_{R} are defined in which these costs increase in a fixed step versus C_{ex}. It should be noted that the minimum value of obtained cost in step 1 is smaller than the defined value for C_{R}. In each stage, the robustness function (7.26) subject to constraints (7.27)–(7.29) is solved and the optimal amount of the confidence level ($\hat{\alpha}_k > \hat{\alpha}_{k-1} > \cdots > \hat{\alpha}_2 > \hat{\alpha}_1$) is obtained. By using of it, can obtain the optimal purchased price ($\left(\hat{\alpha}_k + 1\right) \times \lambda$) and the amount of purchased power in proportion to it is obtained in solving the robustness function.
3. After obtaining the result of the minimized cost from step 1, critical cost (C_{O}) for opportunity function is defined; this cost decreases in a fixed step. It should be noted that the minimum value of cost obtained in step 1 is larger than the defined value of C_{O}. In each stage, the opportunity function (7.35) subject to constraints (7.36)–(7.38) is solved and the optimal amount of the confidence level according to Fig. 7.4 ($\hat{\beta}_k > \hat{\beta}_{k-1} > \cdots > \hat{\beta}_2 > \hat{\beta}_1$) is obtained. By using it, can obtain the optimal purchased price ($\left(1 - \hat{\beta}_k\right) \times \lambda$) and the amount of purchasing power in proportion to obtained from solved the opportunity function.
4. After solving the opportunity and robustness functions, and obtaining the $\hat{\alpha}$ and $\hat{\beta}$ values in each iteration K ($\hat{\alpha}_k > \hat{\alpha}_{k-1} > \cdots > \hat{\alpha}_2 > \hat{\alpha}_1$ and $\hat{\beta}_k > \hat{\beta}_{k-1} > \cdots > \hat{\beta}_2 > \hat{\beta}_1$), the purchased price will be calculated in accordance with Fig. 7.4. Finally, Fig. 7.5 shows the proposed algorithm to create optimal bidding curve.

7.6 Case Study

The proposed scheme intended for 32 periods includes peak times of summer and winter. Each period includes the peak times of 1 week. This scheme includes 12 weeks of January–March, 17 weeks of June–September, and 3 weeks of December. The amount of demand is obtained by averaging the peak times from Monday to Friday each week. Note that the peak time of summer days is from 11 am to 9 pm, while those of winter days are from 6 am–10 am to 4 pm–10 pm. According to the used method in [65, 66], selected peak times come from the Queensland daily curve in 2012 [67].

The selected forward contracts in this chapter include three contracts (F1–F3). F1 covers the first three months of the time horizons. F2 also covers the 17 weeks of winter and finally, F3 has been selected for 3 weeks in December. Each forward contract consists of 6 blocks, which have specific volume and maximum demand.

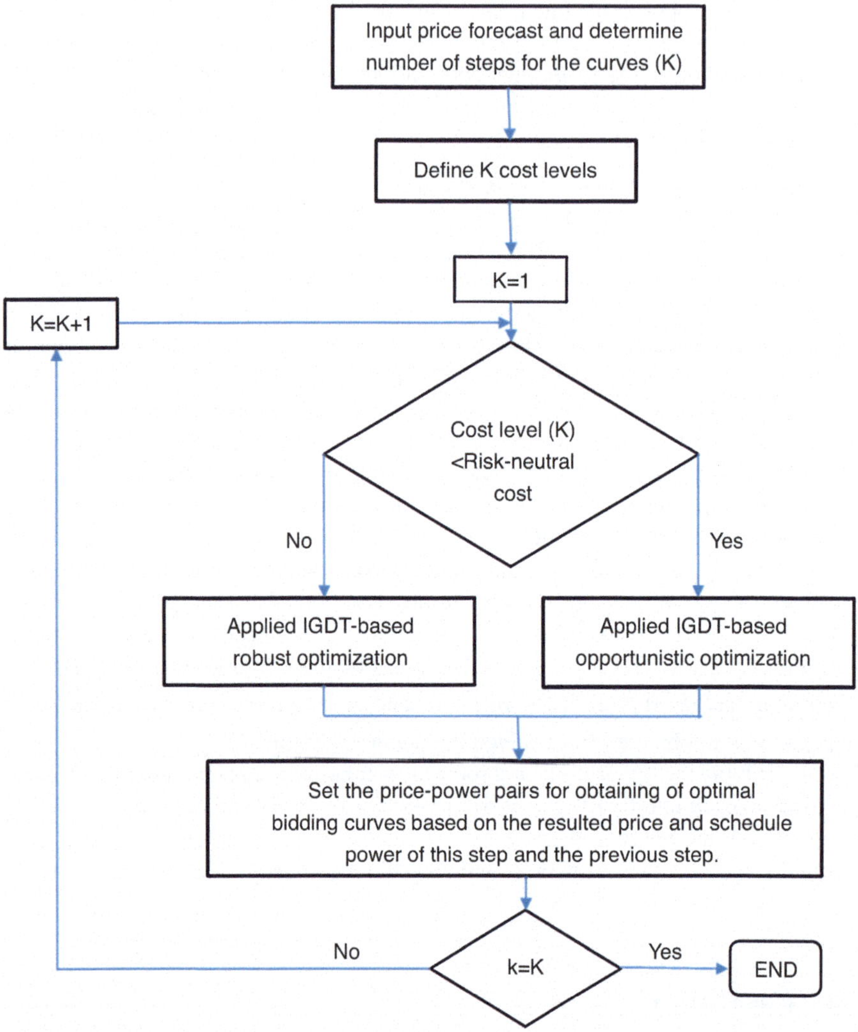

Fig. 7.5 The proposed algorithm to create optimal bidding curve

Forward price for each quarter of Queensland in 2012 are used here [69]. It should be noted that the maximum demand for each block is 450 MW.

Four pool-order options are considered to be used when the market price has been increases. Each contract has a specific volume of demand and a negotiated price for any period. Also, the maximum demand for each pool-order option is 50 MW. If the contract is not profitable, the penalty will be 15% of the total price of the contract.

Forward DR contract is set for a 1-month period. Therefore, eight contracts for the forward DR are set up which is like the forward contract, and each forward DR contract consist of six blocks. Maximum demand for each block of forward DR contract is 75 MW.

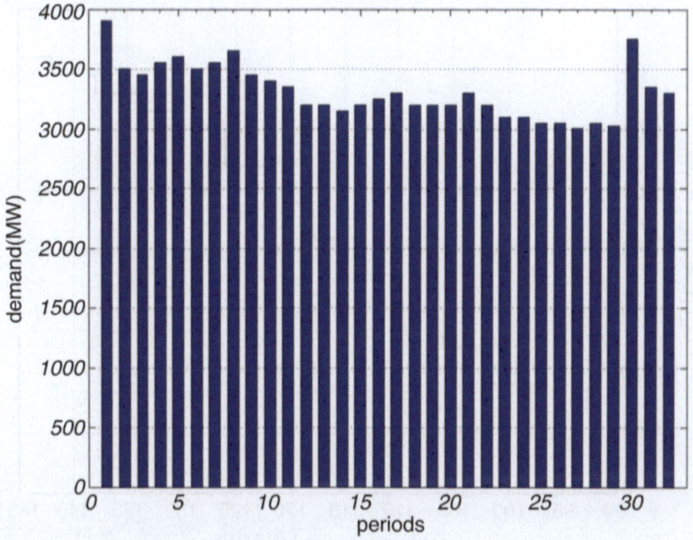

Fig. 7.6 Required demand by the retailers

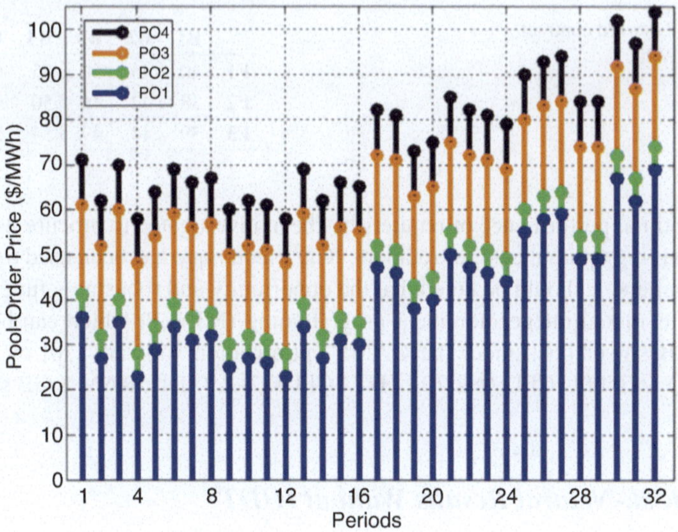

Fig. 7.7 Pool-order option prices

Reward-base DR consists of 14 steps which has a defined reward to reduce demand. Consumers can get more rewards from retailers by increasing the load reduction. Figures 7.6, 7.7, and 7.8 and Tables 7.1 and 7.2 indicate required inputs data. Required demand data are achieved from ref [69].

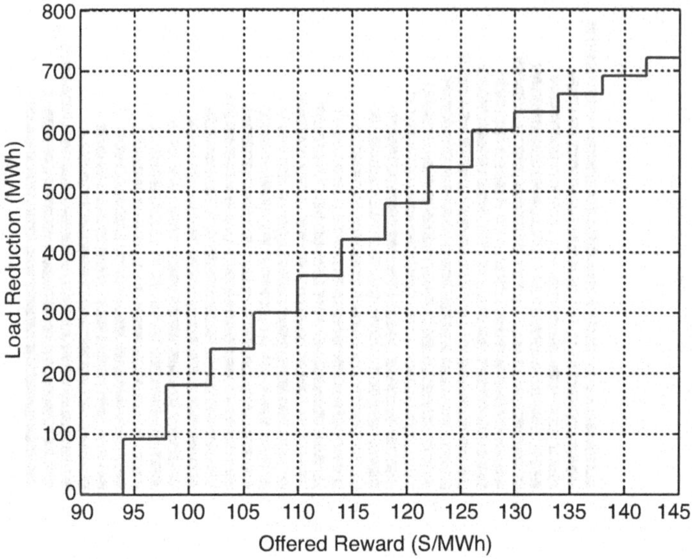

Fig. 7.8 Amount of reward for each reduced demand

Table 7.2 Forward contract prices ($/MWh)

	B1	B2	B3	B4	B5	B6
F1	40	45	50	55	60	65
F2	38	42	46	50	54	58
F3	39	44	49	54	59	64

The optimal performance of retailers has been investigated to procure its energy. After obtaining optimal cost for retailers, IGDT technique has been used to analyze risk of retailers. It should be noted that the opportunity and robustness functions are modeled as mixed-integer nonlinear programming (MINLP) which can be solved using SBB solver [70] under the GAMS optimization software [70] on Intel(R) Core(TM) i7-7500U CPU @ 2.70 GHz (4 CPUs), ~2.9 GHz, RAM 8 GB system.

7.6.1 Risk-Neutral Results Without IGDT

By solving the proposed objective function (7.17) under the constraints (7.2)–(7.17), the retailers cost is obtained for a risk-neutral strategy. This strategy will be solved without considering pool market price uncertainty. Results of considered problems can be considered in two cases as follow:

Case A: Solving the proposed model without considering DRP
Case B: Solving the proposed model with considering DRP

Table 7.3 Forward DR prices ($/MWh)

	B1	B2	B3	B4	B5	B6
FDR1	35	37	39	41	43	45
FDR2	32	34	36	38	40	42
FDR3	29	31	33	35	37	39
FDR4	33	35	37	39	41	43
FDR5	45	47	49	51	53	55
FDR6	51	53	55	57	59	61
FDR7	56	58	60	62	64	66
FDR8	69	71	73	75	77	79

Table 7.4 Results of risk-neutral strategy

	Case A	Case B
The expected cost ($)	4,873,026.5$	4,681,597.4$
The decreased operation cost ($)	0	191447.1$
The decreased operation cost (%)	0	3.93%

Effect of the new DR schemes on the retailers total cost in risk-neutral strategy can be proposed in a table similar to Table 7.3. According to the noted result in risk-neutral strategy, the retailers' costs for case B and case A is 4,681,597.4$ and 4,873,026.5$, respectively. In comparison with cases A and B, beneficial results of DR are applied in case B; therefore, the reduced expected cost in this case is 191,447.1$, which is 3.93% less than case A (Table 7.4).

7.6.2 Robustness and Opportunity Functions

The simulation results of robustness function in IGDT approach are used by retailers when it has a risk-averse strategy. Figure 7.9 indicates the robustness function $\hat{\alpha}\,(C_{\mathrm{r}})$versus the robust cost. Figure 7.9 is obtained by solving the robustness function (7.26) and considering the constraints (7.27)–(7.29). This figure represents the risk-averse strategy of retailers, in which the retailers by paying more money than risk-neutral strategy is more robust against market price rises. In other words according to Fig. 7.9, the robustness function increases as C_{R} increases as expected, and the retailers should pay more cost to have more robust strategy. Fig. 7.9 shows the positive impact of the proposed new DR scheme on retailers costs. For example, the retailers pays 4,903,026.5$ which is 30,000$ more than the risk-neutral strategy, robustness function values are 1.6% in "without DR" mode and 20.4% in "with DR" mode. Therefore, using DR mode, retailers is 18.8% stronger than without DR mode. In other words, retailer by paying 30,000$ more than the risk-neutral strategy can be 1.6% and 20.4% stronger against pool market price increases, in "without" and "with DR" mode. This shows the importance of the proposed DR schemes in this study.

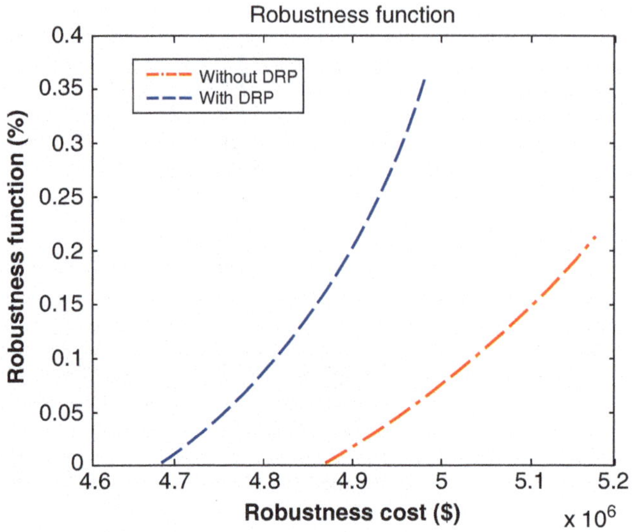

Fig. 7.9 Robustness function

Fig. 7.10 Opportunity function

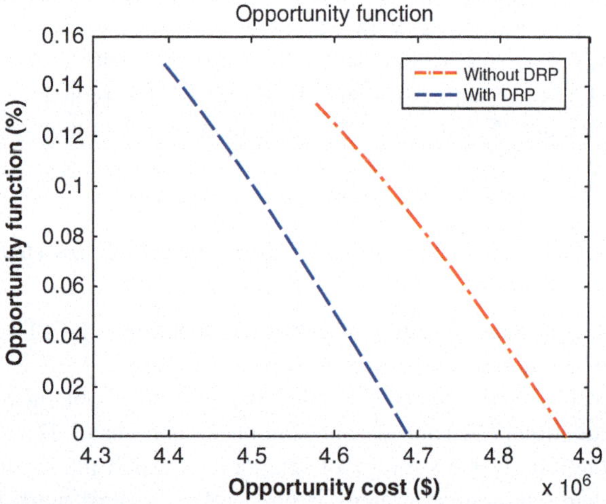

Figure 7.10 is obtained from solving the opportunity function (7.35) and considering the constraints (7.36)–(7.38). This figure represents the risk-taker strategy of retailers, which the opportunity function increases by reducing opportunity cost. In other word, if retailers pays lower cost, it will be risk-taking strategy. For example, with 11% drop in pool market prices, energy purchase costs are 4,471,597.4$ and 4,633,026.5$ for "with DR" and "without DR" modes which indicates that by decreasing 11% in the pool market price, total cost of retailers will decrease 4.6% and 4.1 for "with DR" and "without DR" modes. It can be seen that with the downward prices of the pool market, the retailers achieved more profit by using

DR. The reason is that by pool market price reduction, retailers can use lowest price DRs to supply their consumers energy and achieve more profit than without DR case.

Mentioned results in the above figures indicate the positive effects of proposed new DR schemes. In particular, effects of the DR schemes in the risk-averse strategy is significant but cannot be ignored from positive effects on the risk-taking strategy. Therefore, proposed DR scheme has more effect on the retailers' cost in the presence of pool market price uncertainty.

7.6.3 Optimal Bidding Strategy Result

It should be mentioned that the optimal bidding curves for each time periods are obtained based on the results of solving robustness and opportunity functions. Figure 7.11 illustrates the obtained optimal bidding curves for the 9th, 16th, 17th,

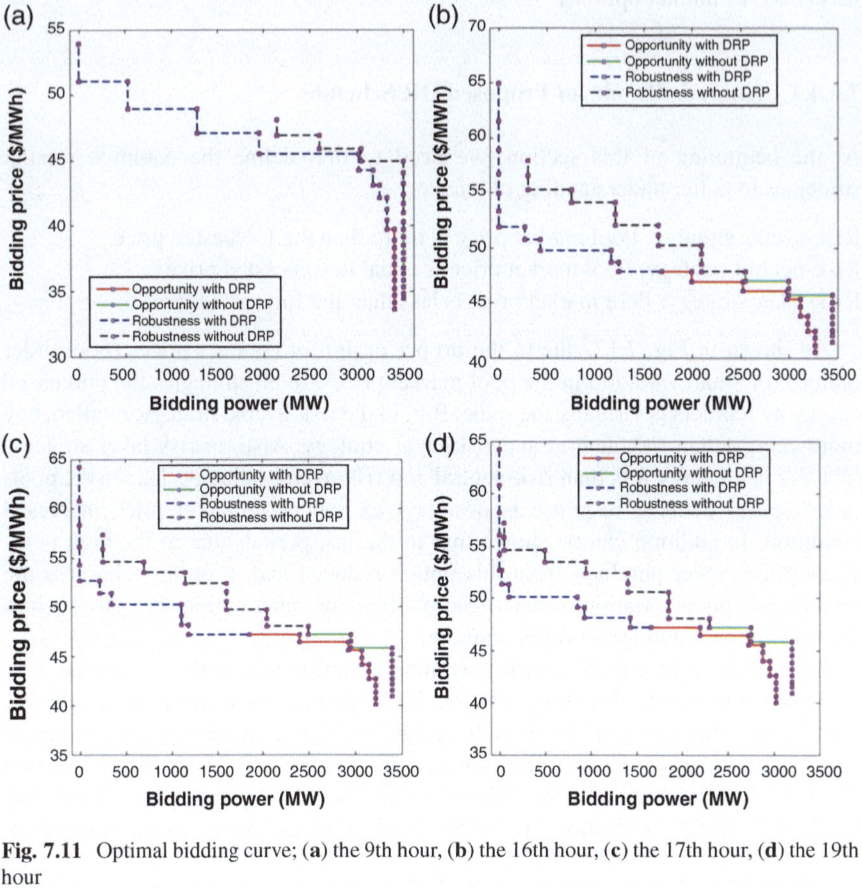

Fig. 7.11 Optimal bidding curve; (**a**) the 9th hour, (**b**) the 16th hour, (**c**) the 17th hour, (**d**) the 19th hour

and 19th periods considering with and without DRP, respectively. These figures show optimal bidding strategy for retailers in with and without DRP modes, which are drawn based on obtained results from robustness and opportunity functions. Optimal bidding curves of the retailers are obtained to bid to the day-ahead market in each period. The bidding price and power are provided in these curves for bidding the day-ahead market to purchase energy for the next day. These curves present the required data for retailers to successfully bid the day-ahead market for the consumer demand considering with and without DRP. The presented results illustrate that the optimal bidding curve considering DRP is more robust than without DRP.

7.6.4 Comparison of Risk-Based Results

In this section, the obtained results are compared in three different strategies as risk-averse, risk-neutral, and risk-taker strategies for retailers performance. This section consists of two parts in which first part analyzes new DR scheme options and second part analyzes market options.

7.6.4.1 Analysis Results of Proposed DR Schemes

At the beginning of this section, we need to first define the common retailer strategies to better understanding of readers:

Risk-averse strategy: Pool market price is more than the forecasted price.
Risk-neutral strategy: Pool market price is equal to forecasted price.
Risk-taker strategy: Pool market price is less than the forecasted price.

As shown in Fig. 7.12, due to the proper design of contract prices (pool-order option) for small variation in the pool market prices, in all strategies the purchased energy by retailers is virtually the same. But, in the risk-averse strategy, retailers buy more energy than risk-neutral and risk-taker strategy. Also, in risk-taker strategy, retailers buy energy less than risk-neutral and risk-averse strategy. Therefore, pool-order option demand response is designed to few pool-market price increased condition. In addition can be shown that in the last periods due to the high pool-order price power purchase from this option reduced and in order to harness the pool market price variations retailers more relying on other option specially such as forward contract and forward DR contract.

In addition, forward DR contracts is another option for retailers to reduce their costs and future risk. According to Fig. 7.13, because forward DR is a safe DR contract for the future, retailers will be able to use more of this option in the risk-averse strategy to reduce their energy procure costs and risk in future. Also this option will decrease in the risk-taker and risk-neutral strategies. In addition, purchasing energy from this contract in risk-neutral strategy is more than a risk-

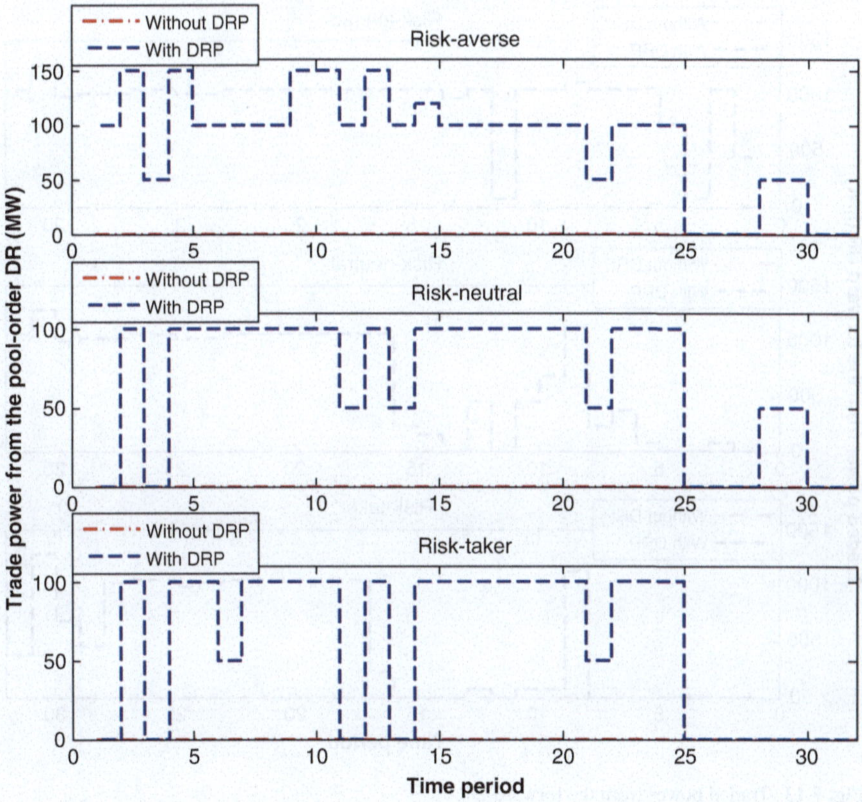

Fig. 7.12 Traded power from the pool-order DR

taker strategy. In addition, it can be shown that this option is an important option to use by retailers to reduce their risk in day-ahead market. Actually, this option is very similar to forward contracts in electricity market, each block contains DR options.

Another option is reward-base DR option, which by encouragement and reward pays, tray to reduce customers' demands. According to Fig. 7.14, purchasing energy from this contract in the risk-taker strategy is less than other strategies. Also, in the risk-aversion strategy, reduced power by cosumer has been increased beacuse of the pool ragket price in this strategy increased and Consumers tend to reduce their consumption. is more than the risk-natural strategy. Because, in the risk-taking strategy, market downturns are considered, retailers are more likely to buy energy from the pool market. Therefore, reward-based DR is different option from used DR schems in this work, which encourages consumers to participate in this DR option in any period (expensive and inexpensive).

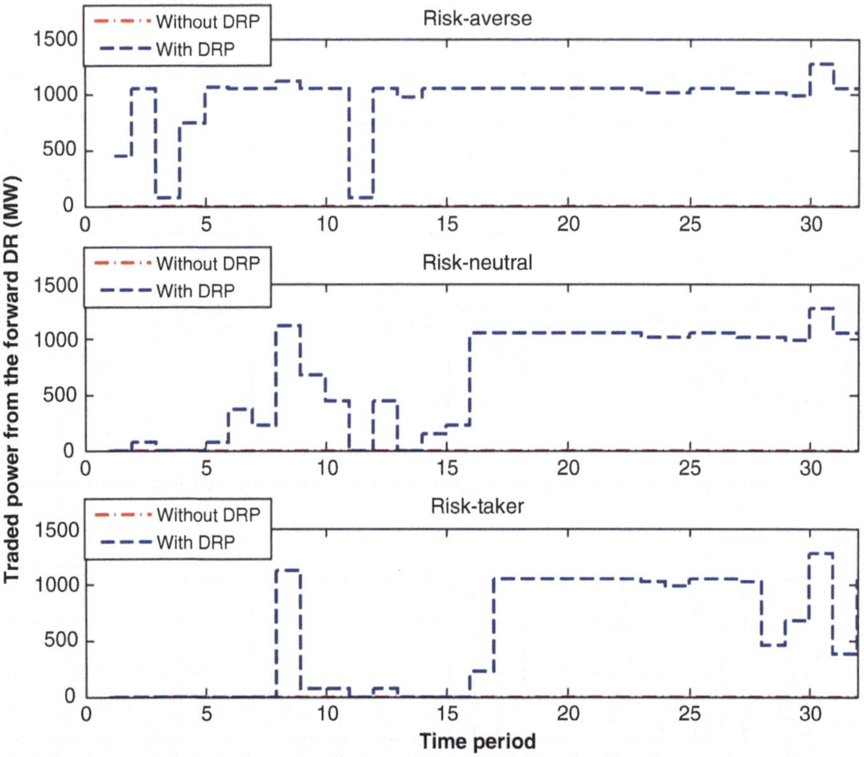

Fig. 7.13 Traded power from the forward DR

7.6.4.2 Analysis Results of Wholesale Market Suppliers

One of the important options in the electricity markets is pool market option, in which electricity power can be exchanged between retailers and wholesalers in the real-time price. As according to Fig. 7.15, purchased power from the pool market by retailer in the risk-taker strategy which considered pool market price decrease, is more than the risk-neutral and risk-aversion strategies which considered pool market price increased. Purchased power from the pool market by retailers in the risk-neural strategy is more than the risk-aversion strategy. It should be noted that purchasing energy from market has reduced in all strategies using the proposed DR programs.

Forward contract is the last option, which is considered in this work. Forward contracts have been created to reduce power procurement risks for power sellers and buyers, which is due to uncertainty in the electricity markets. It can be seen from Fig. 7.16, power purchasing from forward contract in the risk-aversion is more than risk-neutral and risk-taker strategies. Because this contract is for the future, the retailers will buy energy from this contract. It should be noted, purchasing energy from

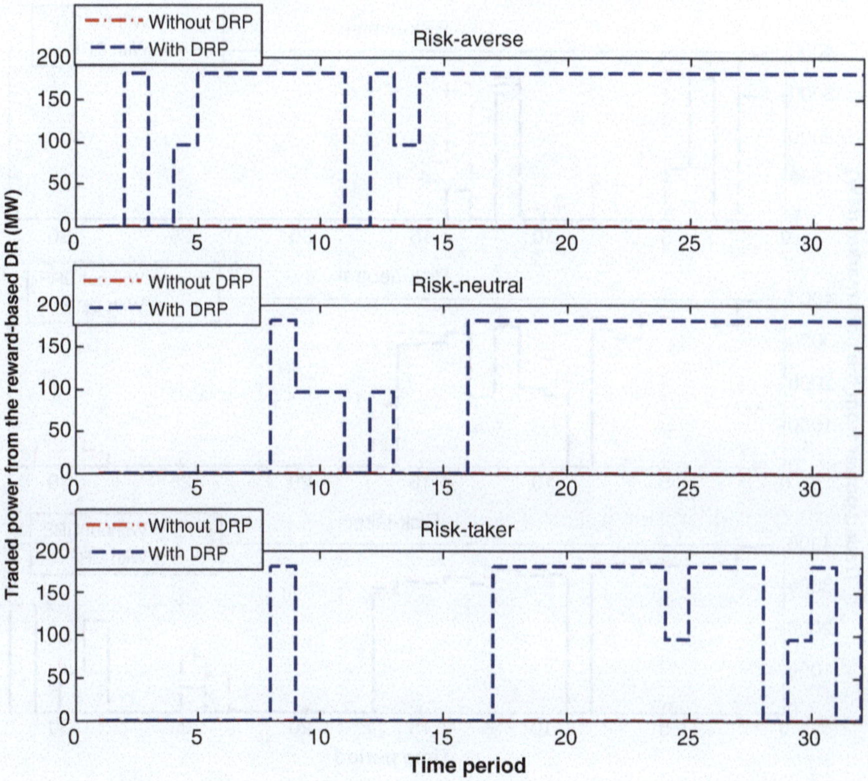

Fig. 7.14 Traded power from the reward-base DR

forward contract has been reduced in all strategies using DR. In addition forward contract has an important role in the peak times due to the possibly exceeded demand from generation capacity.

7.7 Conclusion

In this chapter, new schemes of DR are defined and their impacts on power procurement strategies of retailers are shown. Then, IGDT technique and two important functions of opportunity and robustness are proposed to handle the uncertainty of the market price. Finally, both opportunity and robustness functions are used to obtain optimal bidding curves for the retailers to purchase power in day-ahead market. In risk-neutral strategy, using DR, cost has reduced 3.93% (191,447$). Then, the retailers in the risk-averse strategy can be paid a higher amount of cost to increase their resistance against rising price in pool market which

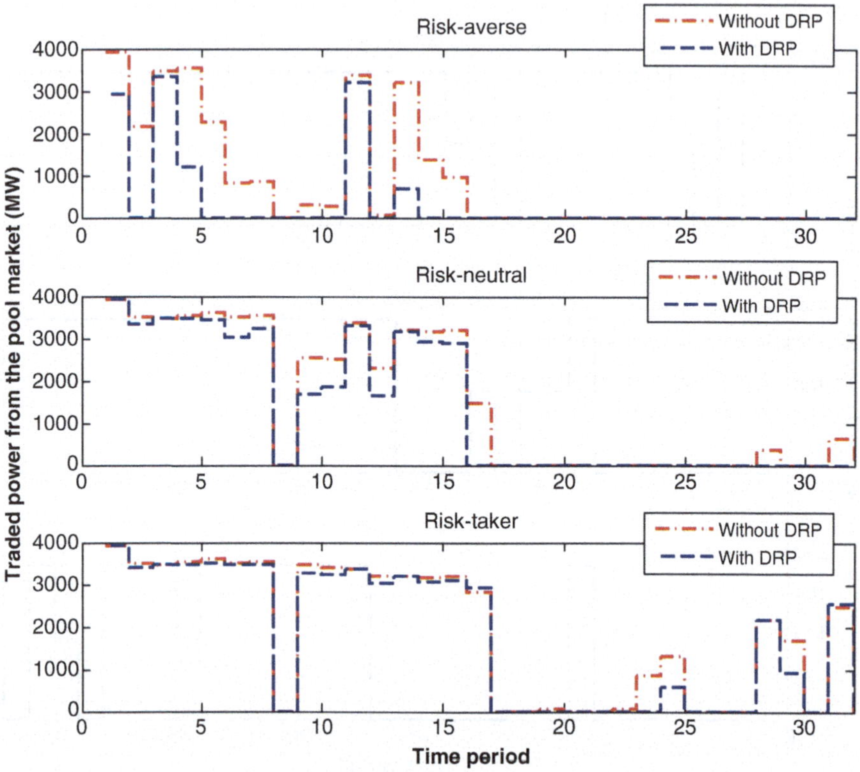

Fig. 7.15 Traded power from the pool market

in using DR has 18.8% higher resistance (lowest loss against pool market price rises) to increasing the market prices. In other words, using DR mode, retailers are 18.8% stronger than without DR mode. This shows the importance of the proposed DR scheme in this study. Also, in the risk-taker strategy, which was determined by the opportunity function, using of DR, the retailers could 161429.1$ increase its profit. Then, it is showed that in the risk-taker strategy of retailers with 11% drop in pool market prices, energy purchase costs are 4,471,597.4$ and 4,633,026.5$ for "with DR" and "without DR" modes, respectively. Retailers in the risk-taking strategy would be able to purchase more power from the market, because this strategy considers market price reduction. Also, the purchased power in the natural risk strategy for all options has been between the two risk-aversion strategy and the risk-taker strategy. Finally, using the two opportunity and opportunity functions, optimal bidding strategy was determined in order to bid to the day-ahead market in each period by electricity retailer.

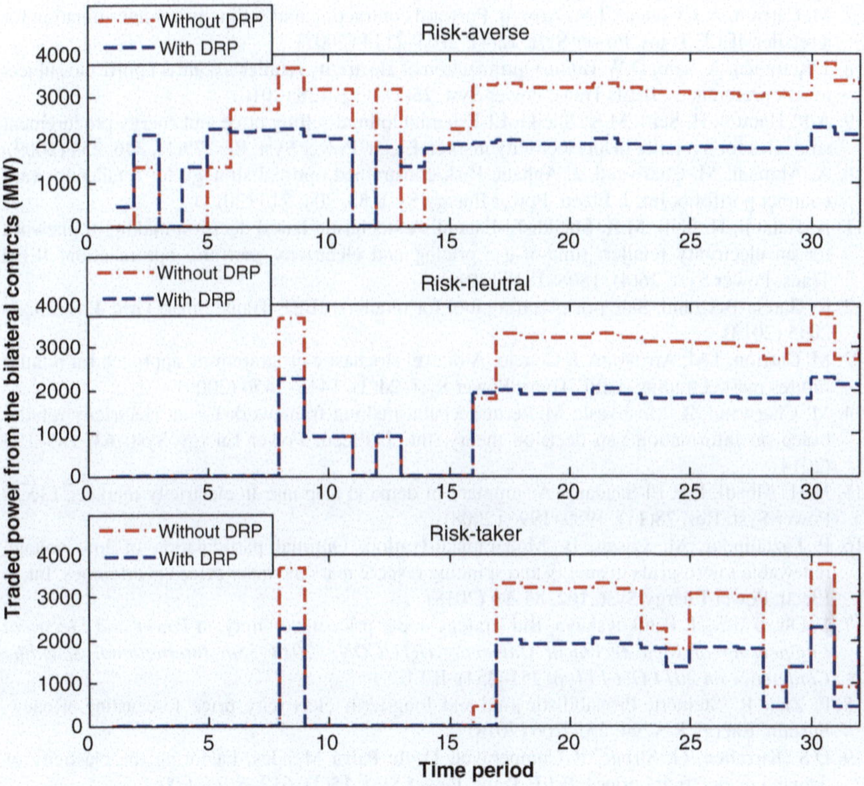

Fig. 7.16 Traded power from the forward contract

References

1. S. Nojavan, K. Zare, B. Mohammadi-Ivatloo, Risk-based framework for supplying electricity from renewable generation-owning retailers to price-sensitive customers using information gap decision theory. Int. J. Electr. Power Energy Syst. **93**, 156–170 (2017)
2. M. Marzband, M.H. Fouladfar, M.F. Akorede, G. Lightbody, E. Pouresmaeil, Framework for smart transactive energy in home-microgrids considering coalition formation and demand side management. Sustain. Cities Soc. **40**, 136–154 (2018)
3. S. Nojavan, K. Zare, B. Mohammadi-Ivatloo, Selling price determination by electricity retailer in the smart grid under demand side management in the presence of the electrolyser and fuel cell as hydrogen storage system. Int. J. Hydrogen Energy **42**(5), 3294–3308 (2017)
4. S. Nojavan, M. Mehdinejad, K. Zare, B. Mohammadi-Ivatloo, Energy procurement management for electricity retailer using new hybrid approach based on combined BICA–BPSO. Int. J. Electr. Power Energy Syst. **73**, 411–419 (2015)
5. M. Nazari, A.A. Foroud, Optimal strategy planning for a retailer considering medium and short-term decisions. Int J. Electr Power Energy Syst. **45**(1), 107–116 (2013)
6. S.A. Gabriel, A.J. Conejo, M.A. Plazas, S. Balakrishnan, Optimal price and quantity determination for retail electric power contracts. IEEE Trans. Power Syst. **21**(1), 180–187 (2006)

7. M. Carrion, A.J. Conejo, J.M. Arroyo, Forward contracting and selling price determination for a retailer. IEEE Trans. Power Syst. **22**(4), 2105–2114 (2007)
8. J. Kettunen, A. Salo, D.W. Bunn, Optimization of electricity retailer's contract portfolio subject to risk preferences. IEEE Trans. Power Syst. **25**(1), 117–128 (2010)
9. A.R. Hatami, H. Seifi, M.K. Sheikh-El-Eslami, Optimal selling price and energy procurement strategies for a retailer in an electricity market. Electr. Power Syst. Res. **79**(1), 246–254 (2009)
10. A. Ahmadi, M. Charwand, J. Aghaei, Risk-constrained optimal strategy for retailer forward contract portfolio. Int. J. Electr. Power Energy Syst. **53**, 704–713 (2013)
11. A. Hatami, H. Seifi, M.K. Sheikh-El-Eslami, A stochastic-based decision-making framework for an electricity retailer: time-of-use pricing and electricity portfolio optimization. IEEE Trans. Power Syst. **26**(4), 1808–1816 (2011)
12. R. Garcia-Bertrand, Sale prices setting tool for retailers. IEEE Trans. Smart Grid **4**(4), 2028–2035 (2013)
13. M. Carrion, J.M. Arroyo, A.J. Conejo, A bilevel stochastic programming approach for retailer futures market trading. IEEE Trans. Power Syst. **24**(3), 1446–1456 (2009)
14. M. Charwand, Z. Moshvash, Midterm decision-making framework for an electricity retailer based on information gap decision theory. Int. J. Electr. Power Energy Syst. **63**, 185–195 (2014)
15. M.H. Albadi, E.F. El-Saadany, A summary of demand response in electricity markets. Electr. Power Syst. Res. **78**(11), 1989–1996 (2008)
16. P. Fazlalipour, M. Ehsan, B. Mohammadi-Ivatloo, Optimal participation of low voltage renewable micro-grids in energy and spinning reserve markets under price uncertainties. Int. J. Electr. Power Energy Syst. **102**, 84–96 (2018)
17. V. Oboskalov, T. Panikovskaya, Bid strategy under price uncertainty, in *Power and Electrical Engineering of Riga Technical University (RTUCON), 2014 55th International Scientific Conference on 2014 Oct 14* (pp. 251–254). IEEE
18. F. Ziel, R. Steinert, Probabilistic mid-and long-term electricity price forecasting. Renew. Sustain. Energy Rev. **94**, 251–266 (2018)
19. D.S. Kirschen, G. Strbac, P. Cumperayot, D. de Paiva Mendes, Factoring the elasticity of demand in electricity prices. IEEE Trans. Power Syst. **15**(2), 612–617 (2000)
20. H.A. Aalami, M.P. Moghaddam, G.R. Yousefi, Modeling and prioritizing demand response programs in power markets. Electr. Power Syst. Res. **80**(4), 426–435 (2010)
21. H. Zhong, L. Xie, Q. Xia, Coupon incentive-based demand response: theory and case study. IEEE Trans. Power Syst. **28**(2), 1266–1276 (2013)
22. P.K. Adom, M. Insaidoo, M.K. Minlah, A.M. Abdallah, Does renewable energy concentration increase the variance/uncertainty in electricity prices in Africa? Renew. Energy **107**, 81–100 (2017)
23. V. Fanelli, L. Maddalena, S. Musti, Asian options pricing in the day-ahead electricity market. Sustain. Cities Soc. **27**, 196–202 (2016)
24. A.H. Mohsenian-Rad, V.W. Wong, J. Jatskevich, R. Schober, A. Leon-Garcia, Autonomous demand-side management based on game-theoretic energy consumption scheduling for the future smart grid. IEEE Trans. Smart Grid **1**(3), 320–331 (2010)
25. L. Gelazanskas, K.A. Gamage, Demand side management in smart grid: a review and proposals for future direction. Sustain. Cities Soc. **11**, 22–30 (2014)
26. A. Sheikhi, M. Rayati, A.M. Ranjbar, Demand side management for a residential customer in multi-energy systems. Sustain. Cities Soc. **22**, 63–77 (2016)
27. F.C. Robert, G.S. Sisodia, S. Gopalan, A critical review on the utilization of storage and demand response for the implementation of renewable energy microgrids. Sustain. Cities Soc. **40**, 735–745 (2018)
28. H. Shakouri, A. Kazemi, Multi-objective cost-load optimization for demand side management of a residential area in smart grids. Sustain. Cities Soc. **32**, 171–180 (2017)
29. S. Nojavan, K. Zare, B. Mohammadi-Ivatloo, Optimal stochastic energy management of retailer based on selling price determination under smart grid environment in the presence of demand response program. Appl. Energy **187**, 449–464 (2017)

30. A.J. Conejo, J.M. Morales, L. Baringo, Real-time demand response model. IEEE Trans. Smart Grid **1**(3), 236–242 (2010)
31. E. Celebi, J.D. Fuller, Time-of-use pricing in electricity markets under different market structures. IEEE Trans. Power Syst. **27**(3), 1170–1181 (2012)
32. L. Kreuder, C. Spataru, Assessing demand response with heat pumps for efficient grid operation in smart grids. Sustain. Cities Soc. **19**, 136–143 (2015)
33. F. Sehar, M. Pipattanasomporn, S. Rahman, Integrated automation for optimal demand management in commercial buildings considering occupant comfort. Sustain. Cities Soc. **28**, 16–29 (2017)
34. C. Eid, E. Koliou, M. Valles, J. Reneses, R. Hakvoort, Time-based pricing and electricity demand response: existing barriers and next steps. Utilities Policy **40**, 15–25 (2016)
35. P. Guo, V.O. Li, J.C. Lam, Smart demand response in China: challenges and drivers. Energy Policy **107**, 1–10 (2017)
36. D.T. Nguyen, M. Negnevitsky, M. De Groot, Pool-based demand response exchange—concept and modeling. IEEE Trans. Power Syst. **26**(3), 1677–1685 (2011)
37. D.T. Nguyen, M. Negnevitsky, M. de Groot, Walrasian market clearing for demand response exchange. IEEE Trans. Power Syst. **27**(1), 535–544 (2012)
38. O. Sezgen, C.A. Goldman, P. Krishnarao, Option value of electricity demand response. Energy **32**(2), 108–119 (2007)
39. J.Y. Joo, S.H. Ahn, Y.T. Yoon, J.W. Choi, Option valuation applied to implementing demand response via critical peak pricing, in *Power Engineering Society General Meeting, 2007. IEEE 2007 Jun 24* (pp. 1–7). IEEE
40. R. Tyagi, J.W. Black, J. Petersen, Optimal scheduling of demand response events using options valuation methods, in *Power and Energy Society General Meeting, 2011 IEEE 2011 Jul 24* (pp. 1–5). IEEE
41. S.C. Oh, J.B. D'Arcy, J.F. Arinez, S.R. Biller, A.J. Hildreth, Assessment of energy demand response options in smart grid utilizing the stochastic programming approach, in *Power and Energy Society General Meeting, 2011 IEEE 2011 Jul 24* (pp. 1–5). IEEE
42. S. Pal, S. Thakur, R. Kumar, B.K. Panigrahi, A strategical game theoretic based demand response model for residential consumers in a fair environment. Int. J. Electr. Power Energy Syst. **97**, 201–210 (2018)
43. M. Jin, W. Feng, C. Marnay, C. Spanos, Microgrid to enable optimal distributed energy retail and end-user demand response. Appl. Energy **210**, 1321–1335 (2018)
44. A.R. Hatami, H. Seifi, M.K. Sheikh-El-Eslami, Hedging risks with interruptible load programs for a load serving entity. Decis. Support Syst. **48**(1), 150–157 (2009)
45. R. Baldick, S. Kolos, S. Tompaidis, Interruptible electricity contracts from an electricity retailer's point of view: valuation and optimal interruption. Oper. Res. **54**(4), 627–642 (2006)
46. M. Carrion, A.B. Philpott, A.J. Conejo, J.M. Arroyo, A stochastic programming approach to electric energy procurement for large consumers. IEEE Trans. Power Syst. **22**(2), 744–754 (2007)
47. A.A. Algarni, K. Bhattacharya, A generic operations framework for discos in retail electricity markets. IEEE Trans. Power Syst. **24**(1), 356–367 (2009)
48. H. Li, Y. Li, Z. Li, A multiperiod energy acquisition model for a distribution company with distributed generation and interruptible load. IEEE Trans. Power Syst. **22**(2), 588–596 (2007)
49. I. Horowitz, C.K. Woo, Designing pareto-superior demand-response rate options. Energy **31**(6), 1040–1051 (2006)
50. S. Nojavan, B. Mohammadi-Ivatloo, K. Zare, Robust optimization based price-taker retailer bidding strategy under pool market price uncertainty. Int. J. Electr. Power Energy Syst. **73**, 955–963 (2015)
51. S. Nojavan, B. Mohammadi-Ivatloo, K. Zare, Optimal bidding strategy of electricity retailers using robust optimisation approach considering time-of-use rate demand response programs under market price uncertainties. IET Gener. Transm. Dis. **9**(4), 328–338 (2015)
52. A.K. David, Competitive bidding in electricity supply, in *IEE Proceedings C (Generation, Transmission and Distribution) 1993 Sep 1* (Vol. 140, No. 5, pp. 421–426). IET Digital Library

53. F. Wen, A.K. David, Optimal bidding strategies and modeling of imperfect information among competitive generators. IEEE Trans. Power Syst. **16**(1), 15–21 (2001)
54. J.V. Kumar, D.V. Kumar, Optimal bidding strategy in an open electricity market using genetic algorithm. Int. J. Adv. Soft Comput. Appl. **3**(1), 55–67 (2011)
55. S.N. Singh, I. Erlich, Strategies for wind power trading in competitive electricity markets. IEEE Trans. Energy Convers. **23**(1), 249–256 (2008)
56. T. Joshi, G. Uberoi, Graphical scheme for determination of market clearing price using quadratic bid functions. Int. J. Adv. Eng. Technol. **1**(2), 144 (2011)
57. L. Ma, W. Fushuan, A.K. David, A preliminary study on strategic bidding in electricity markets with step-wise bidding protocol, in *Transmission and Distribution Conference and Exhibition 2002: Asia Pacific*. IEEE/PES 2002 Oct 6 (Vol. 3, pp. 1960–1965). IEEE
58. N. Mahmoudi, M. Eghbal, T.K. Saha, Employing demand response in energy procurement plans of electricity retailers. Int. J. Electr. Power Energy Syst. **63**, 455–460 (2014)
59. J.C. Hull, S. Basu, *Options, Futures, and Other Derivatives* (Pearson Education India, Noida, 2016)
60. D.S. Kirschen, G. Strbac, *Fundamentals of Power System Economics* (Wiley, New York, 2004), p. 22
61. N. Mahmoudi, T.K. Saha, M. Eghbal, Developing a scenario-based demand response for short-term decisions of electricity retailers, in *Power and Energy Society General Meeting (PES), 2013 IEEE 2013 Jul 21* (pp. 1–5). IEEE
62. Y. Chen, J. Li, Comparison of security constrained economic dispatch formulations to incorporate reliability standards on demand response resources into Midwest ISO co-optimized energy and ancillary service market. Electr. Power Syst. Res. **81**(9), 1786–1795 (2011)
63. C.L. Su, D. Kirschen, Quantifying the effect of demand response on electricity markets. IEEE Trans. Power Syst. **24**(3), 1199–1207 (2009)
64. Y. Ben-Haim, *Information Gap Decision Theory, Designs Under Severe Uncertainty* (Academic, San Diego, 2001)
65. K. Bhattacharya, Competitive framework for procurement of interruptible load services. IEEE Trans. Power Syst. **18**(2), 889–897 (2003)
66. T.F. Lee, M.Y. Cho, Y.C. Hsiao, P.J. Chao, F.M. Fang, Optimization and implementation of a load control scheduler using relaxed dynamic programming for large air conditioner loads. IEEE Trans. Power Syst. **23**(2), 691–702 (2008)
67. AEMO price & demand Data [Online], http://www.aemo.com.au/data/pricedemand.html
68. AER, State of the Energy Market, Melbourn, Australia (2011)
69. Power of choice-giving consumers options in the way they use electricity, Direction Paper, March, 2012
70. The GAMS Software Website; 2016 [Online], <http://www.gams.com/dd/docs/solvers/sbb.pdf>

Chapter 8
Stochastic Cooperative Charging Scheduling of PEV Fleets in Networked Microgrids Considering Price Responsive Demand and Emission Constraints

Mehdi Shamshirband, Farhad Samadi Gazijahani, and Javad Salehi

8.1 Introduction

8.1.1 Motivation

The exponential worries about climate change, CO_2 emission, as well as concerns regarding the available amount of fossil fuel resources have led to an increase in research and study on the using of renewable energy resources (RES). Using RESs not only can lessen air pollutions, but also can diminish the attachment to fossil fuels. On the other hand, a significant increase in energy demand because of growing population, industry, and the advancement of technology are the most important challenges that networked microgrids (NMG) have encountered [1]. In Canada, for example, the public transportation sector is accounted as the second biggest origin of greenhouse gas emissions. According to Canadian transportation statistics, about 35% of all energy demanded by this country is for the public transportation sector [2]. In this context, increasing concerns about reducing the amount of fossil fuels, along with augmenting air pollution, which mainly came from combustion engines of gasoline-based cars, require to increase the using of RESs as well as electric vehicles (EV) in NMGs [3–5]. In fact, governments with replacing EVs instead of cars with combustion engines can play a momentous role in preventing climate change and also postponing the early warming of the earth.

EV is considered as an alternative to existing transportation systems because of their considerable advantages like less pollutant emissions, low energy expenditures, and high energy efficiency. Hence, rapid improvement and development of electric vehicles can simultaneously reduce the amount of pollution, reduce oil dependency,

M. Shamshirband · F. S. Gazijahani (✉) · J. Salehi
Department of Electrical Engineering, Azarbaijan Shahid Madani University, Tabriz, Iran
e-mail: m.shamshirband@azaruniv.ac.ir; f.samadi@azaruniv.ac.ir; j.salahi@azaruniv.ac.ir

S. Nojavan, K. Zare (eds.), *Electricity Markets*,
https://doi.org/10.1007/978-3-030-36979-8_8

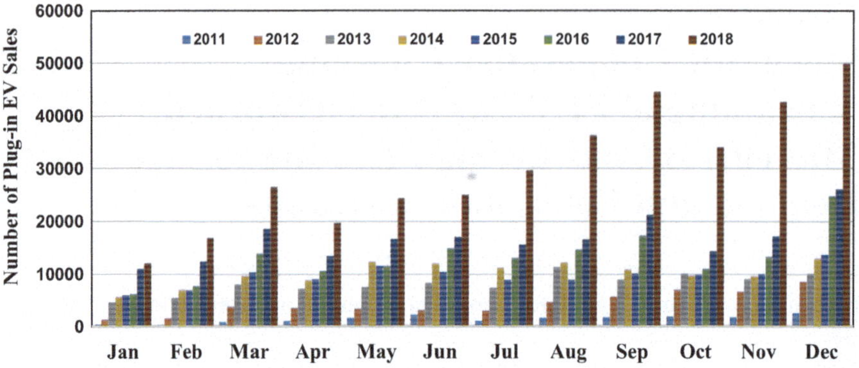

Fig. 8.1 US report for plug-in car sales

and also reduce the effects of greenhouse gases [6]. Therefore, governments and automotive companies made an agreement to replace the EVs instead of cars with combustion engines in the public transport section by 2020 [7]. In addition, by investigating and reviewing the reports from the InsideEVs monthly Plug-In electric vehicle (PEV) sales report, it can be seen that the demand for purchase and use of these vehicles is also increasing day by day. Figure 8.1 shows the monthly sales of PEVs from 2011 to 2018 [8]. With regard to Fig. 8.1 and investigation on reports can realize that the demand for EVs is increasing day by day. Thus, with these reports, concern about the uneven charging demand of these vehicles is one of other major challenges facing the power grids. Considering that the demand for these vehicles is much higher than the residential buildings. Therefore, this excessive demand for electric vehicles can create a separate peak load for the power grids, and causing problems such as reducing system reliability, increasing load demands, violating voltage limits, and increasing the losses in NMGs [9, 10]. For example, the Nissan Leaf vehicle needs an average of 3.3 kilowatts of energy, and this demand is about twice as high as the demand for residential homes [11, 12]. Besides that, Tesla's fully electric car also needs an average of 10 kilowatts of battery charge [13]. As a result, this huge load demand is causing many problems such as mentioned above to the power grids. With regard to the issues mentioned before, there is a public welcome in the using of optimized electric vehicles charging coordination (EVCC) plan for charging and discharging these vehicles in NMGs. However, a large number of EVs will result in detrimental impacts on the efficiency of the distribution grids, such as reducing power quality, increasing power losses, voltage changes, and reversing on the customer's energy costs, but combination of using these vehicles with RESs will have significant benefits for the distribution network (DN).

In recent years, various proceedings have been taken to optimize and change power grids from a traditional structure to a new and privately owned one. During this time, the electric power industry has undergone major changes in terms of management and ownership, due to increased utilization efficiency and encouragement of investors. In order to create a suitable competitive environment, various parts of it including generation, transmission, and distribution, became independent. In a restructured environment of the electricity industry, convincing market players is not easy to invest in multibillion-dollar projects. These changes along with the other factors stated above such as environmental pollution, the problems of the creation of new transmission networks, and the advancement of technology in the field of economics of manufacturing small-scale generation units in comparison with large generation units have increased use of small generation units referred to as distributed generation resources (DGR), which are mainly connected to DNs. Aside from this, increasing the portfolio on the DGRs, associated with environmental issues related to this energy development, especially the carbon dioxide linked "greenhouse effect" and the polluting effects of acid rain, increase public and political attention and also increase the importance of planning to reduce these contaminations and decrease in use of fossil fuel-related DGRs, increases the tendency to using renewable energy resources in distributed generation resources. Figure 8.2 is collected by the US Energy Information Administration. This figure illustrates the amount of generation increase in using of RESs in recent years, as well as growing influence of these energy resources by 2040. Figure 8.3 has also been collected by the US Energy Information Administration, which shows the influence of these RESs in recent years, as well as the increase in the penetration of these resources by 2040 [14].

The increasing use of digital devices, increasing the utilization of EVs, growing of industries and communities, and the requirements to reduce greenhouse gas emissions in power grids, as well as feeding the customers' high load demands, along with increased penetration of DGRs, RESs and uncertainties in the generation of these resources are causing many problems in terms of coordinating their relevance. For this reason, increasing the contribution of RESs instead of fossil fuel-

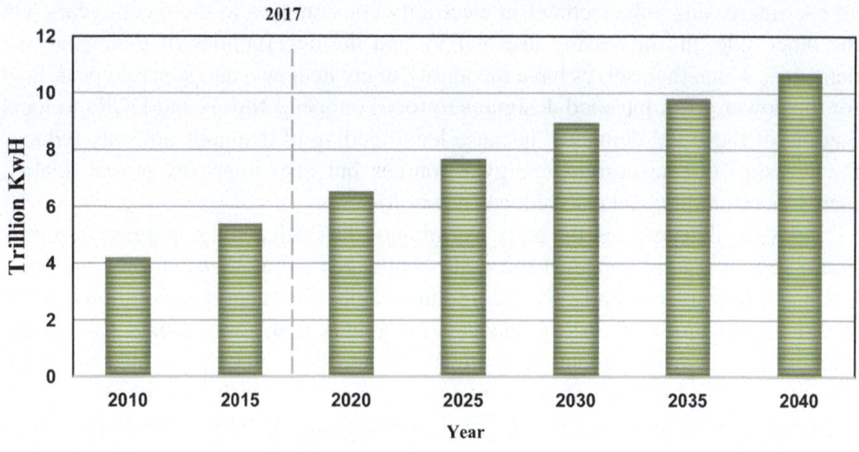

Fig. 8.2 Influence of renewable energy resources by 2040

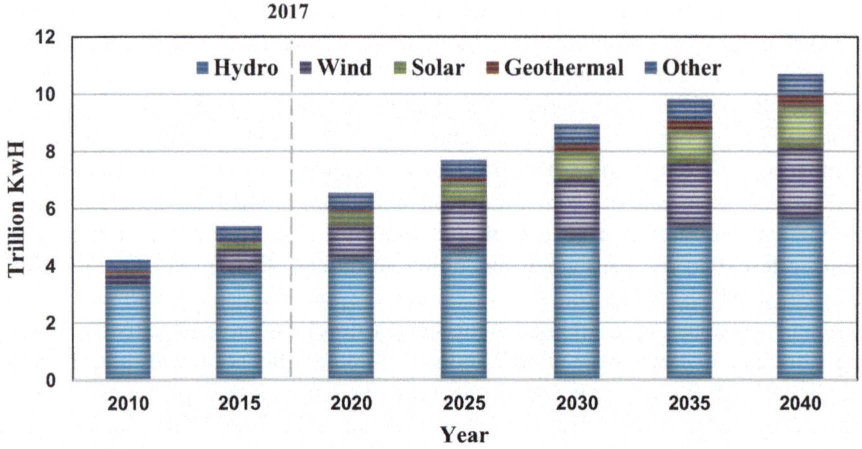

Fig. 8.3 World electricity generation from renewable energy resources by 2040

related resources of energy, as well as increase in the reliability of the system under the conditions above and to provide a relevance between all of these components of the network along with the optimal operation of EVs, it has led designers to use smart grids (SMG).

8.1.2 Literature Review

Ever-increasing growth of the community and the increasing use of electrical equipment, as well as the development of electricity-based industries have led to an ever-increasing enhancement in electricity consumption in the recent years. On the other side, the increasing use of EVs and the uncertainties of their charging demands, which themselves have the ability to create a new and separate peak load for the power grids, have led designers to focus on using NMGs and DGRs to local feeding of these EV demands, because local feeding of demands not only reduces the pressure on the current energy resources but also improves power quality, increases reliability, and also reduces power losses.

Today the debate about the early warming of the earth and the increase in greenhouse gas emissions is one of the main challenges facing distribution companies (DISCO). Because currently, the main source of electrical energy generation is fossil fuel-related sources of energy. Hence, researchers and governments have made many efforts and investments in using RESs and the replacement of these clean energy sources with fossil fuel-related sources of energy. On the other hand, there is uncertainty about the generation of RESs and the unavailability of these resources, as well as the advancement of technology and population growth, which leads to an ever-increasing demand for energy making it difficult to coordinate balance between generation and demands of energy to DISCO in NMGs. In addition, community acceptance of using electric vehicles, which was mentioned before could bring a new and separate peak load demands to the power grids, making the requirement of EVCC programs on NMGs urgent more than ever. These problems of generation and demands, as well as the issue of creating coordination between these components of NMGs, have led DISCO and designers to focus on using SMGs which have the ability to monitor accurately and communicate bilaterally between the consumer and the generator by the distribution network operator (DNO). Thus, by monitoring accurately and bilateral power and information exchange between consumers and manufactures, as well as creating scheduling planning, problems such as uncertainties in the generation of RESs, charging and discharge of EVs, as well as the high demands of energy can be solved, as well as increasing reliability of systems and loss reduction in power systems.

To date, multiple researchers all around the world have already done various extensive studies in this area; some investigators have utilized different models like mixed integer linear programming (MINLP) to precisely handle the energy scheduling of PEVs in unbalanced DGRs. Different layouts have been proposed for energy management of PEVs in the literature consisting energy exchange from grid to vehicle (G2V), vehicle to the grid (V2G), and vehicle to a vehicle (V2V) [15]. In addition, integrated analysis of PEVs in power grids shows that PEVs have the ability to exchange power to the grids. Using this feature allows the DNOs to use energy stored in PEV batteries and by using this technology, the amount of reserve energy required from conventional energy resources has decreased [16]. In addition, [17] proposed water filling algorithm to solve the EVs charging scheduling considering V2G capability in a decentralized manner so that the energy transaction between them has been regarded as bilateral.

Today, using metaheuristic algorithms has been very effective in solving optimization problems. Hence in many studies, using these algorithms was conducted to optimize the load demands in order to optimally charging of EVs. This method has been used aiming to increase the system reliability and DN security in order to minimize voltage, overload, and power losses compared to an uncoordinated charging scheduling program of EVs, assuming there is a two-way information exchange between the DNO and charging points in order to control the recharging process of EVs. This algorithm divides time periods into different distances, to allow users to prioritize their EVs charging program to different distances. Also in each step, the algorithm uses the sensitivity index to identify and use a more suitable electric vehicle with the aim of reducing power loss [18].

In electricity markets, demand response (DR) is a mechanism for managing consumers, under certain conditions of supply. Advantage and purpose of DR are for both the consumer and the DISCO, to benefit from having a smart method to schedule energy consumption. While the rules for using the classical power systems were such that the demand for the load was immediately fed by the existed sources of generation, in the new philosophy, attempts are made to keep demand fluctuations small in order to increase the efficiency of the system. Therefore, the main objective of DR is flattening of the system's load curve, which takes into account changing hours of using energy from peak hours to off-peak hours. Hence, in [19], a scheduling model was introduced along with a series of plug-in hybrid electric vehicles (PHEV), wind turbines (WT), and DR. The above model was implemented on the Illinois grid considering four different charging scenarios including limited charging with 3-h delay and smart charging with executing DR. As a result, the outcomes indicate the useful effect of the DR on the profit of whole system.

As discussed earlier, the increase in greenhouse gas emissions and the limitation of fossil fuels in the future is a worrying issue. Using PHEVs is one of the solutions to this problem. Although the coordination of these vehicles with the DNs is significant, and in case of inconsistency it is possible to create a new demand peak load, or even worse than the current peak in the power grids, on the other hand, if the coordination of these vehicles is done correctly, it can be used to help service in the DN, which is the demand-side management (DSM). For this purpose, this chapter examines the scheduling problem of charging and discharging of PEVs in NMGs with the aim of reducing greenhouse gas emissions as well as reducing the cost of the DISCO.

8.1.3 Contributions

As mentioned earlier, today the use of electric vehicles is growing. On the other hand, concerns about increasing environmental pollution and greenhouse gas emissions are also rising. So far, there have been a lot of works investigated regarding introducing new plans for EVs scheduling in NMGs. But along with these studies, there exist so many problems in the state-of-the-art. In this regard, DSM, as well as the expansion of the influence of RESs and also the uncertainty surrounding the generation of these sources of energy, have caused several new problems for DISCO. Hence, this chapter outlines an optimized method to properly manage the charge and discharge of two different types of EVs including battery electric vehicle (BEV) and PHEV in NMGs aimed at increasing the benefit of operators, decreasing the operation outlays of DGRs and power losses as well as decreasing the emission of greenhouse gases. In addition, in this study different types of DGRs have been used including DGRs with fossil fuels as well as RESs, including the wind turbine (WT) and photovoltaic (PV) to reduce the dependence on fossil fuels. The proposed EVCC program allows the DNO to use energy from EV batteries using V2G and G2V technology to feed the demands of grid consumers during peak load hours. EVs used in this study include various travel patterns such as separate routes, different arrival and departure times to smart parking lots (SPL), as well as the different duration of travel time. This novel EVCC program is presented aimed at reducing power losses, reducing the cost of power bought from the main grid and the operating cost of DGRs, and ultimately increasing the benefit of DISCO along with decreasing greenhouse gas emissions. Further, this chapter develops an efficient scenario-based model to bridge the defects of the state-of-the-art approaches for uncertainty management. On the other side, if the car owners would decide to apply their combustion engines instead of electric ones, the emissions of greenhouse gases emitted by these engines, pollution from the grid as well as greenhouse gases emitted by DGRs will be penalized by grid operator in order to minimize the air pollution. Finally, by modeling the smart parking lots (SPL) such as smart home parking lots and smart workplace parking lots as an aggregator to store the energy of EV batteries, modeling the EVs and managing their charging and discharging plans, the proposed problem with the grey wolf optimizer (GWO) algorithm was solved.

In accordance with the explanations discussed above, the major participations of this chapter can be briefly presented as below:

- Presenting an eco-environmentally friendly approach to optimally assess the various models of EVs with different driving patterns to reduce overall system operation costs, increase DISCO profits, and reduce greenhouse gas emissions in the presence of heterogeneous distributed energy resources units.
- Providing a stochastic model, in order to control the uncertainties related to the generation of RESs. Further, the multi-objective economic-emission cost functions are modeled as a MINLP and converted to single objective one through applying weighted sum method (WSM) and eventually find the best optimal answer with using the GWO algorithm.
- Numerical results indicate a significant reduction in greenhouse gas emissions, power losses, and network operation costs, as well as a significant increase in DISCO profits. Also, using the DR program, V2G and G2V technology in SPLs and energy stored in EV batteries show significant load changes from peak intervals to off-peak ones. Moreover, the results highlighted the effectiveness of the EVCC program in NMG and reducing the cost of power purchased from the upstream network.

8.1.4 Chapter Organization

The rest of the chapter is constructed as follows: in Sect. 8.2, DR program is shown. Section 8.3 is assigned with the categorization of electric vehicles and their explanation. Section 8.4 mathematically formulates the EV scheduling problem. Section 8.5 shows the uncertainty modeling approach by constructing scenarios. Section 8.6 indicates the case studies and simulation results which are specified to show the efficiency of the proposed method and finally, the last section presents the main results of this chapter and the conclusion.

8.2 Incentive-Based DR Programs

8.2.1 Concepts

In recent years, many factors have contributed to increase the penetration of DGRs in power systems. These factors include increased energy consumption, increased fossil fuels prices, technology advances, increased costs associated with the construction of the new power plant, and increased environmental concerns. Hence, using DGRs could potentially reduce the need to expand traditional power systems. Although controlling a large number of these resources, along with increasing the penetration of RESs and uncertainties in the generation of these DGRs, will create a difficult challenge for the secure and efficient operation of power grids, this challenge can be largely managed by establishing two-way communication between these resources and the DNO in the concept of SMGs.

So far, the SMGs have provided significant interests for the power sector, and one of its important components is demand side management (DSM) system. However, in the early days of DSM, peak load demand reduction was a way to increase the power systems capacity. But in recent decades, the DSM program has changed to manage the load profiles. The purpose of these programs is to encourage energy consumers to manage their consumption time, while balancing supply and demand with electrical energy, reducing costs, and increasing the reliability of the power system [20]. DR is one of the superlative important programs of DSM in SMG that can be used to ensure a balance between supply and load demand. Additionally, the DR program has a high potential for reducing high peak load demand. This capability can help to delay the development of generation capacity, as well as decrease the operating costs and air pollutions.

In general, DR program is divided into two major groups: time-based DR (TBDR) and incentive-based DR (IBDR). As shown in Fig. 8.4, TBDR group includes three programs, such as RTP, TOU, and CPP. Moreover, the IBDR group

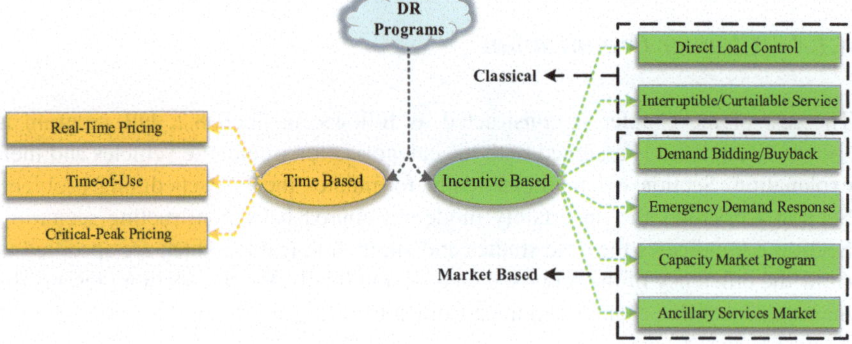

Fig. 8.4 Categories of demand response programs

contains six separate programs including DLC, I/C, DB, EDRP, CAP, and A/S [21]. RTP program gives consumers the incentive to reduce their energy consumption during hours with high electricity price. Since the electricity prices in an electricity market depend on the demands and the availability of energy generation, the high price occurs usually when the demand for energy is high or the grid has a shortage of energy in the generation. In addition, in the opposite situation, customers may shift their energy consumption to the hours when the electricity generation is excessive and the costs are at the lowest levels. This demand shifting could allow higher amount of RESs that could be integrated in the power system without increased need for curtailment [22]. TOU program is a less complicated way to introduce variations in the price of electricity to the customer and it is one of the most commonly used residential consumer programs. In fact, this method can be seen as a very long time lag in the RTP tariff, which is based on two or three different predefined prices of time throughout the day, i.e., tariffs at on-peak and off-peak times. Each of these predefined prices is somehow a reflection of the average price of electricity generated during the hours belonging to that interval. The intention of the CPP program is to offer consumers to reduce their energy consumption under normal circumstances while the retailer has the possibility to increase the energy rates for some hours every year when the total energy demand is high. CPP hours are usually the same for all region, but with the advancement of measuring technology, CPP program could also be based on the peak demand in a local electrical distribution system [23].

IBDR are voluntary incentive-based and usually based on economic motivation and are often not included in the electricity rates. As shown in Fig. 8.4, IBDR program is divided into two categories of classical programs and market-based programs. The classical IBDR program category includes DLC and I/C, while the market-based category includes DB, EDRP, CAP, and A/S. In the classic IBDR programs, customers who want to participate in this program will usually receive the bill credit or rebate in return for their requested actions by operator. In the IBDR, customers participating in this program will be awarded proportion to reduction in their consumption at critical situations. In DLC, the operator is able to remotely turn off the participant's remote control equipment such as air conditioners and water heaters on short notice. These types of schemes are mostly for the benefit of residential customers as well as small commercial customers. Similar to the DLC program, the customers who would participate in I/C program can earn incentive payments or rate discounts. In this program, participants are asked to lessen their energy consumption to a predetermined value. On the other hand, participants who do not respond may be fined according to the program terms and conditions.

DB program (also known as Buyback program) is a program in which consumers bid a specific reduction in their energy consumption in electricity sale market. If the bid offer is lower than the price in the market, it will be accepted. Once the customer bid is accepted, a customer participating in the program should curtail its consumption, otherwise it will be penalized. Additionally, in EDRP, the amount of reward is paid incentives to the customers for measured load reductions during emergency conditions. Besides, in the CAP the customers who are able to reduce their demand when the system is at the risk (i.e., unmoral conditions). Participants in this type of program usually receive a day-ahead notice of possible events and amount of load reductions and if they do not respond to the load reduction call, they will be penalized. A/S program allows the customers to participate in this program with biding consumption reduction in the wholesale market. Then, when the customers bid for participation in this program is accepted, the participants are paid the price cleared in the day-ahead market for perpetrating to be on standby [24].

8.2.2 Modeling

This chapter will cover and discuss IBDR program. To assess consumer engagement in DR programs, it is essential to develop a model that determines load profiles, profit and loss for consumers in terms of electricity prices, incentive payments, and fines. Hence, in order to demonstrate the sensitivity of the load to price changes, we can use a concept called "elastic" or "load elastic." Load elastic is defined as the sensitivity of demand to price changes. Equation (8.1) presents the elastic math expression of the load elastic.

$$\varepsilon = \frac{\rho_0}{d_0} \frac{\partial d}{\partial \rho} \tag{8.1}$$

in which ρ is equal to the price of electricity, d is the demand, ε is equal to elastic, ρ_0 and d_0 respectively represent the price and load quantities before the price changes. With respect to Eq. (8.1), the elasticity of the i period is defined as Eq. (8.2) with the respect to price changes over the j period [25].

$$\varepsilon\,(i, j) = \frac{\rho_0(j)}{d_0(i)} \frac{\partial d(i)}{\partial \rho(j)} \tag{8.2}$$

Figure 8.5 shows that due to the load sensitivity to the price, decrease in the demand side will greatly affect the price of electrical energy [25].

Frequently, the prices of different electrical energy in different periods of time are divided into two categories which include single-period and multi-period. Loads that cannot be displaced in different periods, which only turn on or off (like loads of light), are single-period loads, and the response of these kinds of loads versus price is called single-period sensitivity which is evaluated by self-elasticity. The sign of self-elasticity is always negative, because when the price increases in a period, the amount of demand decreases during the same period or vice versa. On the other hand, loads that move in different periods, so that consumption from peak duration to mid-time load or low-load, is called multi-period load. The response of this type of load versus price is multi-period sensitivity and is evaluated with cross elasticity. The sign of this coefficient is always positive, because when the price increases in a

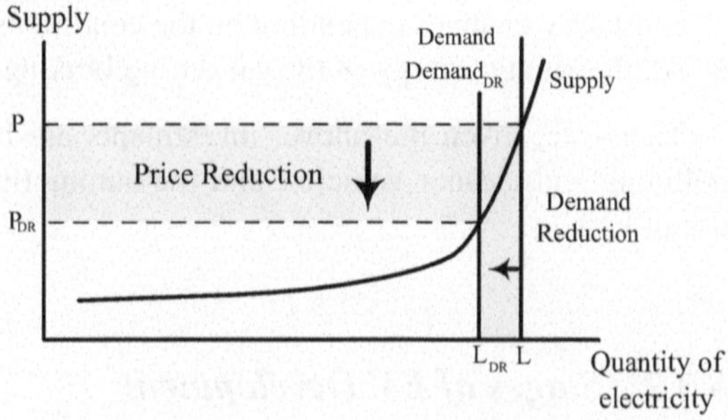

Fig. 8.5 Effect of load elasticity on electricity prices

specific time, the level of consumption will be increased at other times. Eventually, considering the amount of incentive and penalty in the DR program, the economic model of demand can be expressed as Eq. (8.3) [26].

$$d(i) = d_0(i)\left\{1+\sum_{j=1}^{12} \varepsilon\,(i, j)\left[\frac{\rho(j) - \rho_0(j) + \mathrm{Inc}(j) - \mathrm{Pen}(j)}{\rho_0(j)}\right]\right\} \tag{8.3}$$

where Inc(j) is equal to the amount of encouragement and Pen(j) shows the amount of penalizing. Based on the amount of incentive and fines, the position of each demand response program is deployed and affected by its final priority.

8.3 Electric Vehicle

8.3.1 Aim

With increasing concerns about the energy crisis, the issue of using the EVs as a substitute for cars with combustion engines in transport has been seriously considered. One of the main problems of these vehicles is the lack of long-distance travel as well as the need to recharging. On the other hand, the low cost of energy and the compatibility of these vehicles with the environment can be named as their most important advantages. In addition, increasing concerns about rising greenhouse gases is one of the main issues facing industrialized countries. One of the main sources of greenhouse gas emissions in industrialized countries is the use of cars with combustion engines. Also, in addition to over-emission of greenhouse gases, these cars also have several other disadvantages. Disadvantages such as [27]:

- Size of the combustion engines is such that they can provide the necessary power for the desired acceleration for the car. So these engines are usually bulky and heavy.

- The thermal efficiency of these motors depends on their operating point and constantly changes depending on the conditions during the car's motion.
- All the kinetic energy of the car during braking is turned to the heat and wasted.

Therefore, given the above, investments are being made to move toward less polluting and cleaner vehicles, and the automotive industry is leading toward the use of EVs.

8.3.2 *Stages of EV Development*

The stages of EV development are categorized into three generations including EVs, Hybrid electric vehicles (HEV), and PHEVs. EVs are typical electric cars and are made up with a battery pack connected to an electric motor that provides the required power to the wheels. The main drawback of these vehicles is their limited driving time, depending on the type of battery. HEVs are a common type of EV that has two engines including fuel and electric engine. The disadvantages of EVs have been mostly overcome in these vehicles. It can also be said that the disadvantages of cars with internal combustion engines have also been somewhat resolved. The important advantages of these vehicles compared to the car with internal combustion engines are the ability to drive in a steady load and optimum working point. This advantage will increase engine efficiency and reduce pollution as well as lower fuel consumption. In addition, during braking or negative acceleration, kinetic energy is stored electrically in the battery of these vehicles. This function will result in less engine operation and thus reduce pollution and reduce fuel consumption. But on the other hand, the disadvantages of these vehicles include the lack of the ability to charge batteries from the power grid and its dependence on the fuel engine. But ultimately, PHEVs have eliminated the disadvantages of HEVs. PHEVs are able to receive energy through the power grid or an internal combustion engine.

8.3.3 *Classification of EVs*

In general, EVs can be divided into two large groups:

- Autonomous EVs, including HEV and Fuel cell electric vehicle (FCEV)
- Plug-in EVs, including a PHEV and an EV based on an electric battery (BEV)

Figure 8.6 shows the division of these two EVs categories. On the other hand, solar EVs are not considered as an independent group. Because the addition of photovoltaic cells (for battery charging, power supply to electric motors, or power supplies for vehicle systems) can be done in vehicle of each of the above categories. As can be seen in Fig. 8.6, Autonomous vehicles include HEVs and FCEV. HEVs combine an internal combustion engine system with an electric propulsion system.

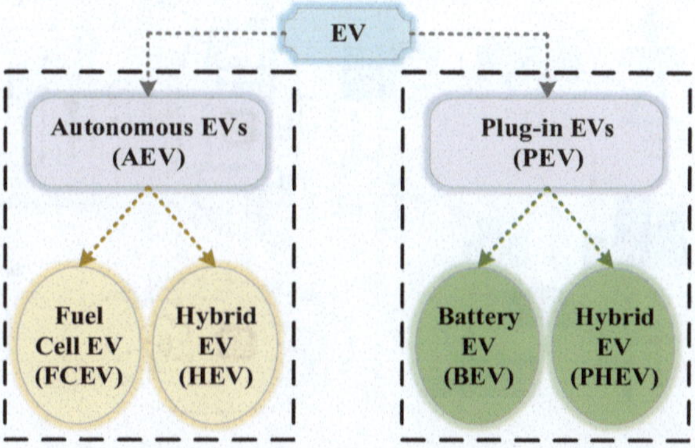

Fig. 8.6 Classification of EVs

FCEVs are also a kind of EVs where the fuel cell is replaced instead of conventional internal combustion gasoline engines. In addition to autonomous EVs, there are plug-in EVs as well. PEVs are also divided into two groups of PHEVs and BEVs. While BEVs are known to use their batteries energy for their electric propulsion system, PHEVs typically have smaller batteries because they are also using an internal combustion engine. This hybrid system makes the driving range of these vehicles more than electric vehicles based on batteries. As a result, there is no need to worry about the completion of battery charge.

Recent research and the growing use of PEVs, along with the advent of smart grids in power systems, which have provided two-way communication capabilities for power grids, has led to a reduction in energy consumption and a reduction in greenhouse gas emissions. PEVs are able to communicate bilaterally with the DNO, while they are in SPLs or smart charging stations. By doing this bilateral communication, they are simultaneously able to play the role of loads or power sources in SMGs. Therefore, using these PEVs can be divided into two categories: PEVs which can only be charged, named G2V, as well as PEVs that include both G2V and V2G technologies [28].

8.3.4 V2G Technology

V2G technology has been much considered in recent years. This technology reduces the dependence on small expensive power generation units, reduces the costs of launching these units, as well as managing load and peak load volatility, increasing spinning reserve and system reliability. In addition, PEVs can exchange energy stored in their batteries with electricity markets and participate in the economic aspects of power systems. In order to exchange this stored energy in markets, aggregators and NMGs can participate together as an interface. Also, with the assumption that there is uncertainty in the DN, V2G technology can create a two-way power flow and increase the interaction between the DN and PEV. Finally,

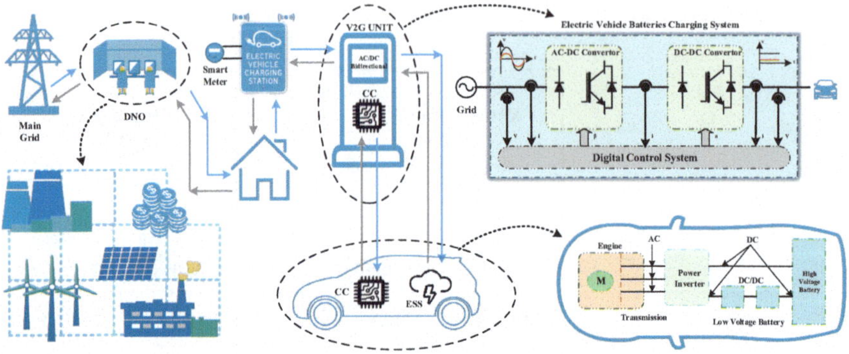

Fig. 8.7 The structure of typical bidirectional charger and the connection of PEV to SPLs

considering two ways of power exchange between PEVs and the power grid, two types of V2G technology are defined as one-way and two-way V2G technologies.

In one-way V2G technology, the PEV battery charge rate is controlled in a one-way power transmission between the EV and grid. One-way V2G technology is possible by adding a simple controller to manage charge rates and also required a low cost. Using this technology will increase the flexibility of network operation. Also, the notion of two-way V2G technology provides the two-way power circulation between the PEVs and power grid to achieve multiple benefits.

Figure 8.7 shows a typical two-way charger within the PEV, consisting both AC/DC and DC/DC converters. When it is charging mode, to control active and reactive power the charger must receive a sinusoidal current with predetermined phase angle from the network. Also in the mode of injection into the grid, the charger must return a sine current to the grid. In this figure, an AC/DC converter is used to rectifying the AC power taken from the power grid and is converted to DC power during charging the PEV. Also in discharging mode, the DC power is converted to AC power before being injected into the network. Moreover, the DC/DC converter is responsible for bidirectional power transactions via different control methods. The DC/DC converter, respectively, behaves as a buck or boost converter during charge and discharge modes [29].

Two-way V2G technology provides more flexibility and capabilities including active and reactive power patronage, and power factor adjustment to ameliorate the operation of power grids. In addition, reactive power support by two-way V2G technology can provide peak-load modification and flattening load profile. This service is available through the charging of PEVs during off-peak hours and the injection of PEVs energy to the power grid during peak hours. Two-way V2G technology has also the capability to provide reactive power to regulate network voltages. This service could be done with the proper measurement of DC Capacitor, charger, and correct control keying. In order to further explain, if at the same time DN needs reactive power and at the same time the PEV would be in the charge mode, the charger should be operated in charging (i.e., capacitive mode). So, the

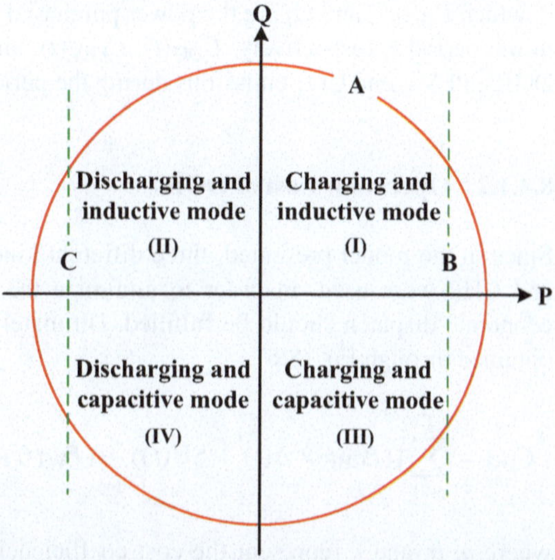

Fig. 8.8 Capability curve of PEV chargers

power grid will furnish the active power for the PEV battery charging and the PEV injects the reactive power to the grid by using V2G two-way technology. The overall structure of the charging and discharge curve of the PEV is shown in Fig. 8.8 [30].

8.4 Problem Formulation

8.4.1 Objective Function

This chapter provides a novel planning for optimal active power management of NMG in the presence of decentralized PEVs in order to decrease the overall cost of DISCO, operating costs of various DGRs including WTs, PVs and combined heat and power (CHP), and greenhouse gas emissions reduction.

$$\text{Minimize}\left\{F^{\text{Cost}}+F^{\text{Emission}}\right\} \tag{8.4}$$

8.4.1.1 Total Cost Function

The total cost of system containing WTs and PVs, CHPs and participation of PEVs associated with the air pollution cost are expressed in Eq. (8.5).

$$\mathrm{F}^{\text{Cost}}=\sum_{t=1}^{T}[(P_{\text{Grid}}(t)\times\Omega_t)+(C_{\text{DG}}(t))+(C_{\text{PEV}}(t))+(C_{\text{EM}}(t))] \tag{8.5}$$

in which $P_{\mathrm{Grid}}(t)$ and Ω_t are the power purchased from the grid and the energy price in the period t, respectively. $C_{\mathrm{DG}}(t)$, $C_{\mathrm{PEV}}(t)$, and $C_{\mathrm{EM}}(t)$ also denote the cost of DGRs, PEVs, and CO_2 emissions during the period of t.

8.4.1.2 Operation Cost of DGRs

Since in the model presented, three different kinds of DGRs including WTs, PVs, and CHP were used, in order to minimize the operating costs of each unit, an economic dispatch should be fulfilled. Ultimately, the total cost of these DGRs is obtained through Eq. (8.6).

$$C_{\mathrm{DG}} = \sum_{t=1}^{T} [(P_{\mathrm{CHP}} \times a(t) + \mathrm{SU}(t)) + (P_{\mathrm{WT}}(t) \times b(t)) + (P_{\mathrm{PV}}(t) \times c(t))] \quad (8.6)$$

where a, b, and c represent the cost coefficiencies of DGRs. $\mathrm{SU}(t)$ expresses the startup cost of CHP over the time t and P_{CHP}, P_{WT} and P_{PV}, respectively, represent the active output power of CHP, WTs, and PVs.

8.4.1.3 Operation Cost of PEVs

The objective from utilization of V2G in PEV was to decrease the costs of power grids and thereby reduce the cost of DISCO as well as reduce greenhouse gas emissions. In fact, PEVs with using an optimal EVCC program can detract losses in NMG, and minimize the costs of whole system and reduce the greenhouse gas emissions. So, the charging/discharging costs of PEVs can be determined through Eq. (8.7).

$$\begin{aligned} C_{\mathrm{PEV}} = \sum_{t=1}^{T} \sum_{n=1}^{N_{\mathrm{PEV}}} \big[& k \times P_{\mathrm{Charge/Discharge}}(n,t) \\ & + \left(P_{\mathrm{Charge}}(n,t) \times C_{\mathrm{Charge}} - P_{\mathrm{Discharge}}(n,t) \times C_{\mathrm{Discharge}}\right)\big] \end{aligned} \quad (8.7)$$

where k represents the operation factor of PEVs in dollars per kilowatt-hour and C_{Charge}, $C_{\mathrm{Discharge}}$ are cost factor for charging and discharging PEVs in period t.

8.4.1.4 Cost of Greenhouse Gas Emission

Equation (8.8) is used to minimize the amount of greenhouse gas emissions. This amount includes greenhouse gas emissions by the main grid and CHP power source. Also, two different types of PEV including BEV and PHEV have been used in the presented model. Therefore, in order to obtain the most accurate amount of greenhouse gas emissions in this equation, the amount of greenhouse gas emitted by PHEV when their owners decided to use their PHEVs combustion engine was taken into account.

$$C_{\mathrm{EM}} = \sum_{t=1}^{T}\left[\left(P_{\mathrm{Grid}}(t) \times \mathrm{Em}_{\mathrm{Grid}}\right) + \left(P_{\mathrm{CHP}}(t) \times \mathrm{Em}_{\mathrm{DG}}\right) + \sum_{n=1}^{N_{\mathrm{PHEV}}} \left(P_{\mathrm{PHEV}}(n,t) \times \mathrm{Em}_{\mathrm{PHEV}}\right) \right] \tag{8.8}$$

where $\mathrm{Em}_{\mathrm{Grid}}$, $\mathrm{Em}_{\mathrm{DG}}$, and $\mathrm{Em}_{\mathrm{PHEV}}$, respectively, represent the greenhouse gas emission cost factor by main grid, CHP and PHEVs when the owners decided to use PHEVs combustion engine.

8.4.2 *Constraints*

In order to minimize the proposed program, it is necessary to consider different constraints as presented here.

8.4.2.1 Power Mismatch Constraint

PEVs can be regarded as portable storage which charge at off-peak periods and inject the stored energy to the NMG at on-peak periods. According to Eq. (8.9), the amount of energy produced by this portable energy storage as well as DGRs should be equal with load and losses.

$$P_{\mathrm{Grid}}(t) + \sum_{i=1}^{I} P_{\mathrm{DG}}(i,t) + \sum_{n=1}^{N_{\mathrm{PEV}}} P_{\mathrm{PEV}}^{\mathrm{Dischare}}(n,t) = D_t + \sum_{n=1}^{N_{\mathrm{PEV}}} P_{\mathrm{PEV}}^{\mathrm{Charge}}(n,t) + \mathrm{Loss}(t) \quad \forall t \in \{1,\ldots,T\} \tag{8.9}$$

where $P_{\mathrm{PEV}}^{\mathrm{Charge}}$, $P_{\mathrm{PEV}}^{\mathrm{Dischare}}$ represent power charge and discharge of nth PEV in period t and D_t and Loss(t), respectively, show the total hourly demand and power loss of the distribution network in period t.

8.4.2.2 Lines Limit

The capacity of lines must transmit the power over a given range due to restrictions such as thermal limits as Eq. (8.10).

$$S_{(i,j)} \leq S_{(i,j)}^{\max} \tag{8.10}$$

where $S_{(i,j)}$ is the power flow of lines between bus i and j, and $S_{(i,j)}^{\max}$ is their maximum capacities.

8.4.2.3 Limit of Power Flow in the Lines

With respect to Eq. (8.11), the power transmission through the lines must not exceed the maximum value.

$$\left|P_{(i,j)}\right| < P_{(i,j)}^{\max}, \quad \forall t \in T, \forall s \in Ns \tag{8.11}$$

where $P_{(i,j)}$ is power flow in the line connected between bus i and j, as well as $P_{(i,j)}^{\max}$ denotes the maximum capacity of line between bus i and bus j.

8.4.2.4 Under/Over Voltage Limits

One of the constraints in the operation of power systems is the limitation of the voltage limit in a given range, which can be seen in Eq. (8.12).

$$V_{\min} \leq V \leq V_{\max}, \quad \forall t \in T, \quad \forall s \in Ns \tag{8.12}$$

where $V_{\min}$ and $V_{\max}$, respectively, show the minimum and maximum values of voltage.

8.4.2.5 PEVs Limitations

In addition, PEVs cannot simultaneously be charged and discharged in each period of time. This limitation is shown in Eq. (8.13).

$$X_{n,t} + Y_{n,t} \leq 1 \quad \forall n \in N_{\mathrm{EV}}, \forall t \in \{1, \ldots, T\} \tag{8.13}$$

where $X_{n,\,t}$ and $Y_{n,\,t}$, respectively, declare the binary variables for charging/discharging modes of nth PEV at tth hour.

Additionally, in the operation of the PEVs, their battery charge balance should be taken into account. Hence, in Eq. (8.14), SOC(n, t) indicates the amount of energy stored in PEVs batteries in EV n and at period t. The traveling energy over hour t $\left(\mathrm{SOC}_{(n,t)}^{\mathrm{Trip}}\right)$ is with the energy remaining in the previous hour and charging/discharging over the time interval.

$$\mathrm{SOC}\,(n,t) = \mathrm{SOC}\,(n,t-1) + \eta_n^C \times P_{\mathrm{PEV}}^{\mathrm{Charge}}\,(n,t) - \mathrm{SOC}_{(n,t)}^{\mathrm{Trip}} - \frac{1}{\eta_n^D} \times P_{\mathrm{PEV}}^{\mathrm{Discharge}}\,(n,t) \quad \forall t \in \{1,\ldots,T\};\ \forall n \in \{1,\ldots,N_{\mathrm{PEV}}\} \tag{8.14}$$

where in this equation, η_n^C and η_n^D are, respectively, efficiency coefficients of G2V and V2G charging. According to Eq. (8.15), charging/discharging of PEV batteries are also limited in a certain range.

$$E_{\mathrm{PEV}}^{\min} \le E_{\mathrm{PEV}} \le E_{\mathrm{PEV}}^{\max} \tag{8.15}$$

where $E_{\mathrm{PEV}}^{\min}$ and $E_{\mathrm{PEV}}^{\max}$, respectively, display the minimum and maximum energy stored in the PEVs.

8.4.2.6 DGRs Constraints

In the proposed problem, various models of DGRs including CHP, WTs, and PVs have been used. According to Eqs. (8.16)–(8.18), the amount of power generated by these DGRs is limited to the minimum and maximum range.

$$P_{\mathrm{CHP}}^{\min} \le P_{\mathrm{CHP}} \le P_{\mathrm{CHP}}^{\max} \tag{8.16}$$

$$P_{\mathrm{WT}}^{\min} \le P_{\mathrm{WT}} \le P_{\mathrm{WT}}^{\max} \tag{8.17}$$

$$P_{\mathrm{PV}}^{\min} \le P_{\mathrm{PV}} \le P_{\mathrm{PV}}^{\max} \tag{8.18}$$

in which $P_{\mathrm{CHP}}^{\min}$, $P_{\mathrm{CHP}}^{\max}$, $P_{\mathrm{WT}}^{\min}$, $P_{\mathrm{WT}}^{\max}$, $P_{\mathrm{PV}}^{\min}$, and $P_{\mathrm{PV}}^{\max}$ are the minimum and maximum generation capacity of CHP, WT, and PV, respectively.

8.5 Scenario Modeling

The scenario generation method is one of the Looking ahead methods, according to which the future conditions are designed by stochastic scenarios. In fact, generating a scenario is not just a prediction of a particular future, it is a description of all probabilities. Therefore, if the scenario includes all future approaches, it can be a powerful tool for power systems planning.

In the operation of power systems, there are some uncertain parameters such as sun irradiation and wind speed. The probable behavior of these parameters can lead to significant changes in the production of RESs, including PVs and WTs. On the other hand, the amount of load demand in DNs by consumers has an unexpected temperament. Hence, in this chapter, a scenario-based model has been used to simulate the stochastic behavior of these parameters. In addition, for decreasing the computational burden of the optimization process, the Kantorovich distance method is used. In the scenario-based approaches, the values of uncertain parameters are determined from their PDF figures. To this end, each PDF is divided into several parts with defined probability for each part. Subsequently, these parts are combined with each other to the scenarios and eventually the optimization process will be implemented for all scenarios [31]:

$$C(s) = [\text{load}(s)\text{solar}(s)\text{wind}(s)] \tag{8.19}$$

$$£_s = £_s^{\text{D}} \times £_s^{\text{PV}} \times £_s^{\text{WT}} \tag{8.20}$$

$$\sum_{s=1}^{S} £_s^{\text{D}} \times £_s^{\text{PV}} \times £_s^{\text{WT}} = 1 \tag{8.21}$$

8.6 Simulation Results

8.6.1 Data and Case Study

As shown in Fig. 8.9, the optimal NMG management planning in the presence of BEV and PHEV and DGRs has been implemented in the 69-bus IEEE test system [32]. This NMG consists of three interconnected MG with separate forecasted daily load profiles including residential, commercial, and industrial microgrids. Each MG along with various forecasted daily load patterns in addition to feeding their own load demands are also able to cover the demanded load in the other MGs for the purpose of flattening the load curve.

Table 8.1 shows the energy price in the day-ahead market [28]. As shown in Fig. 8.9, in the network implemented, various models of DGRs including the source of fossil fuels including CHP, as well as the source of renewable energy generation including WTs and PVs have been used. Also, the location of these DGRs is also specified in this testing network. The CHPs are located at Bus 69, the WTs are located at buses 11, 31, and 40, and also the PVs are located at buses 46, 50, 56, and 63. The operation of these DGRs also includes costs to the electricity company.

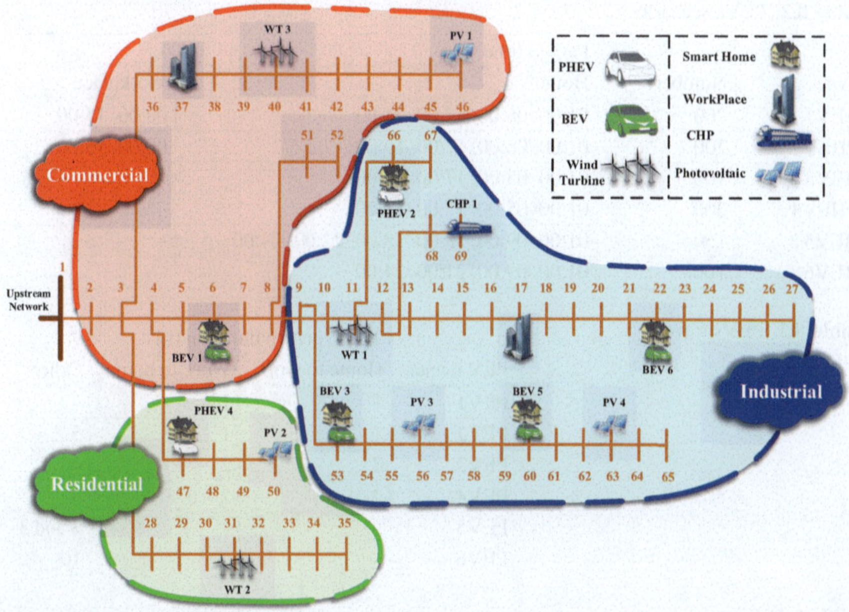

Fig. 8.9 69-bus IEEE test system with three interconnected MGs

Table 8.1 Energy price at the power market

t$/kWh	1	2	3	4	5	6	7	8
	0.033	0.027	0.020	0.017	0017	0.029	0.033	0.054
t$/kWh	9	10	11	12	13	14	15	16
	0.215	0.572	0.572	0.572	0.215	0.572	0.286	0.279
t$/kWh	17	18	19	20	21	22	23	24
	0.086	0.059	0.050	0.061	0.181	0.077	0.043	0.037

Therefore, in order to operate them optimally and to calculate the cost of these DGRs, the start-up and fuel cost factors of CHP are assumed to be 10$ and 4.5$, respectively. Also, in the planning calculations, for the operation of RESs, there are also charges called maintenance and repair cost that are shown for WTs and PVs with the cost coefficient of *b* and *c*, respectively, where both of them are considered to be $ 2.

To implement the proposed EVCC program, the system operator is able to control the amount of generation and demand in NMG by using two-way communication with SPLs in the SMG. As shown in Fig. 8.9, the location of these SPLs includes two categories of residential smart parking lots and workplace smart parking lots are considered. The task of these SPLs is to manage charge and discharge of PEVs by checking the hourly prices of energy, as well as the amount of energy stored in the PEV batteries, as well as the load status of NMG (peak and off-peak) by using two-way communication with the DNO. The site of residential smart parking lots is

Table 8.2 PEVs scenario

Type	Number	Parked time	
		Home	Workplace
BEV1	200	01:00–06:00, 17:00–24:00	08:00–15:00
PHEV2	200	01:00–06:00, 17:00–19:00	08:00–15:00
BEV3	100	01:00–05:00, 17:00–24:00	–
PHEV4	200	01:00–05:00, 17:00–19:00	–
BEV5	150	01:00–09:00, 13:00–18:00, 22:00–24:00	–
BEV6	100	01:00–09:00, 21:00–24:00	–

Table 8.3 PEVs travel duration

PEV owner	Trip type/trip duration (h)		
	Home to work	Work to home	Others
PEV1	1	1	–
PEV2	1	1	5
PEV3	2	2	–
PEV4	2	2	5
PEV5	–	–	3 and 3
PEV6	–	–	10

Table 8.4 Greenhouse gas emissions cost coefficient

Type	Coefficient	Cost
Grid	Em_{Grid}	$ 0.143
CHP	Em_{CHP}	$ 0.127
PHEV	Em_{PHEV}	$ 0.03

considered in buses 6, 22, 47, 53, 60, and 66. Moreover, the location of workplace smart parking lots is also considered in buses 17 and 37.

EVs used in this EVCC planning include two types of PEVs such as PHEV and BEV. These vehicles have a variety of patterns, such as travel time, hours of arrival and departure to SPLs, as well as different charging and discharging periods. Tables 8.2 and 8.3 show the number of each PEVs, duration of their presence in SPLs, as well as the travel time of these vehicles. On the other hand, at some hours of the day, PHEV owners decide to use their own EV combustion engine. Hence, at these hours, these vehicles also emit greenhouse gases. For this reason, in order to study and accurately calculate the emitted greenhouse gases, the amount of greenhouse gas emissions released by these PHEVs during the period of using combustion engines is also calculated. Besides, the average air pollutions released from fossil fuel-based power plants would be calculated at off-peak hours (1–6 and 23–24), average hours (6–20), and peak hours (19–22). The emission coefficients are assumed to be 0.050, 0.562, and 0.985 kg/kW for the main grid [33]. Finally, to determine the greenhouse gases emitted by the main grid, CHP, as well as PHEVs, the cost coefficients of Em_{Grid}, Em_{CHP}, and Em_{PHEV}, respectively, are considered in Table 8.4 [34, 35].

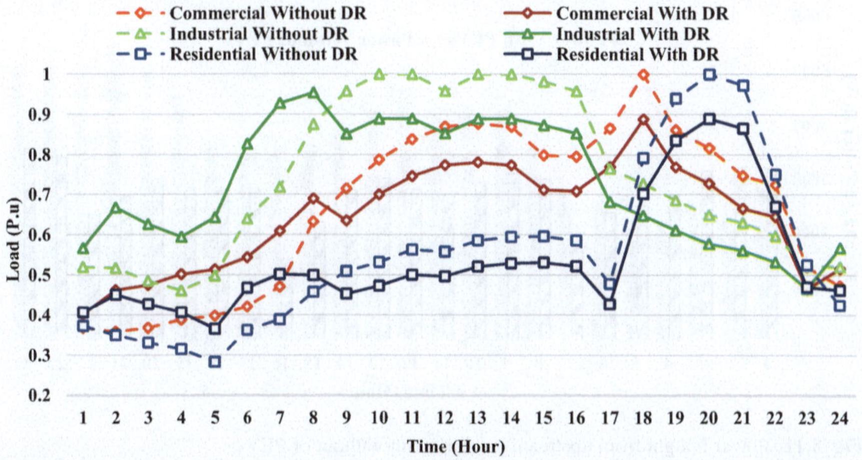

Fig. 8.10 Forecasted load profiles before and after implementation of DR program

8.6.2 Numerical Results

Three linked MGs with different load profiles were examined with the aim of reducing the cost of electricity companies, as well as reducing greenhouse gas emissions. The proposed model reduces the cost of DISCO and greenhouse gas emissions with the EVCC program by using PEVs V2G and G2V capability. Here, the results of proposed method will be shown. In order to implement the EVCC planning in NMG, DR program was used to flatten the system load forecasted profiles. These three interconnected MGs load forecasted profile, such as commercial, industrial, and residential networks, are visible before and after the implementation of the DR program in Fig. 8.10.

The amount of power purchased from the upstream network by the DISCO as well as the power generated by DGRs in the studied network are shown in Figs. 8.11 and 8.12, respectively. Regarding Fig. 8.11, the presence of PEVs, and DR program, have reduced the power purchased by the DISCO from the upstream network. In addition, according to section *b* in Fig. 8.12, the power generated by DGRs has decreased in the presence of PEVs, which reduces the operation cost of these DGRs as well.

Regarding the fact that the planning was implemented to reduce the cost of the DISCO and along with it, reducing greenhouse gas emissions. Hence, after implementation of the DR program as well as the EVCC program, the emission of greenhouse gases by the main grid, the source of CHP, as well as the amount of greenhouse gas emissions by PEVs when the owners decide to using their combustion engines in Fig. 8.11 is visible. Also, in Fig. 8.13, greenhouse gas emission reductions are observed before and after the implementation of the EVCC and DR planning, partially for the main grid, the source of CHP and PEVs.

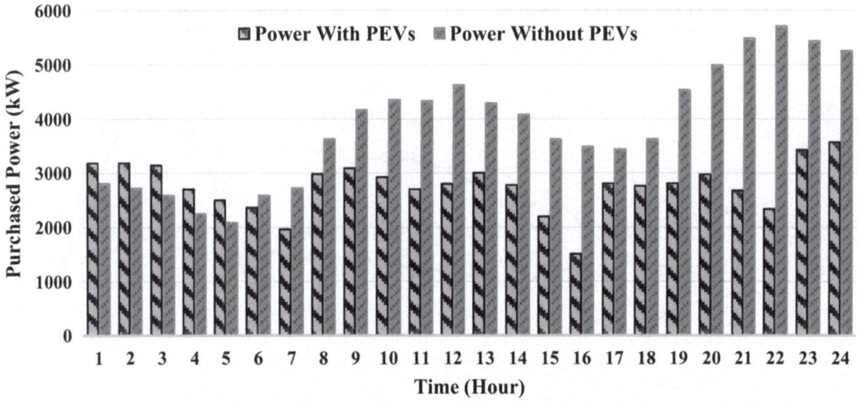

Fig. 8.11 Power bought from upstream network with/without of PEVs

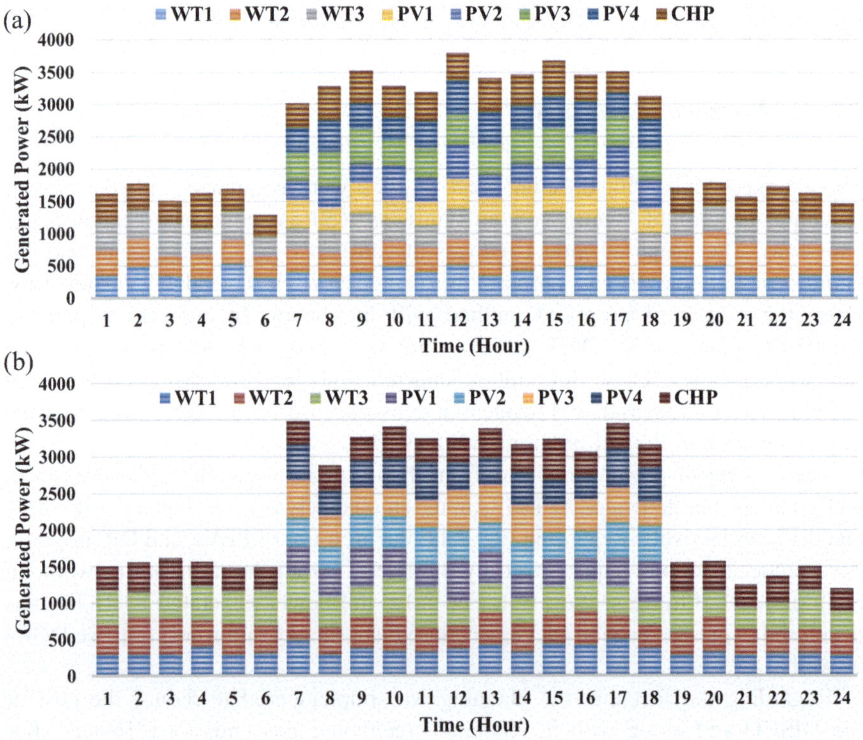

Fig. 8.12 Power generated by each DGRs without (**a**) and with (**b**) presence of PEVs

According to Fig. 8.14, it is obvious that the implementation of the proposed DR program, as well as the optimal use of EVCC program during PEVs presence in SPLs, causes a significant reduction in greenhouse gas emissions. Figure 8.15 illustrates the daily charge and discharge schedule of PEVs in this NMG. As can be

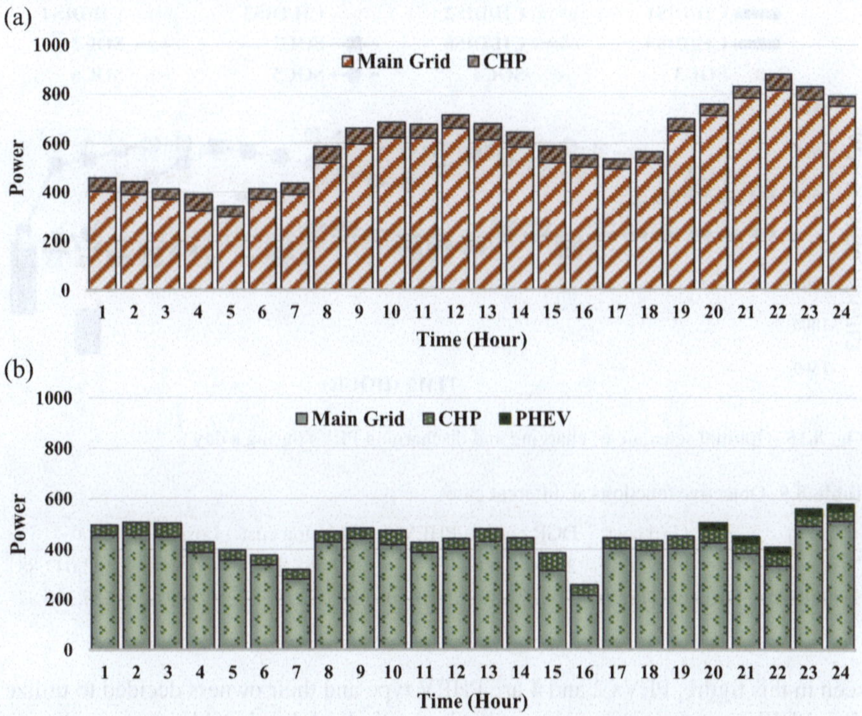

Fig. 8.13 Amount of greenhouse gas emission without (**a**) and with (**b**) presence of PEVs

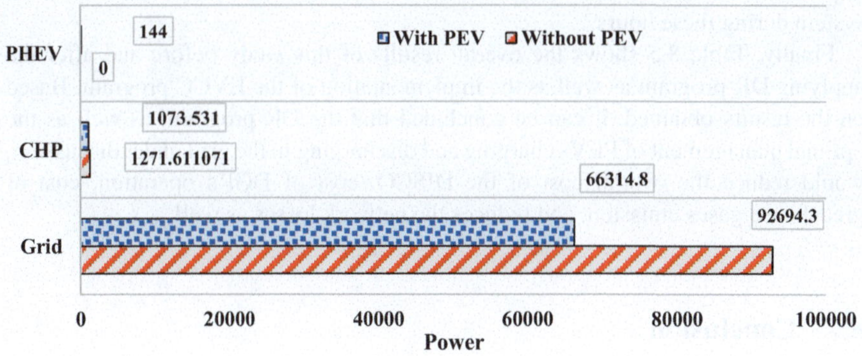

Fig. 8.14 Greenhouse gas emission observed partially for main grid, CHP and PEVs

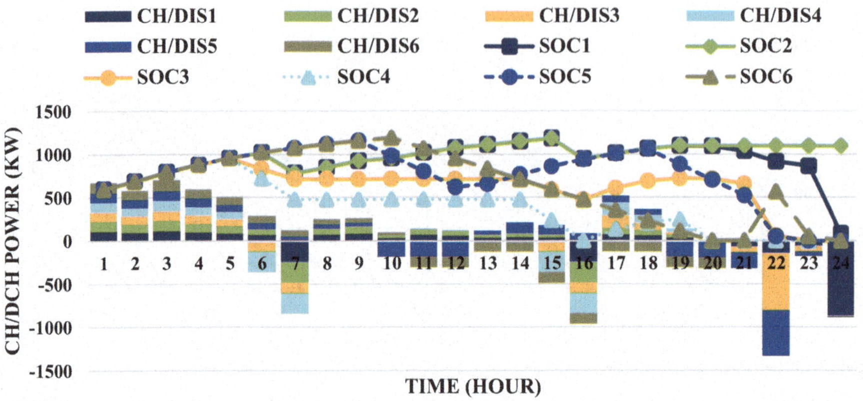

Fig. 8.15 Optimal schedule of charging and discharging PEVs during a day

Table 8.5 Objective functions at different cases

Cost ($)	Grid cost	DGR costs	PHEV	Emission cost	Loss cost	Total
Without PEV	92,694.39	244,440.00	–	117,688.44	4915.66	302,047.88
With PEV + DR	66,315.06	179,201.80	2369.30	83,332.98	3280.01	225,360.87

seen in this figure, PEVs 2 and 4 are PHEV type and their owners decided to utilize their PHEVs combustion engines at the last periods of day. In addition, according to the EVCC planning, during the peak hours of energy consumption (18–21), PEVs in the SPLs use V2G ability to inject the energy stored in their batteries into the grid to preserve the network performance and contribute to increase the reliability of the system during these hours.

Finally, Table 8.5 shows the overall results of this study before and after the applying DR program as well as the implementation of the EVCC program. Based on the results obtained, it can be concluded that the DR program, as well as the optimal management of PEVs charging and discharging in the grid under discussion, would reduce the overall cost of the DISCO, cost of DGRs operation, cost of greenhouse gases emission, and reduces the network losses as well.

8.7 Conclusion

Today, one of the major problems facing power systems is energy management. Increasing demand for energy, increasing the use of RESs and uncertainty about their availability, increasing the use of EVs and their uneven charging demand threaten the reliability of power grids. Hence, in this study, we presented a new methodology for the optimal operation of three interconnected MGs to manage energy in the presence of DGRs, including CHP, WTs, and PVs in the 69-bus IEEE test system. The purpose of this method is to reduce the cost of DISCO, along with the simultaneous reduction of greenhouse gas emissions. Hence, using V2G and G2V technology in PEVs, and the ability to exchange power between SPLs and DNO, we exchanged the power between the consumer and the DISCO.

Additionally, in this planning, each of these MGs studied has a separate daily load forecasted profile, including commercial, industrial, and residential. Therefore, with the implementation of the DR program, we plan to flatten the load profile of these three interconnected MGs. Finally, the intermittencies induced by renewables were modeled by scenarios. The presented multi-objective problem was converted into a single-objective one and minimized by an integrated population-based metaheuristics algorithm namely GWO and Taguchi test method that can acquire a satisfactory solution.

The results obtained from simulations show that the proposed scheme not only significantly reduces the operating cost of system and power losses, but also decreases the greenhouse gas emission pollution. The charging of PEV is done when the price is low and then PEVs discharge when the price is high at the power market. It should be emphasized that the inclusion of environmental objectives into the problem modifies the charging/discharging pattern of PEVs which this issue is subject to an increase in the economic costs of system. Besides, applying DR program could smooth the load profile which leads to decrease the power bought from the main grid.

Some future works developed from this work can be as follows:

- Multi-agent based modeling of EV aggregators within the power markets
- Assessing the impact of price-maker EV aggregators on the revenue adequacy of market
- Enforcing other uncertainty management instruments to curb the market participation risk
- Considering the reserve and frequency regulation services provided by different EVs along with their energy management problem

References

1. F.S. Gazijahani, J. Salehi, Optimal bi-level model for stochastic risk-based planning of microgrids under uncertainty. IEEE Trans. Indus. Inform. **14**(7), 3054–3064 (2018)
2. S. Nojavan, K. Zare, B. Mohammadi-Ivatloo, Application of fuel cell and electrolyzer as hydrogen energy storage system in energy management of electricity energy retailer in the presence of the renewable energy sources and plug-in electric vehicles. Energy Convers. Manag. **136**, 404–417 (2017)
3. A. Abdolahi et al., Probabilistic multi-objective arbitrage of dispersed energy storage systems for optimal congestion management of active distribution networks including solar/wind/CHP hybrid energy system. J. Renew. Sustain Energy **10**(4), 045502 (2018)
4. M. Montazeri-Gh, M. Mahmoodi-K, Optimized predictive energy management of plug-in hybrid electric vehicle based on traffic condition. J. Clean. Prod. **139**, 935–948 (2016)

5. S. Nojavan, K. Zare, Interval optimization based performance of photovoltaic/wind/FC/ electrolyzer/ electric vehicles in energy price determination for customers by electricity retailer. Sol. Energy **171**, 580–592 (2018)
6. B.K. Sovacool, R.F. Hirsh, Beyond batteries: an examination of the benefits and barriers to plug-in hybrid electric vehicles (PHEVs) and a vehicle-to-grid (V2G) transition. Energy Policy **37**(3), 1095–1103 (2009)
7. P. Papadopoulos et al., Predicting electric vehicle impacts on residential distribution networks with distributed generation, in *Vehicle Power and Propulsion Conference, VPPC. IEEE*, (2010), pp. 1–5
8. Inside EVs, Monthly Plug-in Sales Report. [Online]. https://insideevs.com/monthly-plug-in-sales-scorecard/. Accessed January 2019
9. S. Lin, Z. He, T. Zang, Q. Qian, Impact of plug-in hybrid electric vehicles on distribution systems, in *Power System Technology, POWERCON. International Conference on*, (2010), pp. 1–5
10. F.S. Gazijahani, J. Salehi, Robust design of microgrids with reconfigurable topology under severe uncertainty. IEEE Trans. Sustain. Energy **9**(2), 559–569 (2018)
11. National Grid USA. National Grid Residential Load Data. [Online]. http://www.nationalgridus.com/energysupply. Accessed 2018
12. Nissan Motor Company Ltd. Nissan Leaf Specification. [Online]. http://www.nissanusa.com/electric-cars/leaf/versions-specs/. Accessed 2018
13. Tesla Motors, Inc. Tesla Model S Specification. [Online]. http://www.tesla.com/models. Accessed 2018
14. S. Nojavan, M. Majidi, K. Zare, Performance improvement of a battery/PV/fuel cell/grid hybrid energy system considering load uncertainty modeling using IGDT. Energy Convers. Manag. **147**, 29–39 (2017)
15. P. Sanchez-Martin, G. Sanchez, G. Morales-España, Direct load control decision model for aggregated EV charging points. IEEE Trans. Power Syst. **27**(3), 1577–1584 (2012)
16. J.R. Pillai, B. Bak-Jensen, Integration of vehicle-to-grid in the western Danish power system. IEEE Trans. Sustain. Energy **2**(1), 12–19 (2011)
17. H. Xing, M. Fu, Z. Lin, Y. Mou, Decentralized optimal scheduling for charging and discharging of plug-in electric vehicles in smart grids. IEEE Trans. Power Syst. **31**(5), 4118–4127 (2016)
18. A.S. Masoum, S. Deilami, M.A. Masoum, A. Abu-Siada, S. Islam, Online coordination of plug-in electric vehicle charging in smart grid with distributed wind power generation systems, in *PES General Meeting| Conference & Exposition, IEEE*, (2014), pp. 1–5
19. J. Wang, C. Liu, D. Ton, Y. Zhou, J. Kim, A. Vyas, Impact of plug-in hybrid electric vehicles on power systems with demand response and wind power. Energy Policy **39**(7), 4016–4021 (2011)
20. H.A. Aalami, H. Pashaei-Didani, S. Nojavan, Deriving nonlinear models for incentive-based demand response programs. Int. J. Electr. Power Energy Syst. **106**, 223–231 (2019)
21. F.S. Gazijahani, J. Salehi, Integrated DR and reconfiguration scheduling for optimal operation of microgrids using Hong's point estimate method. Int. J. Electr. Power Energy Syst. **99**, 481–492 (2018)
22. K. Saberi et al., Optimal performance of CCHP based microgrid considering environmental issue in the presence of real time demand response. Sustain. Cities Soc. **45**, 596–606 (2019)
23. F.S. Gazijahani, J. Salehi, Game theory based profit maximization model for microgrid aggregators with presence of EDRP using information gap decision theory. IEEE Syst. J. **13**(2), 1767–1775 (2019)
24. H. Pashaei-Didani et al., Optimal economic-emission performance of fuel cell/CHP/storage based microgrid. Int. J. Hydrog. Energy **44**(13), 6896–6908 (2019)
25. T. Khalili, S. Nojavan, K. Zare, Optimal performance of microgrid in the presence of demand response exchange: a stochastic multi-objective model. Comput. Electr. Eng. **74**, 429–450 (2019)

26. F.S. Gazijahani, J. Salehi, Reliability constrained two-stage optimization of multiple renewable-based microgrids incorporating critical energy peak pricing demand response program using robust optimization approach. Energy **161**, 999–1015 (2018)
27. K.Ç. Bayindir, M.A. Gözüküçük, A. Teke, A comprehensive overview of hybrid electric vehicle: powertrain configurations, powertrain control techniques and electronic control units. Energy Convers. Manag. **52**(2), 1305–1313 (2011)
28. M. Shamshirband et al., Decentralized trading of plug-in electric vehicle aggregation agents for optimal energy management of smart renewable penetrated microgrids with the aim of CO2 emission reduction. J. Clean. Prod. **200**, 622–640 (2018)
29. K.M. Tan, V.K. Ramachandaramurthy, J.Y. Yong, Integration of electric vehicles in smart grid: a review on vehicle to grid technologies and optimization techniques. Renew. Sust. Energ. Rev. **53**, 720–732 (2016)
30. M. Shamshirband et al., Look-ahead risk-averse power scheduling of heterogeneous electric vehicles aggregations enabling V2G and G2V systems based on information gap decision theory. Electr. Power Syst. Res. **173**, 56–70 (2019)
31. F.S. Gazijahani, S.N. Ravadanegh, J. Salehi, Stochastic multi-objective model for optimal energy exchange optimization of networked microgrids with presence of renewable generation under risk-based strategies. ISA Trans. **73**, 100–111 (2018)
32. F.S. Gazijahani, J. Salehi, Stochastic multi-objective framework for optimal dynamic planning of interconnected microgrids. IET Renew. Power Gener. **11**(14), 1749–1759 (2017)
33. P. Salyani et al., Chance constrained simultaneous optimization of substations, feeders, renewable and non-renewable distributed generations in distribution network. Electr. Power Syst. Res. **158**, 56–69 (2018)
34. J. Salehi et al., Scenario-based co-optimization of neighboring multi carrier smart buildings under demand response exchange. J. Clean. Prod. **235**, 1483–1498 (2019)
35. F.S. Gazijahani, J. Salehi, IGDT based complementarity approach for dealing with strategic decision making of Price maker VPP considering demand flexibility. IEEE Trans. Indus. Inform., 1–9 (2019)

Chapter 9
Robust Scheduling of Plug-In Electric Vehicles Aggregator in Day-Ahead and Reserve Markets

Amir Farahmand-Zahed, Sayyad Nojavan, and Kazem Zare

Nomenclature

Set

t	Time period (hour)
v	PEV groups

Parameters

λ_t	Electricity market price at time interval t (\$/MWh)
λ_t^{R}	Reserve market price at time interval t (\$/MWh)
λ_{d}	Tariffs of selling energy to the owners of PEVs (\$/MWh)
∂_t	Availability percentage of PEVs
α	Status of delivering energy in the reserve market
c_{d}	Battery degradation cost
$б_v$	Charging efficiency of the battery
ε_v	Degradation coefficient of discharging
$\underline{\mathrm{SOC}}_v$	Minimum capacity limit of the battery (MWh)
$\overline{\mathrm{SOC}}_v$	Maximum capacity limit of the battery (MWh)

A. Farahmand-Zahed · K. Zare
Faculty of Electrical and Computer Engineering, University of Tabriz, Tabriz, Iran
e-mail: amir.farahmand96@ms.tabrizu.ac.ir; kazem.zare@tabrizu.ac.ir

S. Nojavan (✉)
Department of Electrical Engineering, University of Bonab, Bonab, Iran
e-mail: sayyad.nojavan@bonabu.ac.ir

S. Nojavan, K. Zare (eds.), *Electricity Markets*,
https://doi.org/10.1007/978-3-030-36979-8_9

$\underline{E}_v$	Minimum limit of the rate of charge of the battery (MWh)
$\overline{E}_v$	Maximum limit of the rate of charge of the battery (MWh)
c_b	Battery cost for one kilowatt-hour ($/kWh)
c_l	Battery replacement labor cost
E_b	Energy capacity cost of battery
L_c	Cycle life of the battery
DOD	Depth of discharge

Numbers

N_v	Number of PEV groups
n^v	Number of PEVs with the same driving pattern

Variables

$P_{t.v}^{\mathrm{sale}}$	Power sold in the day-ahead market from vth vehicle at time interval t (MW)
$P_{t.v}^{\mathrm{R}}$	Power sold in reserve market from vth vehicle at time interval t (MW)
$P_{t.v}^{\mathrm{buy}}$	Power charged to the vth vehicle at time interval t (MW)
$G_{t.\,v}$	Power required by the vth vehicle at time interval t (MW)
$\mathrm{DE}_{t.\,v}$	Total discharged energy from vth vehicle in energy markets at time interval t (MW)
$\mathrm{SOC}_{t.\,v}$	State of charge of the vth vehicle at time interval t (MWh)
$E_{t.\,v}$	Purchased power from the grid for the vth vehicle at time interval t (MW)

9.1 Introduction

Global warming, energy crisis, and other environmental effects caused by fossil fuels increase the popularity of electric vehicles (EVs) [1]. On the other hand, growing EV technology and government's motivations for using EVs speed up their rapid growth. So in the near future, EVs will have a large share of the transportation system and they can be used as demand-side resources [2]. Participation of a single plug-in electric vehicle (PEV) in the power market is not possible because of market regulation, but a group of PEVs can provide enough energy for participating in electricity markets. Aggregation of PEVs has benefits for both PEV owners and the grid. It is possible for PEV aggregators to participate in electricity markets to maximize its benefit by charging PEVs at time intervals with lower energy price and selling it at time intervals with higher energy price. So it is necessary for a PEV aggregator to have an optimal schedule to participate in electricity markets.

In recent years, the scheduling problem of PEVs is analyzed and modeled in many studies. In [3], an algorithm for unidirectional vehicle-to-grid (V2G) charging regulation is developed, which is able to modulate the charging around the charging rate of each EV. A model of an aggregate battery for an EV fleet is presented in [4]. An aggregate model of a V2G fleet is presented in [5] with the goal of demonstrating the energy and power constraints of the V2G fleet, then this model is used to develop an evaluation method to obtain the V2G capacity. In [6], the PEV aggregators operation is optimized in order to maximize their benefit considering market price variations. A hybrid method based on multi-agents and dynamic game theory is provided to model market players, and also, the effect of the contract on the behavior of EV owners is modeled in [7]. In another framework, an integrated model based on Time of Use and Price-Based Demand Response for the EV aggregator's G2V charge scheduling is studied [8]. The existence of uncertainties makes the scheduling problem of PEVs more complicated. Different methods for handling the uncertain parameter are reviewed, classified, and evaluated in [3]. In [9], an optimal stochastic programming model is studied to maximize the profit of EV aggregator from participating in the regulation market. The participation of EV and energy storage aggregators in frequency regulation markets is studied in [10], their presented formulation is a stochastic mixed integer linear programming model, and the uncertainties of energy and frequency regulation prices are considered in their model, as well as the degradation costs of energy storages. In [11], a mixed-integer stochastic linear program model for coordination of unidirectional V2G services and energy trading is presented. Unidirectional charging pattern is used for the short-term scheduling of aggregators in [12]. A stochastic mixed integer linear programming model for the optimization of the participation of aggregators in real-time and day-ahead electricity markets and power quality services is investigated in [13]. A dynamic stochastic optimization method with a stochastic linear programming formulation is studied in [14] to optimize the charging schedule of electric vehicles. Furthermore, the uncertainties of load, electricity pricing, and renewable energy generation are considered in [14]. Charge and discharge managing tools have been developed in order to track automatic generation control (AGC) signal, also a real-time controller is presented which considers bidirectional charging efficiency in [15]. A linear programming model for real-time charging management of an EV aggregator in order to participate in energy and regulation markets is presented in [16]. The self-scheduling problem of a PEV aggregator to participate in balancing services for a wind power producer is studied in [17], and a scenario-based robust approach is used to consider the uncertainties of wind power generation. An approach based on information gap decision theory is analyzed in [18] in order to manage the risks of EV aggregator's revenue caused by electricity price uncertainty. In [19], a framework for efficient charging of PEVs is studied where multiple aggregators exist in a city and they cooperate with each other to schedule the charging of PEVs in different charging stations, either owned by them or not. Furthermore, a bi-objective charging schedule model is presented to maximize both the aggregators' total profit as well as the total number of charged PEVs [19]. In [20], a stochastic programming model is presented for the optimal scheduling problem of PEV aggregators to participate in day-ahead and reserve markets and autoregressive integrated moving average (ARIMA) technique is used to produce scenarios for market price; furthermore, conditional value at risk (CVaR) is employed in the scheduling process of the PEV aggregator.

In this chapter, the optimal scheduling of PEV aggregator for participation in day-ahead and reserve markets is presented. The impact of pool market price uncertainty is considered. Robust optimization approach is used to specify pool market price uncertainty and find the optimal short-term schedule for the aggregator. The proposed model is applied to a case study which consists of 10,000 PEVs. The simulation is done in two cases: deterministic case and robust optimization case. The results indicate that the aggregator can be robust against 20% changes in the market price with just 2.51% decrease in its total profit in robust optimization case. The contributions of this chapter can be summarized as below:

- The optimal short-term scheduling of PEV aggregator for participation in day-ahead and reserve markets considering the market price uncertainty is proposed.
- Robust optimization approach is used to handle the market price uncertainty.

The remainder of this chapter consists of the following sections: Sect. 9.2 introduces the optimal scheduling problem of PEV aggregator, its objective function, and mathematical formulation. In Sect. 9.3, the robust optimization approach is introduced and formulated. A case study, its input data, and the obtained results are presented in Sect. 9.4. The conclusion of the chapter is presented in Sect. 9.5.

9.2 Deterministic-Based Scheduling of PEV Aggregator

The cost of the PEV aggregator is from the energy bought from the grid and battery degradation cost caused by discharging energy due to participation in energy markets. And, the aggregator's revenue is from selling energy to PEV owners and participating in electricity markets. So an optimal schedule is needed to maximize the total benefit of the PEV aggregator considering the uncertainties of market energy price and PEV availability. This can be obtained by charging the PEVs in time intervals with lower electricity price and selling the electricity at time intervals with higher electricity price. The PEVs charging tariffs is assumed to be constant because determining the energy tariff is a long-term decision and it is an input data for short-term scheduling problem and determined based on market competitions and regulatory constraints. In this chapter, the impacts of pool market price uncertainty and the uncertainty of the availability of PEVs on PEV aggregator scheduling problem are considered. Robust optimization approach is used to specify these uncertainties and find the optimal short-term schedule for the aggregator.

Finding the optimal schedule for the participation of PEV aggregators in the power markets is more complicated due to the pool market price uncertainty. The objective of solving the PEV aggregator scheduling problem is to maximize the aggregator's total profit. The aggregator's revenue comes from two sources: the first one is from PEV owners that would pay for charging their vehicles and the second source is from selling electricity in the day-ahead and reserve markets. And, the aggregator's cost is from the energy bought from the grid and battery degradation cost caused by discharging energy due to participation in energy markets. The formulation of the objective function is represented in (9.1).

$$\text{Max} \sum_{v=1}^{N_v} (n^v \times \partial_t) \left\{ \sum_{t=1}^{T} \lambda_t P_{t.v}^{\text{sale}} + \lambda_t^{\text{R}} P_{t.v}^{\text{R}} + \alpha \lambda_t P_{t.v}^{\text{R}} + \lambda_{\text{d}} G_{t.v} - \lambda_t P_{t.v}^{\text{buy}} - c_{\text{d}} \text{DE}_{t.v} \right\} \tag{9.1}$$

where, n^v and ∂_t are the number of PEVs with the same driving pattern and the availability percentage of PEVs, respectively. λ_t^{R} and λ_t are the reserve market price and pool market price at time interval t ($/MWh), respectively. λ_{d} is the constant tariffs of selling energy to the owners of PEVs. c_{d} is battery degradation cost and $\text{DE}_{t.v}$ shows the total discharged energy from vth vehicle in energy markets at time interval t. The status of delivering energy in the reserve market is modeled by α, $\alpha = 1$ is equal to be called to deliver energy in the reserve market. Total discharged energy from the vth vehicle in day-ahead and reserve markets at time interval t can be stated as:

$$P_{t.v}^{\text{sale}} + P_{t.v}^{\text{R}} = \text{DE}_{t.v} \quad \forall t, \forall v \tag{9.2}$$

The SOC of each battery at each time interval as below:

$$\text{SOC}_{t.v} = \text{SOC}_{t-1.v} + (б_v E_{t.v} - G_{t.v} - \epsilon_v \text{DE}_{t.v}) \quad \forall t, \forall v \tag{9.3}$$

Its elements are the battery's SOC at the previous hour ($\text{SOC}_{t-1.v}$), the power required for driving ($G_{t.v}$), power sold in day-ahead and reserve markets ($\text{DE}_{t.v}$), the power delivered by the grid ($E_{t.v}$), charging efficiency ($б_v$) and the degradation coefficient of discharging (ε_v). The limits of SOC and the rate of charge are applied using (9.4) and (9.5).

$$\underline{\text{SOC}}_v \le \text{SOC}_{t.v} \le \overline{\text{SOC}}_v \quad \forall t, \forall v \tag{9.4}$$

$$\underline{E}_v \le E_{t.v} \le \overline{E}_v \quad \forall t, \forall v \tag{9.5}$$

where, $\underline{\text{SOC}}_v$ and $\overline{\text{SOC}}_v$ are the minimum and maximum limits of the vth PEV's battery. $\underline{E}_v$ and $\overline{E}_v$ are minimum and maximum limits for the rate of charge.

It is ensured that PEV batteries only can be charged or discharged when they are plugged in using (9.6). Charge and discharge of batteries cannot happen at the same time which is constrained by (9.7).

$$\text{If}\left\{\forall t, \forall v | G_{t,v} \neq 0\right\} \quad \text{then}: \quad E_{t.v}, P_{t.v}^{\text{sale}}\ P_{t.v}^{\text{R}} = 0 \tag{9.6}$$

$$\text{If}\left\{\forall t, \forall v | E_{t,v} \neq 0\right\} \quad \text{then}: \quad P_{t.v}^{\text{sale}}\ P_{t.v}^{\text{R}} = 0 \tag{9.7}$$

The following constrain is added to the model to limit the rate of change SOC charging:

$$P_{t.v}^{\text{R}} \leq \text{SOC}_{t.v} - \underline{\text{SOC}}_{v} \quad \forall t, \forall v \tag{9.8}$$

Degradation cost of batteries due to discharging energy is calculated using (9.9).

$$c_{\text{d}} = \frac{c_{\text{b}} E_{\text{b}} + c_{\text{l}}}{L_{\text{c}}\ E_{\text{b}}\ \text{DOD}} \tag{9.9}$$

where, c_{b} and c_{l} are the battery cost for 1 kWh and replacement cost, respectively. Battery's energy capacity cost is shown by E_{b} and L_{c} is the cycle life of the battery at a special depth of discharge (DOD).

9.3 Robust Optimization-Based Scheduling of PEV Aggregator

The existence of uncertainties would increase the complexity of the investigated problems. Many methods have been developed in order to handle the uncertain parameter. One of these methods is Robust Optimization Approach. The basis of Robust Optimization Approach is on representing the effect of the uncertain parameter on the optimal solution and reducing the sensitivity of the optimal solution to the uncertain parameter. The advantages of this method can be summarized as below [21]:

- The Robust Optimization Approach requires less calculation effort than stochastic programming.
- The obtained solutions are reliable since the Robust Optimization Approach considers the worst conditions.
- The Robust Optimization Approach does not exert probabilistic distribution functions.

The standard MILP model is formulated in (9.10)–(9.13).

$$\text{Min} \sum_{t=1}^{n} c_t x_t \tag{9.10}$$

Subjected to

$$\sum_{j=1}^{n} a_{ij}x_j \le b_i \quad \forall i = 1, \ldots, m \tag{9.11}$$

$$x_t \ge 0 \quad \forall t = 1, \ldots, n \tag{9.12}$$

$$x_t \in \{0, 1\} \quad \text{for some } t = 1, \ldots, n \tag{9.13}$$

where c_t is a coefficient with an unknown value and its range is $[c_t - d_t \quad c_t + d_t]$. d_t shows the deviation of c_t from its nominal value. In order to formulate the robust mixed-integer linear programming, an integer control parameter is needed which is named Γ (GAMA). Γ must be a real value in the range of $[0, |T_0|]$. The worst case happens when the value of Γ is equal to $|T_0|$.

And the deterministic case is obtained when $\Gamma = 0$. The robust formulation of (9.10)–(9.13) can be defined as (9.14)–(9.22):

$$\text{Min} \sum_{t=1}^{n} c_t x_t + \beta.\Gamma + \sum_{t=1}^{n} \zeta_t \tag{9.14}$$

$$\sum_{j=1}^{n} a_{ij}x_j \le b_i \quad \forall i = 1, \ldots, m \tag{9.15}$$

$$x_t \ge 0 \quad \forall t = 1, \ldots, n \tag{9.16}$$

$$x_t \in \{0, 1\} \quad \text{for some } t = 1, \ldots, n \tag{9.17}$$

$$\beta + \zeta_t \ge d_t \theta_t \quad t \in T_0 \tag{9.18}$$

$$\beta \ge 0 \tag{9.19}$$

$$\zeta_t \ge 0 \quad \forall t = 1, \ldots, n \tag{9.20}$$

$$\theta_t \ge 0 \quad \forall t = 1, \ldots, n \tag{9.21}$$

$$\theta_t \ge x_t \quad \forall t = 1, \ldots, n \tag{9.22}$$

Pool market price is considered to be the only uncertain parameter in the studied base problem (Eqs. (9.1)–(9.9)). The uncertainty of this parameter is modeled using Robust Optimization Approach. The robust formulation of the main problem is as follows:

$$\text{Max} \quad -\beta.\Gamma - \sum_{t=1}^{n} \zeta_t + \sum_{v=1}^{N_v} \left(n^v \times \partial_t\right)$$

$$\times \left\{ \sum_{t=1}^{T} (\lambda_t)\, P_{t.v}^{\text{sale}} + \left(\lambda_t^{\text{R}}\right) P_{t.v}^{\text{R}} + \alpha\,(\lambda_t)\, P_{t.v}^{\text{R}} + (\lambda_{\text{d}})\, G_{t.v} - (\lambda_t)\, P_{t.v}^{\text{buy}} - (c_{\text{d}})\, \text{DE}_{t.v} \right\} \tag{9.23}$$

$$\beta + \zeta_t \geq \left(\lambda_t^{\max} - \lambda_t^{\min}\right) \theta_t \quad \forall t \tag{9.24}$$

$$\theta_t \geq P_{t.v}^{\text{sale}} + P_{t.v}^{\text{R}} + \alpha.P_{t.v}^{\text{R}} - P_{t.v}^{\text{buy}} \quad \forall t = 1, \ldots, T \tag{9.25}$$

$$\text{Eqs. } (9.2) - (9.9) \tag{9.26}$$

where β and ζ_t are two variables which are used to manage the variations of λ_t, and θ_t is an auxiliary variable used to linearize the problem. For $\left(\lambda_t^{\max} - \lambda_t^{\min}\right) = 0$ the value of Γ will be zero.

9.4 Case Study

The proposed robust optimization approach is used for optimal scheduling problem of PEVs to participate in day-ahead and reserve markets for the 24-h time period. The characteristics of PEV batteries are taken from [22]. In order to simplify the problem, it is assumed that PEVs with the same driving pattern would act in the same way. Driving patterns and the count of vehicles with the same driving pattern are illustrated in Table 9.1. In order to calculate the degradation cost of batteries using (9.9), the values of parameters c_b and c_l are considered 300 \$/kWh and 240\$ [23], respectively. And according to [24], at 100% discharge battery's lifetime is 3000-cycle. Maximum and minimum capacity of each battery is assumed to be 25 kWh and 0 kWh, respectively. Also charging efficiency ($\acute{\sigma}_v$) and the degradation coefficient of discharging (ε_v) of PEV batteries are considered to be 100%. Other parameters of PEVs are represented in Table 9.2. It is assumed that an empty battery can fully charge in 2 h, which means that a battery is able to charge 50% of its total capacity in 1 h with its maximum charging speed. The hourly pool price data is represented in Table 9.1 [25]. The hourly price of reserve market (λ_t^{R}) is assumed to be equal to the day-ahead market prices (λ_t).

Table 9.1 Driving patterns of PEVs

	Power requirements of PEVs ($G_{t,v}$) (kW)										
Hour (h)	1	2	3	4	5	6	7	8	9	10	Market price
1	0	0	0	0	0	0	0	0	0	0	31.3780
2	0	0	0	0	0	0	0	0	0	0	31.0648
3	0	0	0	0	0	0	0	0	0	0	30.7052
4	0	0	0	0	0	0	0	0	0	0	30.7284
5	0	0	0	0	0	0	0	0	0	0	30.8560
6	1.486	1.916	2.049	2.209	1.179	1.720	1.676	1.839	0.856	0.847	31.3316
7	2.264	2.425	2.586	2.528	2.679	1.815	1.700	1.490	1.137	1.307	32.5264
8	2.835	2.812	2.810	3.139	3.919	0	0	0	1.633	1.373	35.2988
9	0	0	0	0	0	0	0	0	0.946	0.831	46.9220
10	0	0	0	0	0	2.232	2.701	3.060	0	0	48.6852
11	0	0	0	0	0	3.803	2.775	4.070	0	0	43.5580
12	1.239	1.183	1.296	1.173	0.906	0	0	0	0	0	41.7832
13	1.051	1.146	1.132	0.944	0.939	0	0	0	0	0	41.2148
14	0	0	0	0	0	0	0	0	0	0	40.9480
15	0	0	0	0	0	1.644	1.930	1.976	0	0	41.5628
16	2.033	1.888	1.466	2.034	1.521	1.888	1.767	2.053	0.769	0.93428	48.7664
17	2.897	3.659	2.480	3.716	3.079	0	0	0	1.191	0.952	49.3348
18	1.892	1.626	2.564	1.870	2.154	0	0	0	0.9781	0.9435	59.6588
19	0	0	0	0	0	0	0	0	0	0	57.9536
20	1.477	2.338	1.503	1.836	1.674	2.178	1.829	1.602	0	0	54.2068
21	2.037	2.143	1.961	1.401	2.027	2.109	1.272	1.624	0	0	48.2212
22	0	0	0	0	0	0	0	0	0	0	43.0244
23	0	0	0	0	0	0	0	0	0	0	37.6884
24	0	0	0	0	0	0	0	0	0	0	32.5380
n^v	900	500	570	600	770	1200	1300	830	1830	1500	

Table 9.2 Characteristics of batteries

Parameters	$\underline{SOC}_v$ (kWh)	$\overline{SOC}_v$ (kWh)	$\underline{E}_v$	$\overline{E}_v$	б_v	ε_v
Values	0	25	0%	50%	100%	100%

The optimal scheduling problem of PEVs is formulated as a mixed-integer linear programming (MILP) and is solved using CPLEX solver under GAMS software.

The scheduling of PEVs is investigated in two cases: deterministic case and robust optimization case. The deterministic case is determined with $\Gamma = 0$ and robust optimization case is the result of $\Gamma = 1$.

Figure 9.1 demonstrates the relation of Γ and the total profit of the aggregator. As parameter Γ increases, the total profit of the aggregator decreases. Total profit of the aggregator at deterministic case ($\Gamma = 0$) is 12,811.16 $ which will be decreased by 2.51% at robust optimization case, bringing the total profit to 12,490.09 $. In other words, with a 2.51% decrease in aggregator's total profit, the aggregator will be robust against 20% changes in the market price in robust optimization case.

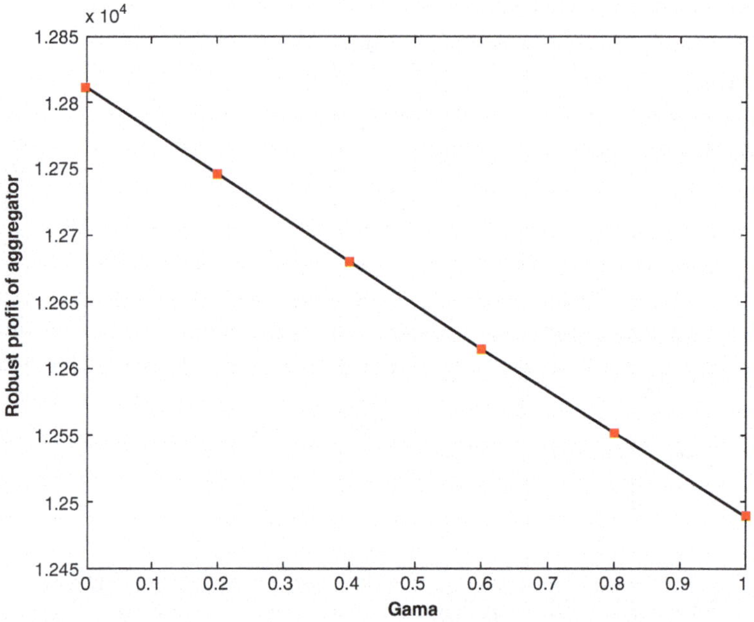

Fig. 9.1 The relation of Γ and the total profit of the aggregator

Total energy sold in the day-ahead market in 1 day is illustrated in Fig. 9.2. According to Fig. 9.2, PEVs are planned to sell energy in the day-ahead market at 9:00, 10:00, 19:00, 20:00, and 22:00. As illustrated in Fig. 9.2, there is an increase of 9.59% in energy sold in the day-ahead market at hour 19:00 in the robust optimization case in comparison to the deterministic case, as well as a small increase at 22:00.

Total energy sold in the reserve markets is illustrated in Fig. 9.3. According to Fig. 9.3, PEVs are scheduled to contribute to the reserve market at 11:00, 12:00, 18:00, and 19:00. The results show that the energy sold in the reserve market in the robust optimization case is 7.63% less than the deterministic case at 18:00, also there is a small increase at 19:00.

Bought energy from the grid is shown in Fig. 9.4. According to Fig. 9.4, the energy bought from the grid is the same in both deterministic and robust optimization cases and PEVs are scheduled to be charged at 8:00, 11:00, 13:00, 14:00, 15:00, and 17:00. The highest amount of bought energy is at 14:00 with 125 MW. According to Figs. 9.2 and 9.4, both charging and discharging of PEVs are planned at 11:00; however, this does not ignore the constraint (9.7), because the same batteries are not scheduled to be charged and discharged at the same time, and they are different batteries.

The state of charge of the batteries is illustrated in Fig. 9.5. According to Fig. 9.5, the SOC of batteries is decreased between 7:00 and 12:00, because of selling energy in reserve and day-ahead market. Also charging PEVs at 13:00 to 16:00 has resulted in an increase of SOC of the batteries at these hours.

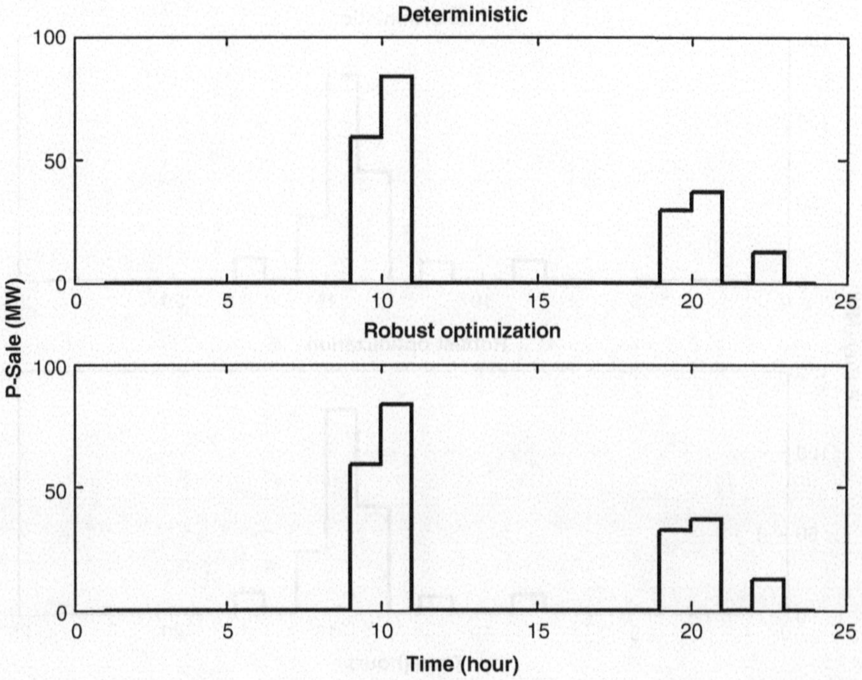

Fig. 9.2 Total energy sold in the day-ahead energy market

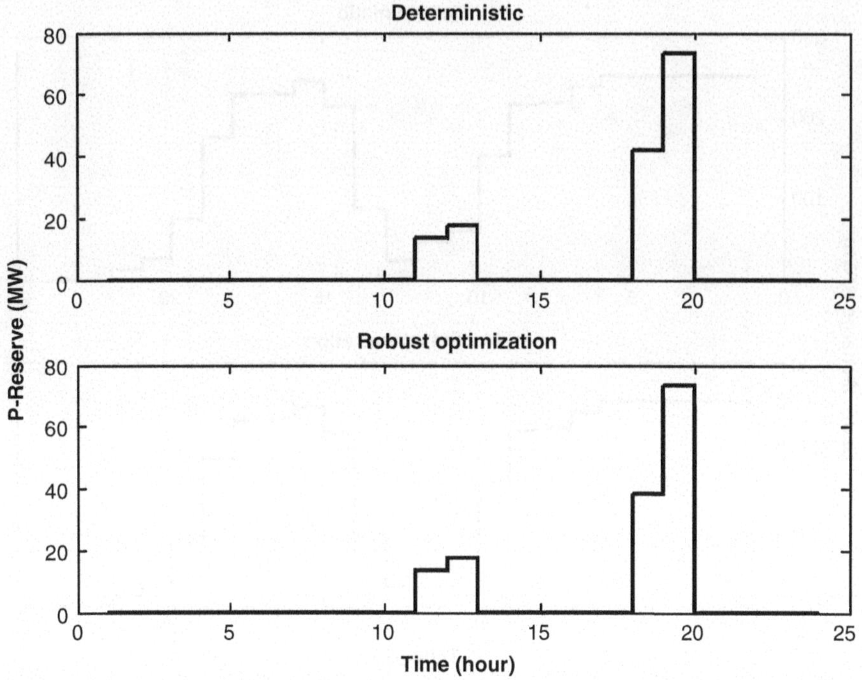

Fig. 9.3 Total energy sold in the reserve market

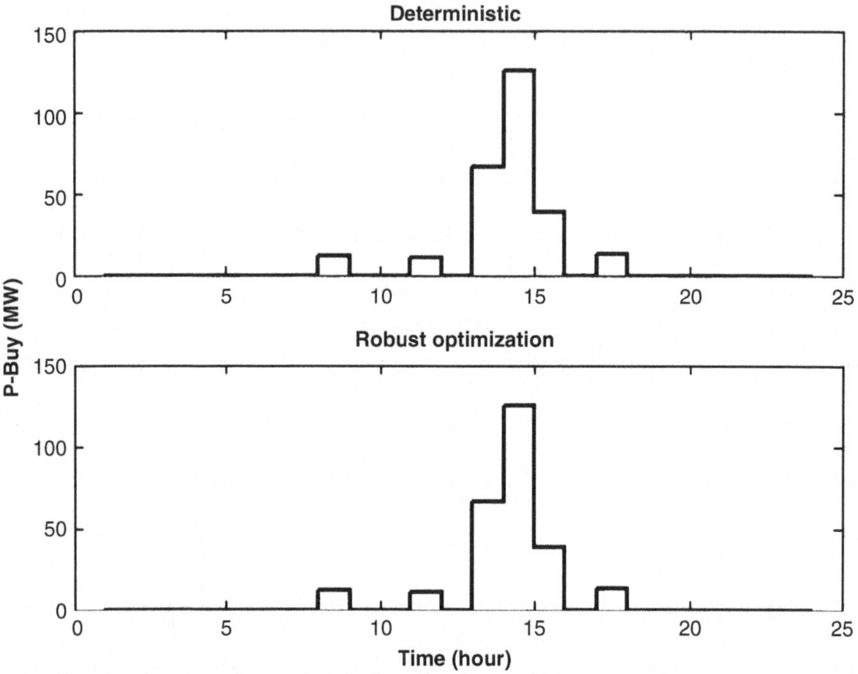

Fig. 9.4 Total energy bought from the grid

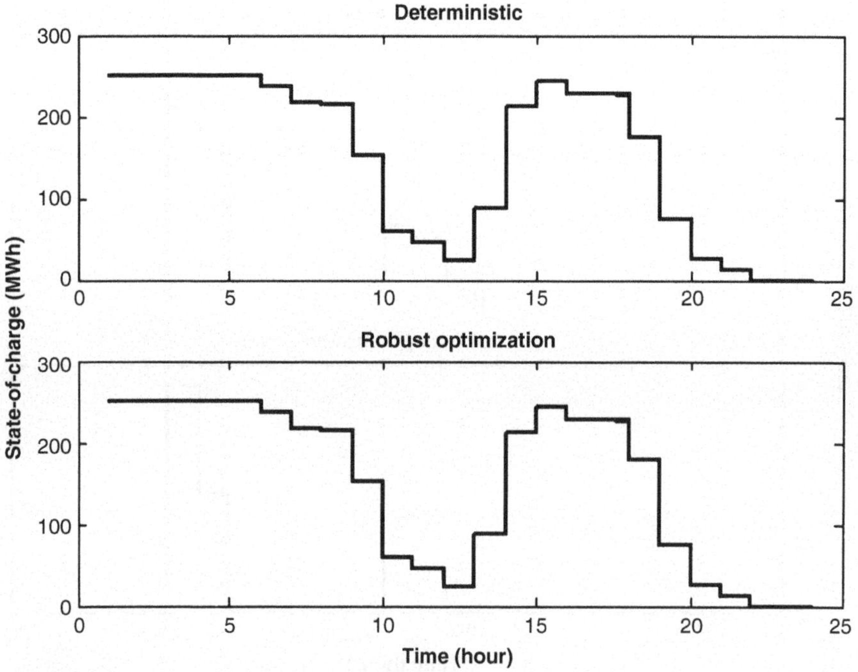

Fig. 9.5 State of charge of the batteries

9.5 Conclusion

In this chapter, the optimal scheduling problem of plug-in electric vehicle (PEV) aggregators participation in day-ahead and reserve markets is studied. The uncertainty of market price is considered using Robust Optimization Approach. Robust optimization model of the scheduling problem of PEVs is presented. The proposed model is tested using a case study consists of 10,000 PEVs, which are divided into 10 groups with the same driving pattern for simplicity. The simulation is done in two cases: deterministic and robust optimization. The results indicate that with a 2.51% decrease of aggregator's total profit, the aggregator will be robust against 20% changes in the market price in robust optimization case.

References

1. B. Amirhosseini, S.M.H. Hosseini, Scheduling charging of hybrid-electric vehicles according to supply and demand based on particle swarm optimization, imperialist competitive and training-learning algorithms. Sustain. Cities Soc. **43**, 339–349 (2018)
2. C. Pang, M. Kezunovic, M. Ehsani, Demand side management by using electric vehicles as distributed energy resources, in *2012 IEEE International Electric Vehicle Conference*, (2012), pp. 1–7
3. E. Sortomme, M.A. El-Sharkawi, Optimal charging strategies for unidirectional vehicle-to-grid. IEEE Trans. Smart Grid **2**(1), 119–126 (2011)
4. B. Škugor, J. Deur, A novel model of electric vehicle fleet aggregate battery for energy planning studies. Energy **92**, 444–455 (2015)
5. H. Zhang, Z. Hu, Z. Xu, Y. Song, Evaluation of achievable vehicle-to-grid capacity using aggregate PEV model. IEEE Trans. Power Syst. **32**(1), 784–794 (2017)
6. N.H. Tehrani, G.B. Shrestha, P. Wang, Vehicle-to-grid service potential with price based PEV charging/discharging, in *The IEEE Power & Energy Society General Meeting*, (2013)
7. M. Shafie-khah, M. Parsa Moghaddam, M.K. Sheikh-El-Eslami, M. Rahmani-Andebili, Modeling of interactions between market regulations and behavior of plug-in electric vehicle aggregators in a virtual power market environment. Energy **40**(1), 139–150 (2012)
8. S. Sharma, P. Jain, Integrated TOU price-based demand response and dynamic grid-to-vehicle charge scheduling of electric vehicle aggregator to support grid stability, in *International Transactions on Electrical Energy Systems 2018*, (2019), pp. 1–22
9. M.W. Khalid, A.T. Al-Awami, E. Sortomme, Stochastic-programming-based bidding strategy for V2G services, in *2013 IEEE PES Innovative Smart Grid Technologies EUROPE (ISGT Europe 2013)*, (2013), pp. 1–5
10. B. Vatandoust, A. Ahmadian, M.A. Golkar, A. Elkamel, A. Almansoori, M. Ghaljehei, Risk-averse optimal bidding of electric vehicles and energy storage aggregator in day-ahead frequency regulation market. IEEE Trans. Power Syst. **34**(3), 2036–2047 (2019)
11. A.T. Al-Awami, E. Sortomme, Coordinating vehicle-to-grid services with energy trading. IEEE Trans. Smart Grid **3**(1), 453–462 (2012)
12. M. Moradi-Dalvand, B. Mohammadi-Ivatloo, Optimal short-term scheduling of aggregators for plug-in electric vehicles. Autonom. Veh., 293–311 (2014)
13. B. Hashemi, M. Shahabi, P. Teimourzadeh-Baboli, Stochastic-based optimal charging strategy for plug-in electric vehicles aggregator under incentive and regulatory policies of DSO. IEEE Trans. Veh. Technol. **68**(4), 3234–3245 (2019)

14. S. Liu, A.H. Etemadi, A dynamic stochastic optimization for recharging plug-in electric vehicles. IEEE Trans. Smart Grid **3053**(c), 1–1 (2017)
15. G. Wenzel, M. Negrete-Pincetic, D.E. Olivares, J. MacDonald, D.S. Callaway, Real-time charging strategies for an electric vehicle aggregator to provide ancillary services. IEEE Trans. Smart Grid **3053**(c), 1–1 (2017)
16. S.I. Vagropoulos, D.K. Kyriazidis, A.G. Bakirtzis, Real-time charging management framework for electric vehicle aggregators in a market environment. IEEE Trans. Smart Grid **7**(2), 948–957 (2016)
17. M.G. Vayá, G. Andersson, Self scheduling of plug-in electric vehicle aggregator to provide balancing services for wind power. IEEE Trans. Sustain. Energy **7**(2), 886–899 (2016)
18. J. Zhao, C. Wan, Z. Xu, J. Wang, Risk-based day-ahead scheduling of electric vehicle aggregator using information gap decision theory. IEEE Trans. Smart Grid **8**(4), 1609–1618 (2017)
19. J.C. Mukherjee, A. Gupta, Distributed charge scheduling of plug-in electric vehicles using inter-aggregator collaboration. IEEE Trans. Smart Grid **8**(1), 331–341 (2017)
20. M. Alipour, B. Mohammadi-Ivatloo, M. Moradi-Dalvand, K. Zare, Stochastic scheduling of aggregators of plug-in electric vehicles for participation in energy and ancillary service markets. Energy **118**, 1168–1179 (2017)
21. E. Köbis, On robust optimization. J. Optim. Theor. Appl. **167**(3), 969–984 (2015)
22. Z. Li, Fundamentals of power system economics [book review]. IEEE Power Energy Mag. **4**(4), 76–78 (2006)
23. C.D. White, K.M. Zhang, Using vehicle-to-grid technology for frequency regulation and peak-load reduction. J. Power Sources **196**(8), 3972–3980 (2011)
24. W. Kempton, J. Tomić, Vehicle-to-grid power fundamentals: calculating capacity and net revenue. J. Power Sources **144**(1), 268–279 (2005)
25. "Historical Market Data." [Online]. https://www.nordpoolgroup.com/historical-market-data. Accessed 28 Sep 2018.

Chapter 10
Optimal Scheduling of Water Distribution Systems' Participation in Demand Response and Frequency Regulation Services

Amir Farahmand-Zahed, Alireza Akbari-Dibavar, Behnam Mohammadi-Ivatloo, and Kazem Zare

10.1 Introduction

In past decades the global demand for electricity has been growing, and in order to face this growth the electricity generation capacity has been increasing as well. But sometimes, at peak times or as a result of the grid contingencies, the generation capacity cannot meet the electricity demand of customers [1]. One possible solution is to create a methodology in which the load responds to the contingencies instead of the generation. In other words, consumers tend to decrease their demand in order to reduce the failure risk of the system. These methodologies are called Demand Response (DR) plans [2]. Demand response is an essential program for restoring the balance between electricity demand and supply in a smart grid.

Water distribution systems (WDSs) are energy-intensive substructures that consume energy to deliver water to consumers. WDS consists of pumps, pipes, reservoirs, valves, and tanks to transfer water from reservoirs to the costumers [3]. In this operation, a great volume of water is being pumped, which requires a high amount of energy. The cost of the energy consumed by pumps can be reduced through scheduling the pumps to consume energy during hours with lower electricity tariffs. This flexibility is possible due to the existence of water tanks, which can store water during hours with cheaper tariffs and deliver the water at high tariff hours [4]. The operation of water storage tanks and pumps have to be optimized to minimize the operation cost of WDS while supplying the water demand and respecting the constraints of WDS. The flexibility caused by pumpsand tanks

A. Farahmand-Zahed (✉) · A. Akbari-Dibavar · B. Mohammadi-Ivatloo · K. Zare
Faculty of Electrical and Computer Engineering, University of Tabriz, Tabriz, Iran
e-mail: amir.farahmand96@ms.tabrizu.ac.ir; a.dibavar96@ms.tabrizu.ac.ir; bmohammadi@tabrizu.ac.ir; kazem.zare@tabrizu.ac.ir

S. Nojavan, K. Zare (eds.), *Electricity Markets*,
https://doi.org/10.1007/978-3-030-36979-8_10

enables WDS to participate in DR and frequency regulation up/down capacity offer services.

In recent years, lots of researches have proposed different methods for minimizing the energy cost of the operation of WDS pumps, but they did not consider the potential of WDS for participating in DR or other services [4–9]. A systematic review of the published models and methodologies for the optimization of the operation of WDS is studied in [10]. A model is provided in [11] that enables power system to use the energy flexibility of WDS. The implementation of demand response programs in the management of water transfer stations is investigated in [12]. A mixed-integer linear programming (MILP) model of WDS is studied in [13], then a robust stochastic programming model considering the uncertainties of rain and water demand is presented for the optimization of WDS. In another framework, a linear optimization model for the pump scheduling problem in drinking water systems while trading demand response in the power market is studied [14], and the uncertainties water demand are considered. A model for the participation of WDS with variable speed pumps in DR is studied in [15, 16], and the comparison of the effects of variable speed pumps in comparison to fixed speed pumps is studied, but the energy flexibility caused by water tanks is not taken into account in these works. A two-step model for the participation of WDS in DR and regulation up and down services is provided in [17], which takes into account the flexibility caused by both variable speed pumps and water tanks. However, these works do not consider the uncertainty of energy market price.

In this chapter, a model for the participation of water distribution system (WDS) in DR and frequency Regulation Up/Down capacity offers is presented, and the uncertainty of wholesale energy price forecast is considered using Robust Optimization Approach. The objective of the optimization is to find the best schedule for operation of water tanks and pumps, in which the WDS's water purchase cost is minimized and the WDS's profit for providing the DR services is maximized.

In the rest of the chapter, at first, the formulation of the proposed method is presented. Then a case study is presented in which the proposed model is implemented on a 15-node test WDS and the obtained robust day-ahead schedule for the operation of water pumps is reported and the results are discussed. And in the last section, a conclusion of the chapter is provided.

10.2 Problem Formulation

In this section, at first robust optimization approach will be introduced briefly. Then the proposed model for the optimal operation of WDS and its participation in DR and frequency regulation services considering the uncertainty of market price forecasts is presented. This model is made of two steps. In the first step, the operation of pumps and water tanks is optimized to minimize the cost of purchasing water and electricity consumption, taking into account the operating constraints of

WDS. Then the optimal result of the first step for the operation of water pumps is used in the second step. In the second step, DR and frequency regulation up/down offers are optimized to maximize the WDS profits, taking into account the operating constraints of WDS. Second step model optimizes the market participation offers by improving the optimal solution of step 1. In order to make sure that the provided offers are practical, a comprehensive model of the water distribution system (WDS) considering its related constraints was combined with the proposed model in both first and second steps.

10.3 Robust Optimization Approach

In recent years many methods have been developed in order to analyze the different uncertainties of studied problems. Recently a novel method has been proposed to model the uncertainty of parameters; this method is named Robust Optimization Approach. The theory of robust optimization method is to show the effect of the uncertain parameter on the optimal result and decrease the optimal result's sensitivity to the uncertain parameter. Dealing with the uncertainty in robust optimization approach is done by the worst case, while stochastic programming applies scenario production. Calculating the probability of the uncertain parameter is the most important part of stochastic programming.

Some advantages of robust optimization approach are [18]:

- It has less computation volume than the stochastic programming, so it gives more effective results.
- Worst conditions are considered so the results are very reliable.
- Probabilistic distribution functions are not used in robust optimization approach.

The standard MILP of the model can be formulated as follows:

$$\text{Min} \sum_{t=1}^{T} c_t x_t \tag{10.1}$$

Subjected to

$$\sum_{j=1}^{T} a_{ij} x_j \leq b_i \quad \forall i = 1, \ldots, m \tag{10.2}$$

$$x_t \geq 0 \quad \forall t = 1, \ldots, T \tag{10.3}$$

$$x_t \in \{0, 1\} \quad \text{for some } t = 1, \ldots, T \tag{10.4}$$

where, c_t is a known coefficient which is considered for the objective function, and its value is unknown, but its bound is known. The value of c_t must be in the range of $[c_t{-}d_t\ c_t{+}d_t]$, where d_t is the deviation of c_t from its nominal value. An integer control parameter must be defined to formulate the robust mixed-integer linear programming. This integer control parameter is named Γ (Gama), and it has a real value in the interval of $[0,|T_0|]$. $\Gamma = 0$ results in the deterministic case and its other values are used for the stochastic case. The worst possible deviations happen for $\Gamma = |T_0|$ and it is used as the most conservative strategy.

The robust formulation of Eqs. (10.1–10.4) is defined as follow:

$$\mathrm{Min} \sum\nolimits_{t=1}^{n} c_t x_t + \beta.\Gamma + \sum\nolimits_{t=1}^{n} \zeta_t \tag{10.5}$$

$$\sum\nolimits_{j=1}^{n} a_{ij} x_j \leq b_i \quad \forall i = 1, \ldots, m \tag{10.6}$$

$$x_t \geq 0 \quad \forall t = 1, \ldots, n \tag{10.7}$$

$$x_t \in \{0, 1\} \quad \text{for some } t = 1, \ldots, n \tag{10.8}$$

$$\beta + \zeta_t \geq d_t \theta_t t \in T_0 \tag{10.9}$$

$$\beta \geq 0 \tag{10.10}$$

$$\zeta_t \geq 0 \quad \forall t = 1, \ldots, n \tag{10.11}$$

$$\theta_t \geq 0 \quad \forall t = 1, \ldots, n \tag{10.12}$$

$$\theta_t \geq x_t \quad \forall t = 1, \ldots, n \tag{10.13}$$

10.4 Water Distribution System Model

The proposed model for market participation of WDS consists of two steps [17]. The objective of step 1 is to minimize the water purchase cost as well as the cost of electricity consumed by pumps, taking into account the WDS hydraulic constraints. A WDS can be modeled using N nodes and J arcs. Each node may be connected to a reservoir (R) or a water tank (T), also each arc consists of a tube (S) and it may contain a water pump. Suppose that W_r^t is the amount of water purchased from the reservoir r at time t and P_p^t is the power consumption of pump p at time t. Also π_r^w and π_t^e represent the water price of reservoir r and Energy market price at time t, respectively. The objective function of step 1 can be modeled as (10.14–10.26), which are the constraints of the objective function of the first step.

$$\min \sum_r \sum_t W_r^t \pi_r^w + \sum_p \sum_t P_p^t \pi_t^e \tag{10.14}$$

The water balance constraint of the water network is modeled in (10.15).

$$\sum_i Q_{in}^t - \sum_j Q_{nj}^t - R_s^t = W_r^t - W_d^{t,n} \quad \forall t \in T, n \in N \tag{10.15}$$

where $W_d^{t,n}$ is the water demand of node n at time t. R_s^t indicates the water inflow of tank s at time t, its positive values demonstrate the inflow of water into the tank, and the negative values demonstrate the outflow of water from the tank. Q_{ij}^t is the volumetric flow rate of arc ij (from node I to node j) at time t. The constraints related to water tanks are formulated in (10.16–10.19).

$$V_s^{t+1} = V_s^t + \tau R_s^t \quad \forall t \in T, s \in Ta \tag{10.16}$$

$$\underline{V_s} \le V_s^t \le \overline{V_s} \quad \forall t \in T, s \in Ta \tag{10.17}$$

$$V_s^0 = V_s^{\text{init}} \quad \forall s \in Ta \tag{10.18}$$

$$-\overline{R_s} \le R_s^t \le \overline{R_s} \quad \forall t \in T, s \in Ta \tag{10.19}$$

The amount of stored water in tanks at each time interval is calculated in (10.16). V_s^t is the water stored in tank s at time t, and it is limited to its upper and lower bounds in (10.17). The number of seconds of the time interval t is demonstrated by τ. $\underline{V_s}$ and $\overline{V_s}$ are the minimum and maximum volume of tank s, respectively. The initial amount of stored water in tanks is set in (10.18). The water inflow rate of tanks is constrained in (10.19), where $\overline{R_s}$ is the maximum water inflow rate of tank s.

The pressure head of node i at time t is shown with H_i^t which depends on the pressure loss of pipes. The empirical Hazen–Williams pressure loss equation [19] is used to calculate H_i^t in (10.20).

$$H_j^t - H_i^t = 10.67 \frac{L_{ij} abs\left(Q_{ij}^t\right)}{C^{1.852} d_{ij}{}^{4.8704}} Q_{ij}^t{}^{0.852} + \alpha_p^{ij} H_p^t \quad \forall t \in T, i, j \in N \tag{10.20}$$

where, L_{ij} and d_{ij} are the length and the diameter of the pipe that connects node i to node j, and C is the Hazen–Williams roughness constant which depends on the type of the pipe. H_p^t is the water pressure increase produced by pump p at time t. And α_p^{ij} is a pump-arc binary parameter which has the value 1 when pump p is located in arc ij. The pressure head of each node is limited to its upper bound $\overline{H_i}$ and lower bound $\underline{H_i}$ in (10.21), and volumetric flow rate of arc ij at time t is bound in (10.22) where $\overline{\overline{Q_{ij}}}$ is the maximum water flow rate of arc ij.

$$\underline{H_i} \leq H_i^t \leq \overline{H_i} \quad \forall t \in T, i \in N \tag{10.21}$$

$$-\overline{Q_{ij}} \leq Q_{ij}^t \leq \overline{Q_{ij}} \quad \forall t \in T, i, j \in N \tag{10.22}$$

The limits mentioned in (10.21) and (10.22) depend on the design characteristics of pipes. The pressure heads of reservoir nodes are fixed to the water pressure of the source which depends on the geographical height of the source.

Water pump constraints are applied in (10.23–10.26). The pressure increase caused by pumps (H_p^t), is calculated in (10.23), also the power consumption of pump p at time t P_p^t is computed in (10.24).

$$H_p^t = {\omega_p^t}^2 \left(a - b \left(\frac{Q_{ij}^t}{\omega_p^t} \right)^c \right) \quad \forall t \in T, p \in P \tag{10.23}$$

$$P_p^t = {\omega_p^t}^3 \left(d - e \frac{Q_{ij}^t}{\omega_p^t} \right) \quad \forall t \in T, p \in P \tag{10.24}$$

where, ω_p^t is the speed of pump p at time t and a, b, c, d, and e are parameters of the pump [19]. The power consumption of pumps is bounded in (10.25), where $\overline{P_p}$ is the maximum capacity of pump p. The direction of water flow in an arc with a pump cannot be negative, which is forced in (10.26).

$$0 \leq P_p^t \leq \overline{P_p} \quad \forall t \in T, p \in P \tag{10.25}$$

$$\alpha_p^{ij} Q_{ij}^t \geq 0 \quad \forall t \in T, p \in P \tag{10.26}$$

The objective function (10.14) is solved subject to constraints (10.15–10.26) and the optimal power consumption of pumps are calculated. P_p^{t*} is the optimal power consumption of pump p at time t which is resulted from the first step and is used to provide optimal DR and frequency regulation offers of the WDS in the second step. The objective function of the second step it is formulated in (10.27) in order to maximize the profit of WDS from participating in electricity markets.

$$\max \sum_{t=1}^{T} \sum_{p=1}^{P} \left(P_{\mathrm{DR}}^{t,p} \pi_t^e + R_u^{t,p} \pi_t^u + R_d^{t,p} \pi_t^d - P_{sh}^{t,p} \lambda_t^e \right) \tag{10.27}$$

where $P_{\mathrm{DR}}^{t,p}$ and $P_{sh}^{t,p}$ are the load reduction and the shifted load of pump p at time interval t, respectively. Also, $R_d^{t,p}$ and $R_u^{t,p}$ are the variables of frequency regulation down and up offers, respectively. The prices of the regulation up, regulation down, and energy markets were denoted by π_t^u, π_t^d, and π_t^e, respectively. λ_t^e is the retail tariff that is used to calculate the cost of the shifted load. The amounts of shifted load and load reduction due to DR are computed from the difference of scheduled power consumption of pumps in step 1 and 2 in (10.28).

$$P_{sh}^{t,p} - P_{\mathrm{DR}}^{t,p} = P_p^t - P_p^{t*} \quad \forall t \in T, p \in P \tag{10.28}$$

The maximum load reduction of each pump is set to their first step power consumption schedule in (10.29), where $\beta_{\mathrm{DR}}^{t,p}$ is a binary variable that indicates the DR participation of pump p at time interval t. When the amount of $\beta_{\mathrm{DR}}^{t,p}$ is equal to one it means that pump p is scheduled to participate in DR at time interval t. $\beta_{\mathrm{DR}}^{t,p}$ is used to separate the time intervals of load reduction and load shifting which cannot occur at the same time. The maximum amount of shifted load cannot exceed the maximum power consumption of pumps which is constrained in (10.30).

$$0 \leq P_{\mathrm{DR}}^{t,p} \leq P_p^{t*} \beta_{\mathrm{DR}}^{t,p} \quad \forall t \in T, p \in P \tag{10.29}$$

$$0 \leq P_{sh}^{t,p} \leq \overline{P_p} \left(1 - \beta_{\mathrm{DR}}^{t,p}\right) \quad \forall t \in T, p \in P \tag{10.30}$$

The limits of the new power consumption of the pumps are constrained in (10.31) and (10.32). These constraints are applied to ensure the deliverability of the regulation up and down offers. The maximum capacity of regulation down and up offers are limited in (10.31) and (10.32), respectively. The upper and lower limits of the power consumption of pumps are constrained in (10.33) and (10.34), respectively.

$$P_p^t \leq P_p^{t*} - P_{\mathrm{DR}}^{t,p} + P_{sh}^{t,p} + R_d^{t,p} \quad \forall t \in T, p \in P \tag{10.31}$$

$$P_p^{t*} - P_{\mathrm{DR}}^{t,p} + P_{sh}^{t,p} - R_u^{t,p} \leq P_p^t \quad \forall t \in T, p \in P \tag{10.32}$$

$$P_p^{t*} - P_{\mathrm{DR}}^{t,p} + P_{sh}^{t,p} + R_d^{t,p} \leq \overline{P_p} \quad \forall t \in T, p \in P \tag{10.33}$$

$$0 \leq P_p^{t*} - P_{\mathrm{DR}}^{t,p} + P_{sh}^{t,p} - R_u^{t,p} \quad \forall t \in T, p \in P \tag{10.34}$$

The frequency regulation market requires changes in the power consumption at a 5-min period, so ramp up (RU) and ramp down (RD) rates of pumps are considered in (10.35) and (10.36) to set the bounds on regulation up and down offers. In (10.36), $\beta_{\mathrm{DR}}^{t,p}$ is used to ensure that DR and regulation down offers do not occur at the same time.

$$0 \leq R_u^{t,p} \leq 5 \times \mathrm{RD} \quad \forall t \in T, p \in P \tag{10.35}$$

$$0 \leq R_d^{t,p} \leq 5 \times \mathrm{RU}\left(1 - \beta_{\mathrm{DR}}^{t,p}\right) \quad \forall t \in T, p \in P \tag{10.36}$$

In the next section, a case study is presented, and the proposed model is implemented on a 15-node test WDS, and the results are presented.

10.5 Case Study

The proposed method was tested on a 15-node test WDS. A schematic diagram of the test WDS is presented in ref. [3]. This test system includes 15 nodes, 14 arcs, 3 pumps in arcs 1, 4, and 7, 2 water tanks at nodes 10 and 13, and 1 reservoir at node 1. The maximum power consumption of each pump is considered to be 4.5 MW. The California ISO prices for electricity and ancillary services markets are used in the simulation [20]. The price data of energy, regulation up, and regulation down markets are illustrated in Fig. 10.1. The water demand provided in [21] is used. The electricity retail tariff and water price of the reservoir are considered to be 0.015 \$/kWh [22] and 0.65 \$/m^3 [23], respectively. The volume of stored water in tanks is assumed to be zero. The results of the simulation are reported in the following.

The results obtained for WDS profits, costs, and revenues from both Risk-Neutral and Risk-Averse cases are summarized in Table 10.1. According to Table 10.1, the total profit of WDS at Risk-Neutral case ($\Gamma = 0$) is 590.558\$ which is reduced to 588.596\$ at Risk-Averse case ($\Gamma = 1$). In other words, with a 0.33% decrease in the total profit of WDS, the schedule will be robust against 30% changes in electricity market price. The revenue of DR at Risk-Neutral case is 289.901\$, but 151.997\$ is spent to supply the shifted load. The profit of WDS from DR is 137.904\$ at Risk-Neutral case which is decreased to 129.05 (6.42% reduction) at Risk-Averse case. The profit of WDS from Regulation Down services is decreased from 166.68\$ to 163.98\$ (1.62%). However, the WDS profit from Regulation Up services is increased by 3.35%, bringing the profit from 285.977\$ (Risk-Neutral case) to 295.57\$ (Risk-Averse case). This increase relieves the reductions of DR and Regulation Up profits.

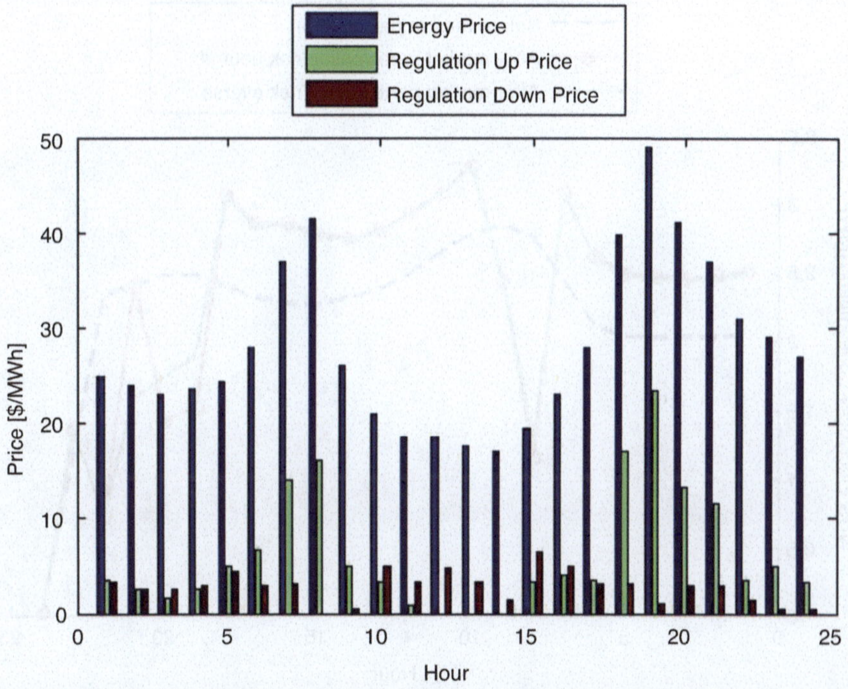

Fig. 10.1 Energy, regulation up, and regulation down price data

Table 10.1 Results obtained for the day-ahead profit of the WDS

	Risk-neutral			Risk-averse		
	Revenue ($)	Cost ($)	Profit ($)	Revenue ($)	Cost ($)	Profit ($)
Demand response	289.901	151.997	137.904	270.536	141.487	129.049
Frequency regulation up	285.977	–	285.977	295.570	–	295.570
Frequency regulation down	166.677	–	166.677	163.977	–	163.977
Total profit	742.555	151.997	590.558	730.083	141.487	588.596

Total electricity consumption of pumps before participation, participation without considering risk (Risk-Neutral), and risk-based participation (Risk-Averse) in markets are illustrated in Fig. 10.2. The results of Risk-Neutral and Risk-Averse cases are almost the same at all hours except 8:00, 19:00, 20:00, and 21:00 as shown in Fig. 10.2, where the maximum power consumptions obtained from Risk-Neutral case are 128.67, 375.36, and 319.15 kW less than Risk-Averse case at hours 8:00, 19:00, and 20:00, respectively, but at 21:00 it is 815.89 kW more than Risk-Averse case. Demand Response offer and load recovery schedule resulted from

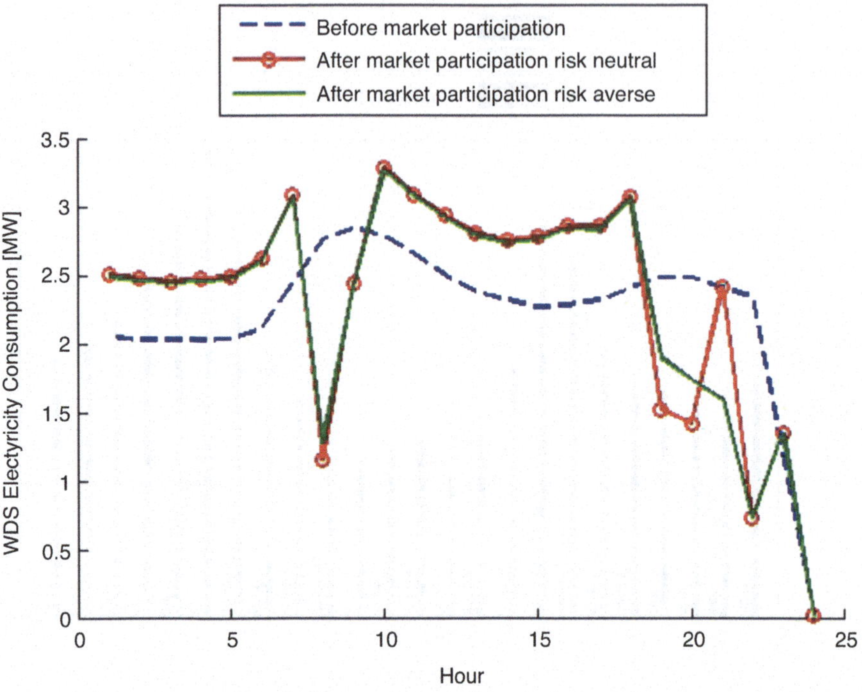

Fig. 10.2 Total electricity consumption of three pumps

Risk-Neutral and Risk-Averse cases are shown in Fig. 10.3. During hours 8–9 and 19–23, WDS has participated in DR because of the high electricity price of these hours and the load recovery is scheduled to be done at lower price hours. This flexibility is possible due to the existence of water tanks. According to Fig. 10.3, the DR offers in Risk-Neutral case are more than Risk-Averse case at hours 8:00, 19:00, and 20:00, with the amounts of 128.671, 375.366, and 319.156 kW, respectively. However, the DR offer in Risk-Averse case is 459.76 kW more than Risk-Neutral case at 21:00. Also, the highest difference in load recovery schedule of Risk-Neutral and Risk-Averse cases happens at 21:00, where the mentioned number for Risk-Neutral case is 356.13 kW more than Risk-Averse case.

The flow rate of water tanks is illustrated in Fig. 10.4. According to Fig. 10.4, in Risk-Neutral case water is stored in the water tank 1 during hours 1–6 and 15–17 and this water is released during hours 7–14 and 18–22. Also during hours 1–6 and 15 water is stored in tank 2 and is released during hours 7–14 and 16–22. The water inflow and outflow of the tank in both Risk-Neutral and Risk-Averse cases is happened at the same time, as shown in Fig. 10.4 and there are just small differences in the amount of flow rate in two cases at hours 19, 20, and 21. The volume of stored water in tank 1 and 2 Risk-Neutral and Risk-Averse cases is illustrated in Fig. 10.5. According to Fig. 10.5, the volume of stored water in tanks in two cases are almost the same.

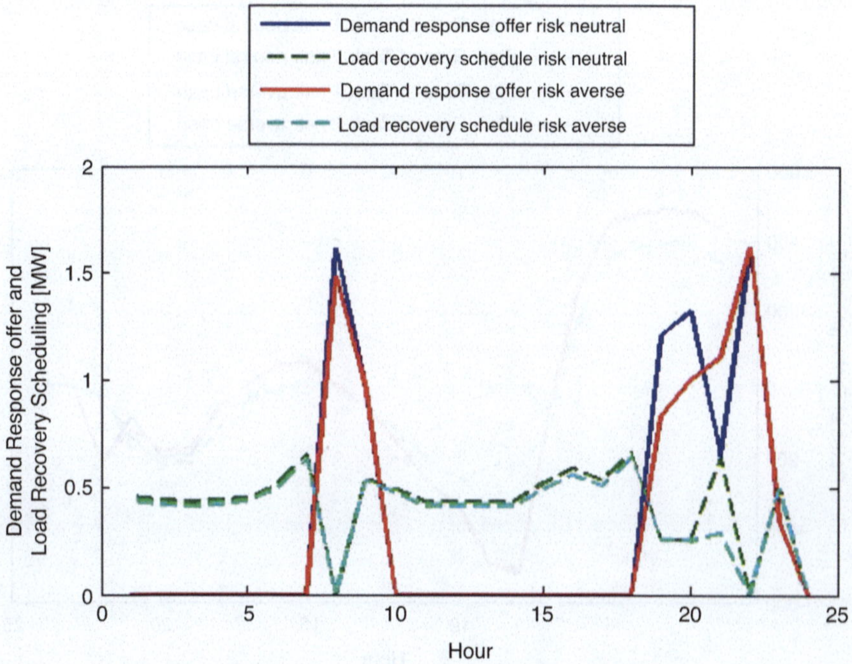

Fig. 10.3 Demand response and load recovery schedule resulted from risk-neutral and risk-averse cases

The WDS frequency regulation up/down offers in both Risk-Neutral and Risk-Averse cases are illustrated in Fig. 10.6. According to Fig. 10.6, the WDS has provided frequency regulation up offer at all hours except 13 and 14 in both Risk-Neutral and Risk-Averse cases. The differences between the results of Risk-Neutral and Risk-Averse cases for regulation up offers are at hours 19, 20, and 21. At the rest of the hours, the results are almost the same. The regulation up offers in Risk-Neutral case at hours 19 and 20 are more than Risk-Averse case, but it is less at 21:00. The frequency regulation down capacity offers are provided at all hours except 8 and 22, and the only difference between the offers of Risk-Neutral and Risk-Averse cases happens at 21:00, where regulation down offer is more in Risk-Neutral case.

The relation of Γ and the total profit of WDS is shown in Fig. 10.7. The relation of WDS Demand Response profit and Recovery Load Cost and Γ is illustrated in Fig. 10.8. Figure 10.9 shows the relation of WDS profit from Regulation Up and Down services and Γ. According to Fig. 10.7, there is a small decrease in the total profit of WDS in the Risk-Averse case. The reason for the smallness of the reduction of WDS total profit can be concluded from Figs. 10.8 and 10.9. According to Figs. 10.8 and 10.9, with the increase of Γ, WDS Demand Response profit decreases but WDS profit from Regulation Up services increases. And this relieves the reduction of WDS total profit.

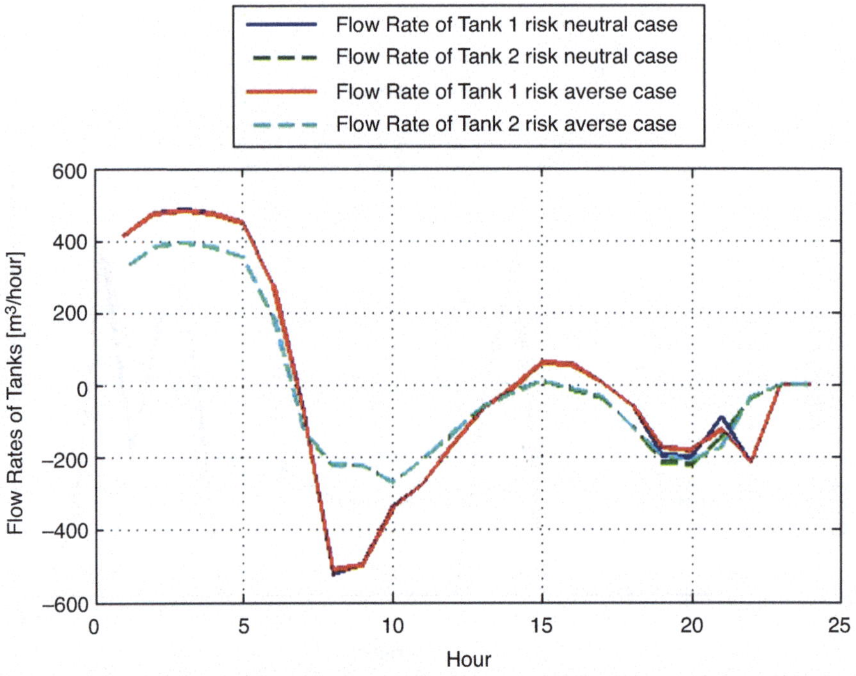

Fig. 10.4 Inflow/outflow rate of water tank 1 and 2 in two cases

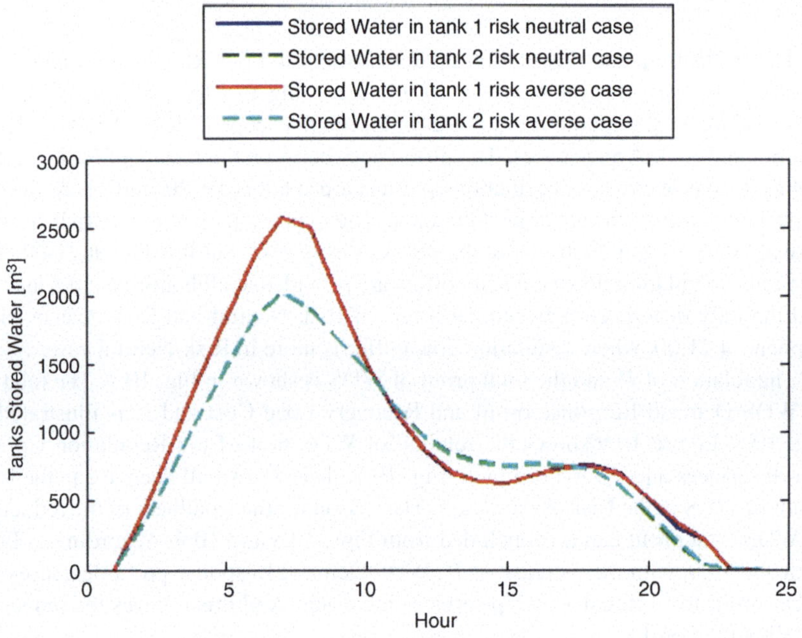

Fig. 10.5 The volume of stored water in tank 1 and 2 in two cases

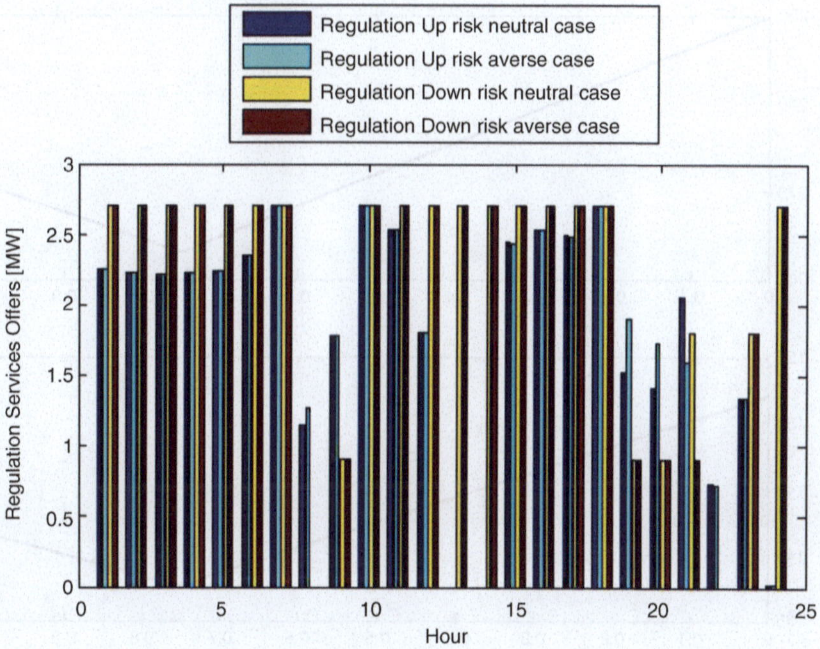

Fig. 10.6 Frequency regulation up/down offers of WDS

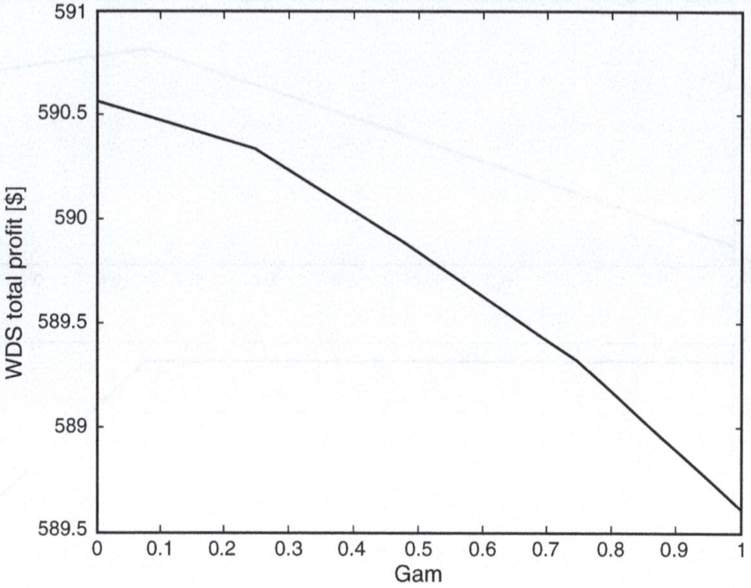

Fig. 10.7 The relation of the total profit of WDS and Γ

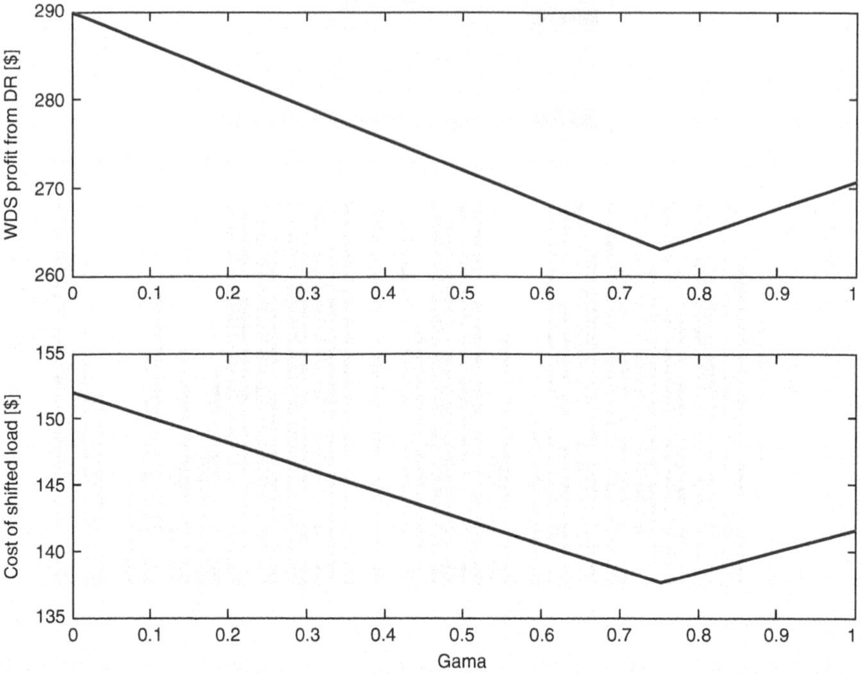

Fig. 10.8 The relation of WDS demand response profit and recovery load cost and Γ

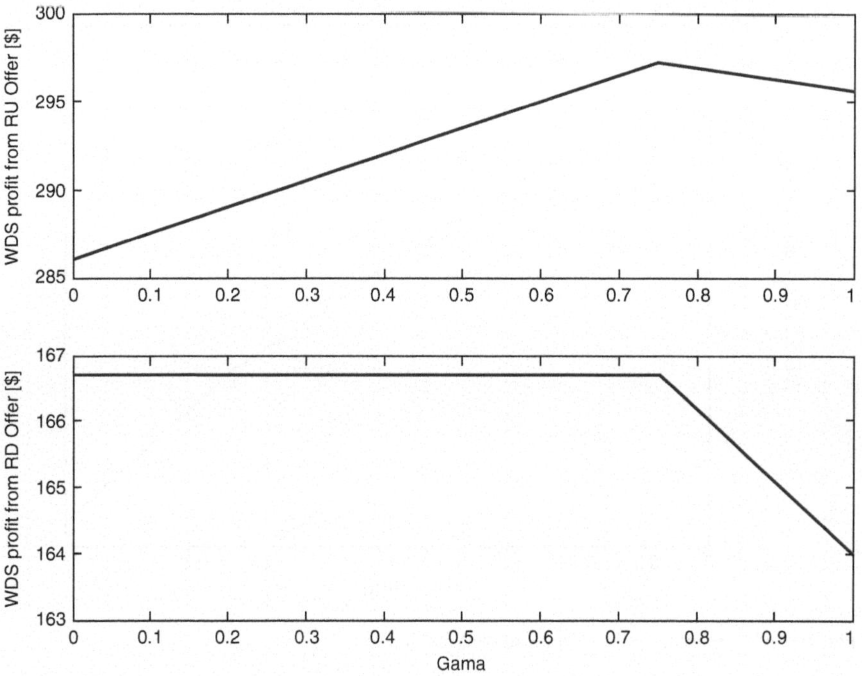

Fig. 10.9 The relation of WDS profit from regulation up and down services and Γ

10.6 Conclusion

A model for optimizing the participation of WDSs in Demand Response market and frequency regulation capacity services considering the uncertainty of wholesale energy price forecast is presented in this chapter. Robust Optimization Approach is used to handle the uncertainty of energy price. This model optimizes the operation schedule of water tanks and pumps in order to minimize the WDS's water procurement cost and maximize the WDS's profit for providing the DR services. The model is tested on a WDS consisting of 15 nodes. The results indicate that there are possibilities for the WDS to gain more profit through participating in Demand Response and frequency regulation up/down capacity offers. On the other hand, this profit can be robust against 30% changes of market price, with only a 0.33% decrease in the total profit of WDS, and that is because with the increase of risk, WDS Demand Response profit decreases but WDS profit from regulation up services increases, and relieves the reduction of WDS total profit.

References

1. C. Diaz, F. Ruiz, D. Patino, Modeling and control of water booster pressure systems as flexible loads for demand response. Appl. Energy **204**, 106–116 (2017)
2. N. Oconnell, P. Pinson, H. Madsen, M. Omalley, Benefits and challenges of electrical demand response: A critical review. Renew. Sust. Energ. Rev. **39**, 686–699 (2014)
3. K. Oikonomou, M. Parvania, Optimal coordination of water distribution energy flexibility with power systems operation. IEEE Trans. Smart Grid **3053**(c), 1–10 (2018)
4. B. Ulanicki, J. Kahler, H. See, Dynamic optimization approach for solving an optimal scheduling problem in water distribution systems. J. Water Resour. Plan. Manag. **133**(1), 23–32 (2006)
5. S.M.S. Abkenar, S.D. Stanley, C.J. Miller, D.V. Chase, S.P. McElmurry, Evaluation of genetic algorithms using discrete and continuous methods for pump optimization of water distribution systems. Sustain. Comput. Inform. Syst. **8**, 18–23 (2015)
6. G. Bonvin, S. Demassey, C. Le Pape, N. Maïzi, V. Mazauric, A. Samperio, A convex mathematical program for pump scheduling in a class of branched water networks. Appl. Energy **185**, 1702–1711 (2017)
7. L.H.M. Costa, B. de Athayde Prata, H.M. Ramos, M.A.H. de Castro, A branch-and-bound algorithm for optimal pump scheduling in water distribution networks. Water Resour. Manag. **30**(3), 1037–1052 (2016)
8. M. López-Ibáñez, T.D. Prasad, B. Paechter, Ant colony optimization for optimal control of pumps in water distribution networks. J. Water Resour. Plan. Manag. **134**(4), 337–346 (2008)
9. R. Menke, E. Abraham, P. Parpas, I. Stoianov, Exploring optimal pump scheduling in water distribution networks with branch and bound methods. Water Resour. Manag. **30**(14), 5333–5349 (2016)
10. H. Mala-Jetmarova, N. Sultanova, D. Savic, Lost in optimisation of water distribution systems? A literature review of system operation. Environ. Model. Softw. **93**, 209–254 (2017)
11. Z. Ghelichi, J. Tajik, M.S. Pishvaee, A novel robust optimization approach for an integrated municipal water distribution system design under uncertainty: A case study of Mashhad. Comput. Chem. Eng. **110**, 13–34 (2018)

12. Y. Yalcin, K. Yigit, B. Acarkan, Energy management of water transfer stations by using demand response programs. 2018 6th international conference on control engineering & information technology (2018), pp. 1–5
13. K. Oikonomou, P. Masood, S. Burian, Integrating water distribution energy flexibility in power systems operation (2017)
14. C. Mkireb, A. Dembele, A. Jouglet, T. Denoeux, A linear programming approach to optimize demand response for water systems under water demand uncertainties. 2018 international conference on smart grid and clean energy technologies (ICSGCE 2018) (2018), pp. 206–211
15. R. Menke, E. Abraham, P. Parpas, I. Stoianov, Demonstrating demand response from water distribution system through pump scheduling. Appl. Energy **170**, 377–387 (2016)
16. R. Menke, E. Abraham, P. Parpas, I. Stoianov, Extending the envelope of demand response provision though variable speed pumps. Procedia Eng. **186**, 584–591 (2017)
17. K. Oikonomou, M. Parvania, R. Khatami, Optimal demand response scheduling for water distribution systems. IEEE Trans. Ind. Informatics **3203**(c), 1–11 (2018)
18. E. Köbis, On robust optimization. J. Optim. Theory Appl. **167**(3), 969–984 (2015)
19. L.W. Mays, *Water Resources Engineering*, Vol. ث ققتق, No. ثق تقتقتق; (Wiley, 2010)
20. OASIS—OASIS Prod—PUBLIC—0. [Online]. http://oasis.caiso.com/mrioasis/logon.do. Accessed 5 Apr 2019
21. L.W. House, *Water Supply Related Electricity Demand in California* (California Energy Commission, no. 500, 2006)
22. Electric Power Monthly—U.S. Energy Information Administration. [Online]. https://www.eia.gov/electricity/monthly/. Accessed 5 Apr 2019
23. Dr. Stephen and S. Marie, *What Will Be the Cost of Future Sources of Water for California?* (2016)

Chapter 11
Optimal Power Scheduling of a GenCo Using Two-Point Estimate Method

Kittisak Jermsittiparsert

Nomenclature

Set

t	Index of time interval
i	Index of generation units
j	Auxiliary index for linear modeling of minimum up-time and minimum down-time constraints

Known Parameters

a_i, b_i, c_i	Quadratic, linear, and fixed coefficients of operation cost function for generation unit
$\mathrm{SD}_i,\mathrm{SU}_i$	Shut-down and start-up costs of generation unit
$P_i^{\mathrm{G,min}},P_i^{\mathrm{G,max}}$	Minimum and maximum powers of generation unit
$\mathrm{UR}_i,\mathrm{DR}_i$	Ramp-up and ramp-down limits of generation unit
$\mathrm{MDT}_i,\mathrm{MUT}_i$	Minimum down/up time limits of generation unit
$\mathrm{Dn}_{i,j},\mathrm{Up}_{i,j}$	Auxiliary parameters for the MDT and MUT constraints
λ_t^D	Electricity price

K. Jermsittiparsert (✉)
Social Research Institute, Chulalongkorn University, Bangkok, Thailand
e-mail: kittisak.j@chula.ac.th

S. Nojavan, K. Zare (eds.), *Electricity Markets*,
https://doi.org/10.1007/978-3-030-36979-8_11

Decision Variables

$U_{i,t}$	Binary variable {0,1}, It is 1 if generation unit is ON; otherwise, it is 0.
$P^{G}_{i,t}$	Generation power of unit
$\Delta P_{i,t}$	An auxiliary variable for using the power generation of unit
$SDC_{i,t}$,$SUC_{i,t}$	Shut-down and start-up costs of generation unit

Functions

Cost (t)	Total cost of GenCo at each time
$F(P^{G}, \lambda^{D})$	Total profit of GenCo

11.1 Introduction

Price-taker GenCo participates in the electricity market to maximize its profit [1]. It has information about forecasted market price to obtain scheduling in order to offer to the electricity market [2]. Therefore, market price uncertainty should be modeled by GenCo to maximize profit [3, 4].

Optimal scheduling of GenCo is studied in many works. A GenCo maximizes the own profit with participation in the energy and reserve markets in [5]. In [6], a scenario-based stochastic framework is provided to obtain scheduling of GenCo considering environment issue. In [7], offering strategy of GenCo is obtained to offer to reserve and energy markets. Optimal offering strategies of GenCo and Distribution Company are obtained using max–min fuzzy method and genetic algorithm in [8]. In [9], a bi-level model is provided to obtain the offering curves of GenCo in the spinning reserve and energy markets. Offering and bidding strategies of a virtual power plant in the energy and reserve markets are provided in [10]. In [11], a genetic algorithm is provided to maximize profit function of a GenCo in the day-ahead market. A bi-level model considering emission issue is presented in [12] to model offering strategy of GenCo. In [13], optimal scheduling of a pumped-storage based GenCo is obtained with a multistage algorithm. Optimal scheduling of GenCo in the oligopolistic electricity market is studied in [14]. In [15], the risk, profit, and emission issues of a GenCo as a multi-objective model are improved by differential evolution algorithm which the risk and emission are minimized while the profit is maximized. Optimal offering strategy of a GenCo is modeled via the supply function equilibrium approach in [16]. In [17], a bat-inspired method is used to obtain supply function of a GenCo in the electricity market. Optimal strategy of a GenCo is determined using a modified reinforcement learning algorithm in the energy market in [18]. In [19], optimal offering curves of a GenCo are developed in the day-ahead electricity market. Time-varying acceleration coefficients and self-organizing hierarchical particle swarm optimization is used to obtain risk-based offering strategy of a GenCo in [20]. In [21], optimal scheduling of a GenCo is provided in the pay-as-bid market. Finally, future contracts and scheduling of a GenCo is studied using the supply function model in [22].

From the viewpoint of uncertainty modeling of electricity market price, self-scheduling of a GenCo is studied in [23] considering the effects of inaccurate price forecasts. Monte Carlo simulation (MCS) [24], probabilistic model [25], Markov decision process [26], and scenario-based stochastic framework [27, 28] are used to model electricity market price uncertainty to obtain robust scheduling of a GenCo in the uncertain environment.

In this chapter, two-point estimate method (TPEM) is applied to model uncertainty of electricity market price in order to obtain robust scheduling of GenCo in the presence of uncertainty in order to maximize its profit. It should be noted that the TPEM is used to model uncertainty in the power system problems such as power flow [29], load flow [30], energy management of microgrid [31], power flow problem considering wind power uncertainty [32], and renewable energy sources-based distribution systems [33].

TPEM in comparison with MCS only requires resolving $2 \times m$ deterministic problem to obtain the behavior of m random variable. Mixed-Integer Quadratic Constrained Programming (MIQCP) is used for solving the deterministic scheduling problem using the SBB MIQCP solver of GAMS software. The objective function is to maximize the expected profit. The main novelty and contributions of this chapter are summarily presented below.

1. Two-point estimated method is proposed to model uncertainty parameter of market price to obtain robust scheduling of a GenCo in the uncertain environment.
2. Appling the Monte Carlo Simulation to model uncertainty parameter of market price.
3. Obtained results from proposed method are compared such as Monte Carlo Simulation and deterministic approach.

The rest of the chapter is categorized as below. Deterministic-based scheduling of a price-taker GenCo is presented in Sect. 11.2. Section 11.3 provides two-point estimate method to model uncertainty parameter of energy market price. The numerical study is studied in Sect. 11.4 in which obtained results are compared with deterministic approach and MCS. Finally, the conclusion of this study is presented in Sect. 11.5.

11.2 Deterministic-Based Scheduling of a GenCo

Deterministic-based scheduling of a price-taker GenCo in the day-ahead electricity market is presented in this section. The profit function of GenCo is presented in Eq. (11.1) in order to obtain optimal scheduling considering forecasted market price [34].

$$F\left(P^{\mathrm{G}}, \lambda^{\mathrm{D}}\right) = \sum_{t=1}^{T} \left\{ \sum_{i=1}^{N_{\mathrm{G}}} \lambda_t^{\mathrm{D}} \times P_{i,t}^{\mathrm{G}} \right\} - \sum_{t=1}^{T} \mathrm{Cost}\,(t) \tag{11.1}$$

The profit function (11.1) is equal to the revenue minus total operation cost. The revenue is obtained with selling power to the day-ahead market. Also, total operation cost of thermal generation unit for GenCo includes the fuel cost as a quadratic function, start-up, and shut-down costs which are provided in Eq. (11.2).

$$\mathrm{Cost}\,(t) = \sum_{i=1}^{N_{\mathrm{G}}} \left[\left\{ a_i \times \left(P_{i,t}^{\mathrm{G}}\right)^2 + b_i \times P_{i,t}^{\mathrm{G}} + c_i \times U_{i,t} \right\} + \mathrm{SUC}_{i,t} + \mathrm{SDC}_{i,t} \right] \tag{11.2}$$

To obtain mixed-integer quadratic constrained program (MIQCP) for self-scheduling of a GenCo, an auxiliary variable $\Delta P_{i,\,t}$ is defined to avoid the complications of multiplied variables in (11.2). This auxiliary variable is limited by Eq. (11.3).

$$0 \le \Delta P_{i,t} \le \left[P_i^{\mathrm{G,max}} - P_i^{\mathrm{G,min}} \right] \times U_{i,t} \tag{11.3}$$

and thus,

$$P_{i,t}^{\mathrm{G}} = U_{i,t} \times P_i^{\mathrm{G,max}} + \Delta P_{i,t} \tag{11.4}$$

By substituting Eq. (11.4) in Eq. (11.2) [35] and bearing in mind $\Delta P_{i,\,t} \times U_{i,\,t} = \Delta P_{i,\,t}$ and $U_{i,t}^2 = U_{i,t}$, one can write:

$$\mathrm{Cost}\,(t) = \sum_{i=1}^{N_{\mathrm{G}}} \left\{ \begin{array}{l} \left[b_i + 2a_i \times P_i^{\mathrm{G,min}} \right] \times \Delta P_{i,t} \\ + \left[c_i + b_i P_i^{\mathrm{G,min}} + a_i \left(P_i^{\mathrm{G,min}} \right)^2 \right] \times U_{i,t} \\ + a_i \Delta P_{i,t}{}^2 + \mathrm{SDC}_{i,t} + \mathrm{SUC}_{i,t} \end{array} \right\} \tag{11.5}$$

Constraints pertaining to the objective function are modeled as follows: Start-up and shut-down costs of thermal generation units are presented in Eqs. (11.6) and (11.7), respectively [36].

$$\begin{array}{l} \mathrm{SUC}_{i,t} \ge \mathrm{SU}_i \left[U_{i,t} - U_{i,t-1} \right] \\ \mathrm{SUC}_{i,t} \ge 0; \quad \forall t, \forall i \end{array} \tag{11.6}$$

$$\begin{array}{l} \mathrm{SDC}_{i,t} \ge \mathrm{SD}_i \left[U_{i,t-1} - U_{i,t} \right] \\ \mathrm{SDC}_{i,t} \ge 0; \quad \forall t, \forall i \end{array} \tag{11.7}$$

To avoid inappropriate stresses on the combustion and boiler devices, ramp-down and ramp-up constraints of a thermal generation units are stated in Eqs. (11.8) and (11.9), respectively.

$$\Delta P_{i,t-1} - \mathrm{DR}_i \leq \Delta P_{i,t}; \quad \forall t, \forall i \tag{11.8}$$

$$\Delta P_{i,t} \leq \Delta P_{i,t-1} + \mathrm{UR}_i; \quad \forall t, \forall i \tag{11.9}$$

Finally, the constraints (11.10) and (11.12) express the minimum up and down time limits, respectively. In these constraints, the auxiliary parameters $\mathrm{Up}_{i,j}$ and $\mathrm{Dn}_{i,j}$ are defined in (11.11) and (11.13) in order to linear model of minimum up and down time constraints, respectively [37].

$$U_{i,t} - U_{i,t-1} \leq U_{i,t+\mathrm{Up}_{i,j}}; \quad \forall i, \forall t, \forall j \tag{11.10}$$

$$\mathrm{Up}_{i,j} = \begin{Bmatrix} j & j \leq \mathrm{MUT}_i \\ 0 & j > \mathrm{MUT}_i \end{Bmatrix} \tag{11.11}$$

$$U_{i,t-1} - U_{i,t} \leq 1 - U_{i,t+\mathrm{Dn}_{i,j}}; \quad \forall i, \forall t, \forall j. \tag{11.12}$$

$$\mathrm{Dn}_{i,j} = \begin{Bmatrix} j & j \leq \mathrm{MDT}_i \\ 0 & j > \mathrm{MDT}_i \end{Bmatrix} \tag{11.13}$$

In this study, the GenCo is assumed as price-taker player in the electricity market in line with [3, 4, 36–38]. TPEM is applied to model market price uncertainty. The implementation of TPEM on optimal scheduling problem is presented in the next section.

11.3 Background of TPEM

The background of TPEM is provided in this section to show the uncertainty model of market price in order to obtain robust scheduling of a GenCo in the uncertain environment. TPEM focuses on the statistical data ready by the first few central moments of random input parameters at s points for each parameter, named concentration. Output random variables of the optimization problem and objective function F in the presence of market price uncertainty are resulted according to these points [39].

Assume that $X\{x_1, x_2, \ldots, x_l, \ldots, x_m\}$ is a parameter with a mean amount μ_{x_l} and standard deviation amount σ_{x_l}. Also, Z is a the objective function of X (i.e., $Z = F(X)$). Each of the s concentrations of parameters x_l can be defined as a pair composed of a weight $w_{l,s}$ and a location $x_{l,s}$. In the proposed TPEM, objective function F has to be optimized only s times for each input parameter x_l at the points made up of the sth location of input parameter x_l and the mean amount (μ_{x_l}) of remaining input parameters. Therefore, the total number of optimizations is $2 \times m$. Equation (11.14) presents the location $x_{l,s}$in TPEM.

$$x_{l,s} = \mu_{x_l} + \xi_{l,s} \cdot \sigma_{x_l} \tag{11.14}$$

In Eq. (11.14), $\xi_{l,s}$ is the standard location of parameter x_l. The weights and standard locations of parameter of x_l are calculated by Eqs. (11.15) and (11.16), respectively.

$$w_{l,1} = -\frac{\xi_{l,2}}{m\left(\xi_{l,1} - \xi_{l,2}\right)}, \quad w_{l,2} = \frac{\xi_{l,1}}{m\left(\xi_{l,1} - \xi_{l,2}\right)} \tag{11.15}$$

and,

$$\xi_{l,1} = \frac{\lambda_{l,3}}{2} + \sqrt{m + \left(\frac{\lambda_{l,3}}{2}\right)^2}, \quad \xi_{l,2} = \frac{\lambda_{l,3}}{2} - \sqrt{m + \left(\frac{\lambda_{l,3}}{2}\right)^2} \tag{11.16}$$

where $\lambda_{l,3}$ defines the skewness of the parameter x_l which is computed by Eq. (11.17).

$$\lambda_{l,3} = \frac{E\left[\left(x_l - \mu_{x_l}\right)^3\right]}{\left(\sigma_{x_l}\right)^3} \tag{11.17}$$

Figure 11.1 illustrated the flowchart algorithm based on the proposed TPEM to obtain robust scheduling of a GenCo in the presence of market price uncertainty. Deterministic-based scheduling of a GenCo problem is formulated in the presence of input data parameters and decision variables. In the proposed optimization problem, electricity market prices as uncertainty parameters have the known probability distribution in which the weights and locations can be calculated as described previously. A deterministic-based scheduling problem should be optimized for each concentration. The procedure of optimal scheduling problem of a GenCo via proposed TPEM is illustrated in Fig. 11.1.

$$Z_{l,s} = F\left\{x_{l,1}, x_{l,2}, \ldots, x_{l,s}, \ldots, x_{m,s}\right\} \tag{11.18}$$

where $Z_{l,s}$ as objective function shows the nonlinear model between the output variables and input parameters in the sth concentration in the optimal scheduling problem. The raw moments of output random variables are obtained as:

$$E(Z) \cong E(Z) + \sum_{s} w_{l,s} \cdot Z_{l,s} \tag{11.19}$$

The solution steps of the proposed algorithm based on TPEM are clearly presented below.

Step 1: Set the second and first moments of *s*th output random variables to zero; E(Z) = 0.
Step 2: Choose the input parameter x_l.
Step 3: Calculate $\lambda_{l,\,3}$, $\xi_{l,\,s}$, $w_{l,\,s}$ according to Eqs. (11.15)–(11.17).
Step 4: The two locations of $x_{l,\,s}$ is estimated.
Step 5: Solve the deterministic-based scheduling problem of a GenCo for each concentration.
Step 6: Update the raw moments of output variables.
Step 7: Repeat the steps 2–6 until all concentrations of all input uncertainty parameters are taken into account. Finally, calculate the statistical data of output random variables.

11.4 Numerical Study

This section provides the numerical simulations to show the capability of proposed approach. A price-taker GenCo including five thermal units is assumed to participate in electricity market. Therefore, 24 time periods as 1 day is considered as the time horizon in this chapter. The GenCo's data and technical constraints are presented in Table 11.1. Also, 54 thermal generation units from IEEE 118-bus test system as large system are used in the second case study in order to show capability of proposed TPEM in comparison with MCS and deterministic methods. All data of 85 thermal units are adopted from [40].

The problem formulation is modeled as a Mixed-Integer Quadratic Constrained Program (MIQCP) which can be solved via GAMS [41] under SBB solver [42].

Table 11.1 Parameters of the generation 5-units

Parameter	Unit 1	Unit 2	Unit 3	Unit 4	Unit 5
a_i[\$/MW^2]	0.018825	0.039	0.00890	0.00000425	0.02265
b_i[\$/MW]	6.7495	2.3215	9.565	19.80375	6.10875
c_i(\$)	250	250	350	300	225
$P_i^{G,\max}$ (MW)	335	232	260	440	250
$P_i^{G,\min}$ (MW)	125	150	50	160	130
DR$_i$(MW/h)	335	232	260	440	250
UR$_i$(MW/h)	335	232	260	440	250
MUT$_i$(h)	10	10	10	10	10
MDT$_i$(h)	1	1	1	1	1
SU$_i$(\$)	500	250	500	507.5	375

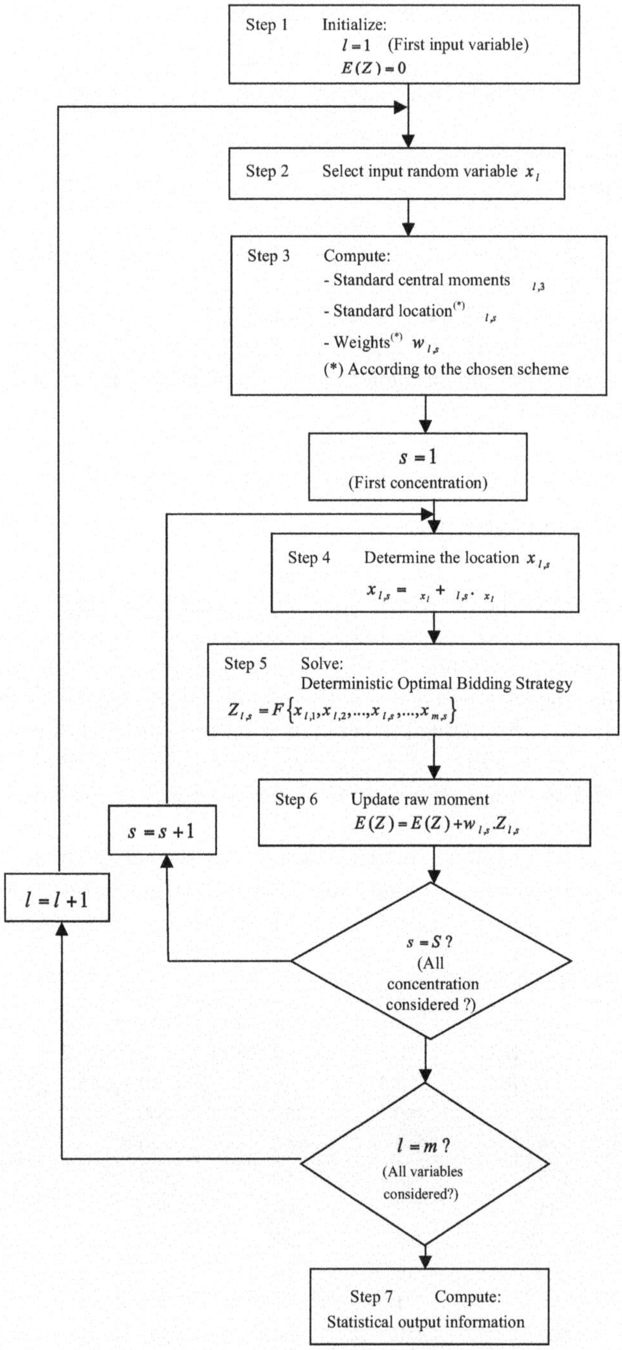

Fig. 11.1 The procedure of optimal scheduling problem of a GenCo via proposed TPEM

Three case studies are clearly categorized to show the capability of proposed TPEM in comparison with MCS and deterministic methods as follow:

Case I: Deterministic optimization (Without uncertainty modeling)
Case II: Uncertainty modeling via TPEM
Case III: Uncertainty modeling via MCS

Case I: Deterministic Optimization
In case I as deterministic optimization, only the mean amount of day-ahead electricity market price is considered in the optimal scheduling problem. The standard deviation and mean amounts of day-ahead electricity market price are presented in Table 11.2. In other words, the deterministic optimization consists of maximizing the objective function (11.1) under the constraints (11.2)–(11.13) with considering the mean amounts of day-ahead energy market price. The expected profit for 5 and 54 thermal units are equal to $119,680.6 and 481,304.0, respectively.

The expected profit may be reduced in the deterministic optimization which can be expected in a self-scheduling of the price-taker GenCo. Note that the value of deterministic expected profit which corresponds to the maximum profit is obtained by ignoring the stochastic behavior of day-ahead energy markets price. The optimal

Table 11.2 Mean and standard deviation amounts of the electricity market price

	Day-ahead market price	
Hour	Mean ($/MW)	St. Dev. ($/MW)
1	14.92	1.865
2	12.5	1.5625
3	12.5	1.5625
4	12.5	1.5625
5	12.5	1.5625
6	12.5	1.5625
7	12.5	1.5625
8	13	1.625
9	13	1.625
10	19.7	2.4625
11	20	2.5
12	20	2.5
13	20	2.5
14	20.01	2.50125
15	13.28	1.66
16	14.78	1.8475
17	14	1.75
18	14	1.75
19	20	2.5
20	27.5	3.4375
21	32.5	4.0625
22	32.5	4.0625
23	19.5	2.4375
24	13.28	1.66

dispatch of 5 thermal units and some units of 54 thermal units for offering to the day-ahead energy market are presented in Tables 11.3 and 11.4, respectively. Moreover, the profits, revenues, costs, and hourly offers to the day-ahead energy market for the 5 and 54 thermal units of GenCo are also summarized in Tables 11.3 and 11.4.

Case II: Uncertainty Characterization Using Two-Point Estimate Method
Due to the stochastic nature of the day-ahead market price, the market analysis requires a probabilistic technique. In this subsection, the uncertainty of electricity market price is modeled via proposed TPEM which the standard deviation and mean amounts of price are considered. In this case, the expected profit for 5 and 54 thermal units are $120,746.1 and $492,961.4 which are higher than the deterministic expected profit obtained from Case I.

The profits, revenues, costs, and the mean amounts of hourly offers to the day-ahead energy market for the five thermal units of GenCo are presented in Table 11.5. Also, the results of this case for 54 thermal units are presented in Table 11.6.

According to the results of Tables 11.5 and 11.6, it is shown that the generations of thermal units differ from case I. In other words, the total profit of GenCo is higher than case I because of considering the electricity market price uncertainty.

Case III: Uncertainty Characterization Using Monte Carlo Simulation
In this subsection, an MCS-based technique is applied to deal the electricity market price uncertainty. If uncertainty parameters have known probability distribution function (PDF) then the MCS technique as a numerical simulation method is applied to the optimization problem. In the MCS technique, a set of amounts for the random parameters as uncertainty parameters are generated based on PDF and many deterministic-based optimization problems are solved in each simulation. The result of MCS is similar to a sample of an experimental observation. By obtaining results from many simulations in MCS technique, the inference to the data set and statistical estimation are possibly obtained.

In optimal scheduling problem of a GenCo, the random parameters take into account by MCS technique via electricity market price uncertainty modeling. It should be noted that the electricity market price uncertainty is distributed with a known mean amount (corresponding to the estimated amount) and a known standard deviation.

In the MCS approach, 5000 random variables are generally achieved by the MATLAB function RANDN for considering the stochastic nature of the electricity market price. Then, the deterministic-based profit function (11.1) considering the constraints (11.2)–(11.13) is solved using 5000 random variables for electricity market price. In this case the expected profit for 5 and 54 thermal units are $120,949.2 and $493,228.6, respectively.

Table 11.3 Hourly offers to the day-ahead energy market, total costs, revenues, and profits in case I for five thermal units

Time	Unit 1 MW	Unit 2 MW	Unit 3 MW	Unit 4 MW	Unit 5 MW	Total revenue ($)	Total cost ($)	Total profit ($)
1	217.012	161.5192	0	0	194.5088	8549.757	7638.826	910.9311
2	152.7357	150	0	0	141.0872	5547.787	4733.496	814.2902
3	152.7357	150	0	0	141.0872	5547.787	4733.496	814.2902
4	152.7357	150	0	0	141.0872	5547.787	4733.496	814.2902
5	152.7357	150	0	0	141.0872	5547.787	4733.496	814.2902
6	152.7357	150	0	0	141.0872	5547.787	4733.496	814.2902
7	152.7357	150	0	0	141.0872	5547.787	4733.496	814.2902
8	166.0159	150	0	0	152.1247	6085.829	5043.548	1042.281
9	166.0159	150	0	0	152.1247	6085.829	5043.548	1042.281
10	335	222.8013	227.7528	0	250	20400.42	14677.33	5723.083
11	335	226.6474	234.4944	0	250	20922.84	14387.5	6535.337
12	335	226.6474	234.4944	0	250	20922.84	14387.5	6535.337
13	335	226.6474	234.4944	0	250	20922.84	14387.5	6535.337
14	335	226.7756	234.7191	440	250	29744.76	23916.53	5828.228
15	173.4529	150	83.48315	160	158.3057	9631.21	9994.782	−363.572
16	213.2935	159.7244	117.191	160	191.4183	12439.25	11631.28	807.9745
17	192.5764	150	99.66292	160	174.1998	10870.15	10693.11	177.0332
18	192.5764	150	99.66292	160	174.1998	10870.15	10693.11	177.0332
19	335	226.6474	234.4944	440	250	29722.84	23401.97	6320.864
20	335	232	260	440	250	41717.5	24034.73	17682.77
21	335	232	260	440	250	49302.5	24034.73	25267.77
22	335	232	260	440	250	49302.5	24034.73	25267.77
23	335	220.2372	223.2584	160	250	23175.66	17507.7	5667.969
24	173.4529	150	83.48315	160	158.3057	9631.21	9994.782	−363.572
Total						413584.8	293904.2	**119680.6**

Table 11.4 Hourly offers to the day-ahead energy market, total costs, revenues, and profits in case I for 54 thermal units

Time	Unit 5 MW	Unit 20 MW	Unit 29 MW	Unit 40 MW	Unit 54 MW	Total revenue ($)	Total cost ($)	Total profit ($)
1	0	175	0	0	0	18575.4	15873.18	2702.218
2	0	50	0	0	0	9781.722	8925.037	856.6851
3	0	50	0	0	0	9781.722	8925.037	856.6851
4	0	50	0	0	0	9781.722	8925.037	856.6851
5	0	50	0	0	0	9781.722	8925.037	856.6851
6	0	50	0	0	0	9781.722	8925.037	856.6851
7	0	50	0	0	0	9781.722	8925.037	856.6851
8	0	139.6042	0	0	0	13895.91	12578.42	1317.485
9	0	139.6042	0	0	0	13895.91	12578.42	1317.485
10	250	250	230	150	0	86069.3	65655.74	20413.56
11	300	250	300	200	0	96,880	71560.25	25319.75
12	300	250	300	200	0	96,880	71560.25	25319.75
13	300	250	300	200	0	96,880	71560.25	25319.75
14	250	250	247.3624	200	0	90,876	65850.55	25025.45
15	100	197.9375	97.36239	100	0	33324.3	32853.83	470.472
16	100	250	86.81193	86.81193	0	42417.01	37887.79	4529.214
17	100	250	80	51.03211	0	36070.2	33662.91	2407.295
18	113.6468	250	113.6468	100	0	38722.99	36461.69	2261.296
19	263.6468	250	263.6468	200	0	116004.4	90442.28	25562.09
20	300	250	300	200	50	184,085	108643.9	75441.08
21	300	250	300	200	50	218,400	109,219	109,181
22	300	250	300	200	50	218,400	109,219	109,181
23	250	250	235.6651	200	25	109764.2	86833.31	22930.85
24	100	197.9375	85.66514	100	25	39476.96	42012.82	−2535.86
Total						1,609,308	1,128,004	**481,304**

Table 11.5 Hourly expected offers to the day-ahead energy market, total costs, revenues, and profits in case II for five thermal units

Time	Unit 1 MW	Unit 2 MW	Unit 3 MW	Unit 4 MW	Unit 5 MW	Expected revenue ($)	Expected cost ($)	Expected profit ($)
1	155.6153	151.5757	0	0	142.9311	5690.99	5962.24	−271.25
2	155.969	151.6908	0	0	143.1332	5702.168	4847.572	854.5952
3	156.0933	151.6905	0	0	143.2042	5704.596	4850.38	854.2165
4	156.0132	151.6228	0	0	143.1585	5699.501	4847.204	852.2971
5	155.9834	151.5867	0	0	143.1414	5697.041	4845.803	851.2377
6	155.9985	151.6444	0	0	143.15	5700.328	4847.302	853.0255
7	155.7912	151.6444	0	0	143.0316	5698.703	4842.612	856.0909
8	168.4835	151.6308	0	0	153.5894	6228.328	5138.686	1089.642
9	168.6215	151.6847	0	0	153.6683	6233.352	5142.903	1090.449
10	330.7451	221.5231	224.8415	0	247.5686	20350.79	14512.46	5838.333
11	330.8446	225.2479	231.4018	0	247.6255	20868.92	14218.54	6650.373
12	330.8549	225.2519	231.412	0	247.6314	20868.95	14219.03	6649.927
13	330.8699	225.2578	231.4271	0	247.64	20869.76	14219.73	6650.025
14	330.9084	225.3957	231.6809	434.5445	247.6619	29661.73	23642.23	6019.501
15	176.0285	151.7941	86.68258	166.1261	159.7518	9996.721	10274.72	−277.997
16	214.1315	161.0829	118.9083	166.0225	191.4403	12769.45	11837.32	932.1305
17	194.271	151.7612	102.1039	166.0137	174.9354	11221.01	10937.77	283.2481
18	194.3566	151.7946	102.1896	166.128	174.9844	11,225	10944.06	280.9444
19	330.7607	225.2151	231.3178	434.3475	247.5775	29627.92	23117.12	6510.805
20	330.7548	230.3424	255.7548	434.3397	247.5742	41550.3	23723.08	17827.23
21	330.8526	230.3805	255.8526	434.4702	247.6301	49120.57	23730.26	25390.32
22	330.7857	230.3544	255.7857	434.381	247.5918	49107.08	23725.35	25381.73
23	330.7945	219.0851	220.5838	166.0593	247.5968	23329.8	17470.92	5858.882
24	175.5502	151.6073	86.20424	165.4883	159.4785	9959.949	10239.6	−279.653
Total						412,883	292136.9	**120746.1**

Table 11.6 Hourly expected offers to the day-ahead energy market, total costs, revenues, and profits in case II for 54 thermal units

Time	Unit 5 MW	Unit 20 MW	Unit 29 MW	Unit 40 MW	Unit 54 MW	Expected revenue ($)	Expected cost ($)	Expected profit ($)
1	0	5.975181	4.419594	2.882344	0	8330.503	7820.237	510.2662
2	0	8.872115	6.344152	4.095717	0	9032.669	7884.631	1148.038
3	5.372881	11.5849	8.152266	5.229488	0	9870.7	8593.27	1277.43
4	7.426742	12.5908	9.78392	6.24689	0	10107.74	8906.405	1201.334
5	9.502201	13.59116	11.44002	7.281062	0	10672.26	9453.689	1218.572
6	11.61703	14.52354	13.13405	8.34028	0	11136.1	9945.569	1190.528
7	13.47429	15.31358	14.59025	11.23305	0	11506.85	10336.3	1170.545
8	15.43836	103.3784	16.15655	13.27111	0	16022.59	14467.67	1554.919
9	20.80233	136.0951	21.74032	47.58436	0	20640.4	19606.57	1033.83
10	247.0322	243.3506	228.9071	169.7257	0	93748.57	70920.1	22828.47
11	294.0425	246.007	292.8468	195.0895	0	102828.2	76292.73	26535.44
12	294.0796	247.5327	292.8957	195.1241	0	103,035	76498.13	26536.83
13	294.1055	246.957	292.9277	195.1464	0	102922.1	76386.95	26535.12
14	250.356	245.4958	246.868	195.1047	0	96998.24	70680.3	26317.93
15	106.5604	195.6868	103.0724	99.24096	0	36223.77	35797.4	426.3707
16	106.4696	242.9702	93.99967	89.07622	0	45253.82	40409.39	4844.43
17	106.4652	242.9668	87.76269	56.33568	0	39000.57	36432.32	2568.243
18	118.7414	244.5296	117.942	100.211	0	41594.25	39212.37	2381.881
19	262.7638	245.993	262.3681	196.9922	1.073966	116210.4	89645.7	26564.66
20	296.2916	247.4731	296.2916	197.9785	46.99838	183434.5	107433.8	76000.69
21	297.0376	247.5313	297.0376	198.0251	48.00913	218737.8	108815.7	109922.1
22	296.9898	247.4915	296.9898	197.9932	48.00913	218488.9	108639.8	109849.1
23	250.0594	245.8418	236.4778	196.8941	24.54558	110,339	86742.1	23596.95
24	105.0022	195.6937	91.02011	99.85551	24.00457	41136.34	43388.65	−2252.31
Total						1,657,271	1,164,310	**492961.4**

11.5 Comparison and Discussion

Table 11.7 shows the result of expected and standard division of profit using TPEM in comparison with the MCS and deterministic approaches for 5 and 54 thermal units. It should be mentioned that the total expected profits of GenCo in cases II and III are higher than Case I because the electricity market price uncertainty are modeled in cases II and III via TPEM and MCS, respectively. Moreover, the total expected profit of GenCo obtained by the Monte Carlo simulation is better than two-point estimate method. However, the execution time and number of runs of two-point estimate method are much less than Monte Carlo simulation.

Therefore, from the view point of saving the time and with a reasonable approximation for the total expected profit of GenCo, the performance of two-point estimate method is acceptable.

MCS is a criterion usually used to test the accuracy and efficiency of algorithms. This chapter uses the MCS with 5000 samples. The calculated equations of the average value (μ_{MCS}) and the standard deviation (σ_{MCS}) are shown as follows:

$$\begin{aligned} \mu_{\mathrm{MCS}} &= \frac{1}{N}\sum_{i=1}^{N} X_i \\ \sigma_{\mathrm{MCS}} &= \sqrt{\frac{1}{N}\sum_{i=1}^{N} (X_i - \mu_{\mathrm{MCS}})^2} \end{aligned} \tag{11.20}$$

Let F_{TPEM} and F_{MCS} denote the obtained results from TPEM and MCS, respectively. The relative error therefore is defined as:

$$\varepsilon = \frac{|F_{\mathrm{MCS}} - F_{\mathrm{TPEM}}|}{F_{\mathrm{MCS}}} \times 100\% \tag{11.21}$$

The relative error of mean value and standard deviation of TPEM are, respectively, less than %0.2 and %5 for 5 thermal units, and %0.1 and %2 for 54 thermal units in case II which shows that the result of TPEM is credible. Finally, the computation burden of the problem along with the number of binary variables, real variables, and constraints are presented in Table 11.8.

Table 11.7 Comparison between the results of two-point estimate method and other methods

GenCo	Methods	Expected profit ($)	St. Dev. profit ($)	Number of run (no)	Computing time (s)
5 thermal units	Deterministic approach	119,680.6	0	1	0.067
54 thermal units	Two-point estimate method	120,746.1	12,702.7	48	5.628
	Monte Carlo simulation	120,949.2	12,685.3	5000	334.99
	Deterministic approach	481,304.0	0	1	34.14
	Two-point estimate method	492,961.4	56,870.6	48	1639.39
	Monte Carlo simulation	493,228.6	56,787.2	5000	170729.16

Table 11.8 Computational size of the problem

Parameter	5-unit	54-unit
Number of binary variables	120	1296
Number of real variables	601	6481
Number of constraints	3076	35,627

11.6 Conclusion

This chapter proposes the deterministic-based and probabilistic-based scheduling for a price-taker GenCo in a day-ahead electricity market. In the probabilistic-based scheduling, market price uncertainty has been modeled using TPEM and MCS. Mixed-Integer Quadratic Constrained Program is used for solving the deterministic-based and probabilistic-based scheduling problem by the GAMS software. Three case studies are used to show capability of proposed approach. The obtained results via proposed TPEM are compared with MCS and deterministic methods. The compared results show that expected profit of GenCo via TPEM is higher than the deterministic method and less than the MCS. But, the execution time and the number of runs of TPEM are much less than MCS. Therefore, TPEM is preferred in comparison with MCS to model uncertainty of market price in the scheduling of a GenCo in order to decrease the calculation time. Also, the relative error of mean amount and standard deviation of the profit in the TPEM are less than the MCS, which shows that the results of TPEM are credible.

References

1. P.M. Joshi, H.K. Verma, An improved TLBO based economic dispatch of power generation through distributed energy resources considering environmental constraints. Sust. Energ. Grids Netw. **18**, 100207 (2019)
2. K.O. Adu-Kankam, L.M. Camarinha-Matos, Towards collaborative virtual power plants: Trends and convergence. Sust. Energ. Grids Netw. **16**, 217–230 (2018)
3. S. Nojavan, K. Zare, M.A. Ashpazi, A hybrid approach based on IGDT–MPSO method for optimal bidding strategy of price-taker generation station in day-ahead electricity market. Int. J. Electr. Power Energy Syst. **69**, 335–343 (2015)
4. S. Nojavan, K. Zare, Risk-based optimal bidding strategy of generation company in day-ahead electricity market using information gap decision theory. Int. J. Electr. Power Energy Syst. **48**, 83–92 (2013)
5. P. Attaviriyanupap, H. Kita, E. Tanaka, J. Hasegawa, New bidding strategy formulation for day-ahead energy and reserve markets based on evolutionary programming. Int. J. Electr. Power Energy Syst. **27**, 157–167 (2005)
6. V. Vahidinasab, S. Jadid, Stochastic multi-objective self-scheduling of a power producer in joint energy and reserves markets. Electr. Power Syst. Res. **80**, 760–769 (2010)
7. J.S. Derk, Simultaneous bidding in day-ahead auctions for spot energy and power systems reserve. Int. J. Electr. Power Energy Syst. **29**, 470–479 (2007)
8. A. Akbari Foroud, M. Amir ahmadi, M. Bahmanzadeh, A. Akbar Abdoos, Optimal bidding strategy for all market players in a wholesale power market considering demand response programs. Eur. T. Electr. Power **21**, 293–311 (2011)

9. H. Haghighat, H. Seifi, R.K. Ashkan, Gaming analysis in joint energy and spinning reserve markets. IEEE Trans. Power Syst. **22**(4), 2074–2085 (2007)
10. E. Mashhour, S.M. Moghaddas-Tafreshi, Bidding strategy of virtual power Plant for Participating in energy and spinning reserve markets—Part I: Problem formulation. IEEE Trans. Power Syst. **26**(2), 949–956 (2011)
11. A. Azadeh, S.F. Ghaderi, B. Pourvalikhan Nokhandan, M. Sheikhalishahi, A new genetic algorithm approach for optimizing bidding strategy viewpoint of profit maximization of a generation company. Expert Syst. Appl. **39**, 1565–1574 (2012)
12. V. Vahidinasab, S. Jadid, Multi objective environmental/techno-economic approach for strategic bidding in energy markets. Appl. Energy **86**, 496–504 (2009)
13. P. Kanakasabapathy, K. Shanti Swarup, Bidding strategy for pumped-storage plant in pool-based electricity market. Energy Convers. Manag. **51**, 572–579 (2010)
14. V. Vahidinasab, S. Jadid, Normal boundary intersection method for suppliers' strategic bidding in electricity markets: An environmental/economic approach. Energy Convers. Manag. **51**, 1111–1119 (2010)
15. C. Peng, H. Sun, J. Guo, G. Liu, Multi-objective optimal strategy for generating and bidding in the power market. Energy Convers. Manag. **57**, 13–22 (2012)
16. S. Soleymani, Nash equilibrium strategies of generating companies (Gencos) in the simultaneous operation of active and reactive power market, with considering voltage stability margin. Energy Convers. Manag. **65**, 292–298 (2013)
17. T. Niknam, S. Sharifinia, R. Azizipanah-Abarghooee, A new enhanced bat-inspired algorithm for finding linear supply function equilibrium of GENCOs in the competitive electricity market. Energy Convers. Manag. **76**, 1015–1028 (2013)
18. M.B. Naghibi-Sistani, M.R. Akbarzadeh-Tootoonchi, M.H. Javidi-Dashte Bayaz, H. Rajabi-Mashhadi, Application of Q-learning with temperature variation for bidding strategies in market based power systems. Energy Convers. Manag. **47**, 1529–1538 (2006)
19. S. Soleymani, A.M. Ranjbar, A.R. Shirani, New approach to bidding strategies of generating companies in day ahead energy market. Energy Convers. Manag. **49**, 1493–1499 (2008)
20. C. Boonchuay, W. Ongsakul, Optimal risky bidding strategy for a generating company by self-organizing hierarchical particle swarm optimization. Energy Convers. Manag. **52**, 1047–1053 (2011)
21. N. Bigdeli, K. Afshar, M. Fotuhi-Firuzabad, Bidding strategy in pay-as-bid markets based on supplier-market interaction analysis. Energy Convers. Manag. **51**, 2419–2430 (2010)
22. Z. Yuan, D. Liu, C. Jiang, Analysis of equilibrium about bidding strategy of suppliers with future contracts. Energy Convers. Manag. **48**, 1016–1020 (2007)
23. B. Mohammadi-Ivatloo, H. Zareipour, M. Ehsan, N. Amjady, Economic impact of price forecasting inaccuracies on self-scheduling of generation companies. Electr. Power Syst. Res. **81**(2), 617–624 (2011)
24. F. Wen, A.K. David, A genetic algorithm based method for bidding strategy coordination in energy and reserve markets. Artif. Intell. Eng. **15**, 71–79 (2001)
25. M. Rahimiyan, H. Rajabi Mashhadi, Risk analysis of bidding strategies in an electricity pay as bid auction: A new theorem. Energy Convers. Manag. **48**, 131–137 (2007)
26. H. Song, C.C. Liu, J. Lawarrée, R.W. Dahlgren, Optimal electricity supply bidding by Markov decision process. IEEE Trans. Power Syst. **15**(2), 618–624 (2000)
27. T. Li, M. Shahidehpour, Z. Li, Risk-constrained bidding strategy with stochastic unit commitment. IEEE Trans. Power Syst. **22**(1), 449–458 (2007)
28. C.L. Tseng, W. Zhu, Optimal self-scheduling and bidding strategy of a thermal unit subject to ramp constraints and price uncertainty. IET Gener. Trans. Distrib. **4**(2), 125–137 (2010)
29. G. Verbic, C.A. Cañizares, Probabilistic optimal power flow in electricity markets based on a two-point estimate method. IEEE Trans. Power Syst. **21**(4), 1883–1893 (2006)
30. M. Mohammadi, A. Shayegani, H. Adaminejad, A new approach of point estimate method for probabilistic load flow. Int. J. Electr. Power Energy Syst. **51**, 54–60 (2013)

31. T. Niknam, F. Golestaneh, A.R. Malekpour, Probabilistic energy and operation management of a microgrid containing wind/photovoltaic/fuel cell generation and energy storage devices based on point estimate method and self-adaptive gravitational search algorithm. Energy J. **43**, 427–437 (2012)
32. X. Li, J. Cao, D. Du, Probabilistic optimal power flow for power systems considering wind uncertainty and load correlation. Neurocomputing **148**, 240–247 (2015)
33. A.R. Soroudi, M. Aien, M. Ehsan, A probabilistic modeling of photo voltaic modules and wind power generation impact on distribution networks. IEEE Syst. J. **6**(2), 254–259 (2012)
34. G. Shrestha, B. Pokharel, T. Lie, S. Fleten, Price-based unit commitment for bidding under price uncertainty. IET Gener. Transm. Distrib. **1**(4), 663–669 (2007)
35. B. Mohammadi-Ivatloo, H. Zareipour, N. Amjady, M. Ehsan, Application of information-gap decision theory to risk-constrained self-scheduling of GenCos. IEEE Trans. Power Syst. **28**(2), 1093–1102
36. J.M. Arroyo, A.J. Conejo, Optimal response of a thermal unit to an electricity spot market. IEEE Trans. Power Syst. **15**(3), 1098–1104 (2000)
37. M. Kazemi, B. Mohammadi-Ivatloo, M. Ehsan, Risk-based bidding of large electric utilities using information gap decision theory considering demand response. Electr. Power Syst. Res. **114**, 86–92 (2014)
38. L. Baringo, A.J. Conejo, Offering strategy via robust optimization. IEEE Trans. Power Syst. **26**(3), 1418–1425 (2011)
39. J.M. Morales, J. Perez-Ruiz, Point estimate schemes to solve the probabilistic power flow. IEEE Trans. Power Syst. **22**, 1594–1601 (2007)
40. [Online]. motor.ece.iit.edu/data/SCUC−118 test.xls. Accessed Feburary 2011
41. A. Brooke, D. Kendrick, A. Meeraus, *GAMS User's Guide*, (The Scientific Press, Redwood City, CA, 1990). [Online]. http://www.gams.com/docs/gams/GAMSUsersGuide.pdf
42. The GAMS Software Website (2019). [Online] http://www.gams.com/dd/docs/solvers/sbb.pdf>

Chapter 12
Bidding and Offering Strategies for Integration of Battery Storage System and Wind Turbine

Kittisak Jermsittiparsert

Nomenclature

Indices

s, t	Scenario, time

Input

P_r	The rated power of WT
P^{ch}_{min}	Minimum amount of charging power of the BSS
P^{ch}_{max}	Maximum amount of charging power of the BSS
P^{disc}_{min}	Minimum amount of discharging power of the BSS
P^{disc}_{max}	Maximum amount of discharging power of the BSS
P^{proc}_{max}	Maximum limit of power procurement from the grid
P^{sell}_{max}	Maximum limit of sold power to the grid
SOC^B_{max}, SOC^B_{min}	Maximum and minimum limits of the BSS's SOC
$V_{cut-out}, V_{cut-in}, V_r$	The cut-out, cut-in, and rated speeds of WT
η_{disc}, η_{ch}	Discharging and charging efficiencies of the BSS
ρ_s	Probability of the each scenario
$\lambda_{t, s}$	Power price
$V_{t, s}$	The predicated wind speed

K. Jermsittiparsert (✉)
Social Research Institute, Chulalongkorn University, Bangkok, Thailand
e-mail: kittisak.j@chula.ac.th

S. Nojavan, K. Zare (eds.), *Electricity Markets*,
https://doi.org/10.1007/978-3-030-36979-8_12

Variables

$P_{t,s}^{\text{sell}}$	Sold power to the market
$P_{t,s}^{\text{pro}}$	Procured power from the market
$P_{t,s}^{\text{WT}}$	Total produced power by the WT
$P_{t,s}^{\text{WT-G}}$	Injected power from WT to the grid
$P_{t,s}^{\text{WT-B}}$	Injected power from WT to the BSS
$P_{t,s}^{\text{B-G}}$	Injected power from the BSS to the grid
$P_{t,s}^{\text{purchase}}$	Purchased power from the upstream grid
$P_{t,s}^{\text{G-B}}$	Procured power by the BSS from upstream grid
$\text{SOC}_{t,s}^{\text{B}}$	State of charge of the BSS
$U_{t,s}^{\text{disc}}, U_{t,s}^{\text{ch}}$	Binary variables of the discharging and charging states of the BSS

12.1 Introduction

By rising consumed electricity and depletion of conventional fuels [1] which are used to produce electric power, using renewable energy sources is a vital issue [2]. Considering environmental challenges such as greenhouse gas emissions and climate change imposes more necessity on this issue [3]. According to extensive availability [4] and mature technology [5], wind power generation is considered as one of the most promising renewable energies [6]. Despite the conventional power plant, wind power is non-dispatchable [7]. Because of that, there is different uncertainty in planning and operation of wind power systems [8].

To cope with intermittent nature and uncertainty of wind power, integration of wind farms and Energy Storage System (ESS) is proposed by [9, 10]. In restructured power market, which provides a competitive environment [11], electricity prices have high volatility [12]. So, presentation of the optimal offering and bidding curves is very important and vital [13]. To obtain optimal offering and bidding strategies different methods have been reported in the literature. In [14], a bi-level stochastic programming is presented to make optimal offering curves for a wind generation unit in the power market. In this work, the profit maximization function of the wind generation unit is carried out in the upper level of the bi-level model, while the market clearing proceedings of real-time and day-ahead markets are considered in the lower level. In addition, wind speed and load uncertainties are modeled by a set of scenarios. Optimal bids are driven for a wind generation under uncertainties in [15]. To evaluate the proposed method in [15], different strategies for day-ahead bidding are compared from a hypothetical wind site in the USA. According to obtained results, the optimal bid is highly depended to the risk preferences of the wind farm owner, and real-time and day-ahead prices. A novel bidding model for large-scale BSS is proposed in [16] to increase its profitability. In this work, a profit maximization of BSS is provided in the presence of a battery cycle life to develop the optimal bids strategy in spinning reserve, regulation, and day-ahead energy markets. A two-stage stochastic model is presented in [17] to get optimal offering strategy for renewable generation-based micro-grids. To capture various system uncertainties, the Monte Carlo simulation method is utilized in [17]. In addition, it is aimed to balance between maximizing the expected profit of the MG and minimizing the MG performance cost under thermal comfort requirements of the consumers. The same problem is studied in [18] by using hybrid robust optimization and stochastic model. To do so, the uncertainties of the electricity price and power output of intermittent generation units are modeled via generating scenarios and the uncertainty of real-time price is considered by the robust optimization. The same goal is pursued in [19] by using game theoretic approach for electric vehicle aggregators in ancillary services and day-ahead energy markets with variable wind generation. Robust bidding curve for arbitrage is provided in [20] by considering uncertainties in electricity price and wind generation using a bi-level stochastic model. In the presented method, the profit maximization of the wind generation is pursued in the first level, while the market clearing is carried out in the second level. To model the uncertainties of the wind speed and load, a set of scenarios is used. IGDT for determining the optimal bidding curves for large consumer is presented in [21]. In this work, different power procurement sources as distributed generation, bilateral contracts, and the pool market are considered for the large consumer. In addition, the impacts of load management are investigated. By considering uncertainty, a stochastic decision-making model for a wind generation is proposed in [22] by taking participation of demand response aggregators into account. A new approach is proposed in [23] for determining the bidding curves for a large-scale hybrid electric energy company by considering demand response program. Hourly offering and bidding strategies is developed in [24] to sell and purchase power for a BSS unit by using MIP approach. In order to guarantee the increasing and decreasing nature of the offering and bidding curves, the sequential constraints are implemented, and then, offering and bidding curves are developed considering the optimal scheduling of the system. An offering/bidding strategy for a hybrid VPP including a storage unit, wind-power unit, flexible demands, and conventional power plant are developed by using a risk-constrained, stochastic-based robust optimization (RO) formulation to model the problem in [25]. In this reference, the uncertainties of power prices and wind speed in the market are modeled considering confidence scenarios and bounds, respectively. An approach for the offering curve of a VPP that participates which includes a wind-power unit, conventional power plant, a storage facility as well as flexible demands is presented in [26]. The bidding strategy problem is modeled via an RO and stochastic model in which market prices uncertainties are taken into account in [27]. To maximize profitability of the integrated wind turbine with BSS, MIP method is used to develop the optimal offering and bidding curves considering wind speed and market price uncertainties. Finally, summary of researches on bidding/offering strategy are presented in Table 12.1.

According to [15], the renewable-based power plants in USA, especially the wind farms, sell their power generation based on long-term power purchase contracts

Table 12.1 The summary of proposed works on optimal bidding and offering strategies

						Uncertainty	
Refs.	Method	Bidding	offering	RES	Storage	Price	Wind speed
[14]	Bi-level stochastic optimization	✓		✓			✓
[15]	Analytical methods	✓		✓		✓	✓
[16]	Analytical methods	✓			✓		
[17]	Two-stage stochastic program	✓		✓			✓
[18]	Hybrid stochastic/robust optimization	✓		✓		✓	
[19]	Stochastic optimization	✓					✓
[20]	Bi-level stochastic optimization		✓				
[21]	Robust optimization		✓	✓	✓	✓	
[22]	Stochastic decision	✓		✓		✓	
[23]	Max–min bi-level mathematical programming		✓			✓	
[24]	Mixed-integer linear programming	✓	✓		✓		
[25]	Stochastic adaptive robust	✓	✓			✓	
[26]	Analytical methods		✓		✓	✓	
[27]	Stochastic robust optimization	✓		✓		✓	
[28]	Proposed chapter	✓	✓	✓	✓	✓	✓

which prevents the plant owners to benefit high power prices in short term. To overcome this problem and to be immune against the imbalance penalties in the day-ahead and real-time markets, a novel structure is proposed for the WT by integrating with the BSS and developing optimal offering and bidding strategies. In this chapter, first a new scheme for integration of WT and BSS is proposed in which the BSS can get power from WT or the upstream grid and sell power to the market. Second optimal offering and bidding strategies are provided by considering uncertainty of electricity price.

The rest of this work is organized as follows: The proposed model is formulated in Sect. 12.2. Required information beside the results analyzing are provided in Sect. 12.3. Finally, Sect. 12.4 provided the conclusion of this work.

12.2 Problem Formulation

In this part, a new scheme is introduced for integration of WT and BSS. As shown in Fig. 12.1, according to market price, generated electrical power can be injected to the grid or be stored in the BSS. On the other hand, the BSS can be charged by WT or procure power from the upstream grid in off peak periods (low price) in which charged or procured power can be sold in high price periods. Formulation of proposed scheme is presented in the following section.

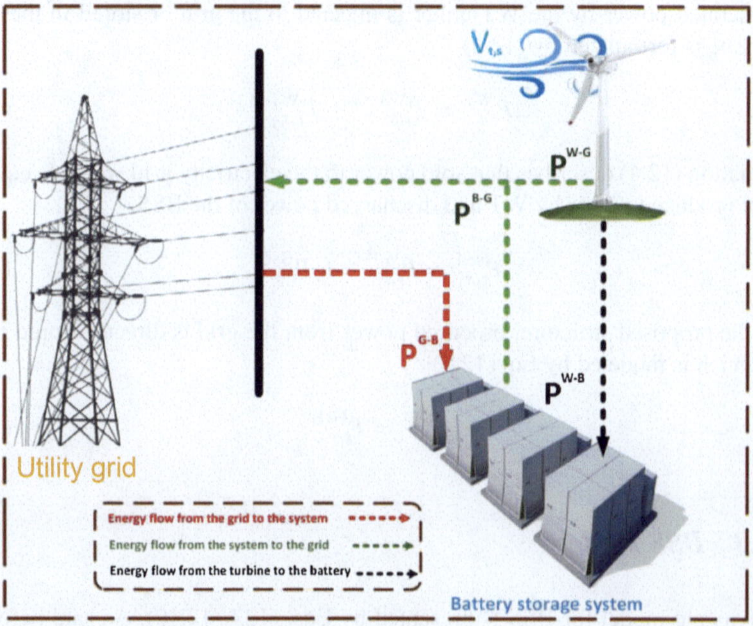

Fig. 12.1 Configuration of proposed model

12.2.1 Objective Function

In order to get maximum amount of profit for proposed structure in previous section, the objective function is presented by Eq. (12.1).

$$\max \ \text{profit} = \sum_{s=1}^{N_s} \rho_s \sum_{t=1}^{N_t} \lambda_{t,s} \left[P_{t,s}^{\text{sell}} - P_{t,s}^{\text{pro}} \right] \tag{12.1}$$

12.2.2 WT Model

The wind turbine is modeled by using Eqs. (12.2–12.5) [28]. The available power of WT based on the wind speed is formulated as Eq. (12.2) [29].

$$P_{t,\text{s}}^{\text{WT}} = \begin{cases} 0, & V_{t,s} \leq V_{\text{cut-in}} \\ P_r \left[\frac{V_{t,s} - V_{\text{cut-in}}}{V_r - V_{\text{cut-in}}} \right]^3, & V_{\text{cut-in}} \leq V_{t,\text{s}} \leq V_r \\ P_r, & V_r \leq V_{t,s} \leq V_{\text{cut-out}} \\ 0, & , V_{t,s} > V_{\text{cut-out}} \end{cases} \tag{12.2}$$

Generated power by the WT either is injected to the grid or stored in the BSS. This issue is formulated by (12.3).

$$P_{t,s}^{\text{WT}} = P_{t,s}^{\text{WT-G}} + P_{t,s}^{\text{WT-B}} \tag{12.3}$$

Equation (12.4) describes that sold power to the electricity grid which is equal to sum of produced power by WT and discharged power of the BSS.

$$P_{t,s}^{\text{sell}} = P_{t,s}^{\text{WT-G}} + P_{t,s}^{\text{B-G}} \tag{12.4}$$

In the proposed structure, procured power from the grid is directly stored in the BSS which is modeled by Eq. (12.5).

$$P_{t,s}^{\text{pro}} = P_{t,s}^{\text{G-B}} \tag{12.5}$$

12.2.3 BSS Model

The dynamic model for BSS is presented by Eqs. (12.6–12.10). As said before, it assumed that the BSS can be charged by WT or power purchased from the upstream grid in low price times and sells the procured power to the upstream grid in high price times. SOC of the BSS is modeled by using Eq. (12.6). Equation (12.7) is applied to limit the SOC of the BSS.

$$\text{SOC}_{t,s}^{\text{B}} = \text{SOC}_{t-1,s}^{\text{B}} + \eta_{ch}\left(P_{t,s}^{\text{G-B}} + P_{t,s}^{\text{W-B}}\right) - \frac{P_{t,s}^{\text{B-G}}}{\eta_{\text{disc}}} \tag{12.6}$$

$$\text{SOC}_{\min}^{\text{B}} \le \text{SOC}_{t,s}^{\text{B}} \le \text{SOC}_{\max}^{\text{B}} \tag{12.7}$$

In order to limit the charged and discharged power of the BSS on their maximum discharge and charge capacity, constraints (12.8) and (12.9) are considered, respectively.

$$P_{\min}^{\text{ch}} \cdot U_{t,s}^{\text{ch}} \le P_{t,s}^{\text{WT-B}} + P_{t,\text{s}}^{\text{G-B}} \le P_{\max}^{\text{ch}} \cdot U_{t,s}^{\text{ch}} \tag{12.8}$$

$$P_{\min}^{\text{disc}} \cdot U_{t,s}^{\text{disc}} \le P_{t,\text{s}}^{\text{B-G}} \le P_{\max}^{\text{disc}} \cdot U_{t,s}^{\text{disc}} \tag{12.9}$$

Also, at each time, the BSS can either charge or discharge. Therefore, Eq. (12.10) is implemented to prevent the simultaneous discharging and charging of the BSS.

$$U_{t,s}^{\text{ch}} + U_{t,s}^{\text{disc}} \le 1 \tag{12.10}$$

In order to obtain optimal bidding and offering curves, MIP method is applied. To ensure that bidding/offering curve is continuously increasing/decreasing, which is a common requirement in power market constraints (12.11) and (12.12) are applied.

$$P_{t,s}^{\text{sell}} \geq P_{t,s'}^{\text{sell}} \mid \lambda_{t,s} \geq \lambda_{t,s'} \tag{12.11}$$

$$P_{t,s}^{\text{pro}} \leq P_{t,s'}^{\text{pro}} \mid \lambda_{t,\text{s}} \geq \lambda_{t,s'} \tag{12.12}$$

The maximum amounts of sold and procured power to or from the grid are limited on by Eqs. (12.13) and (12.14), respectively.

$$P_{t,s}^{\text{sell}} \leq P_{\max}^{\text{sell}} \cdot U_{t,s}^{\text{sell}} \tag{12.13}$$

$$P_{t,s}^{\text{pro}} \leq P_{\max}^{\text{proc}} \cdot U_{t,s}^{\text{pro}} \tag{12.14}$$

To prevent simultaneously selling and purchasing power, Eq. (12.15) is applied.

$$U_{t,s}^{\text{pro}} + U_{t,s}^{\text{sell}} \leq 1 \tag{12.15}$$

12.3 Numerical Simulation

The uncertainties of wind speed and market price are considered by using a set of ten scenarios which are obtained with Weibull and normal distribution functions, respectively. Required information of the WT and BSS are presented in Tables 12.2 and 12.3, respectively. Figures 12.2 and 12.3 present the power price and wind speed scenarios, respectively. To obtain offering and bidding curves proposed model in (12.1–12.15), the model is implemented by using CPLEX solver [30] under GAMS [31].

Results show that by implementing proposed model, total profit is equal to $1433. Discharging and charging powers of the BSS system are shown in Fig. 12.4 which is shown by negative and positive numbers, respectively. According to Fig. 12.4, it

Table 12.2 Coefficients of the WT [32]

$V_{\text{cut - in}}$ (m/s)	V_r (m/s)	$V_{\text{cut - out}}$ (m/s)	P_{rated} (MW)
5	14	25	2.05

Table 12.3 Coefficients of the BSS [32]

$\text{SOC}_{\min}^{\text{B}}$ (MW)	$\text{SOC}_{\max}^{\text{B}}$ (MW)	$P_{\min}^{\text{ch}}$ (MW)
2	10	1
$P_{\max}^{\text{ch}}$ (MW)	$P_{\min}^{\text{disc}}$ (MW)	$P_{\min}^{\text{disc}}$ (MW)
5	1	5

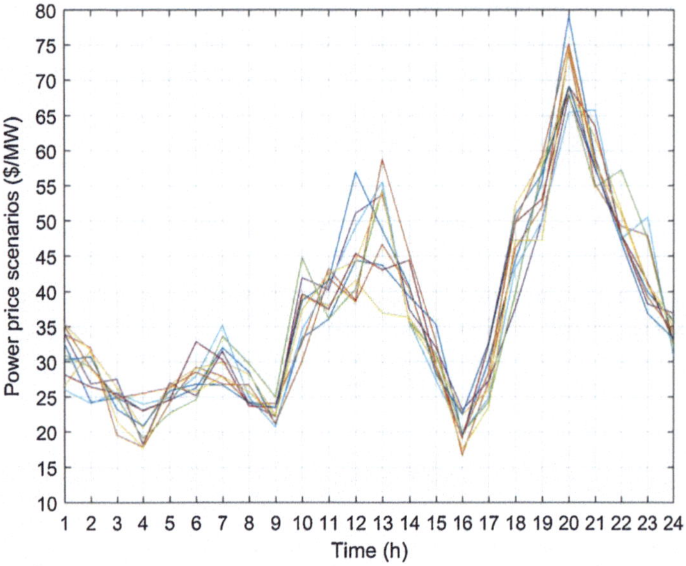

Fig. 12.2 Power price scenarios

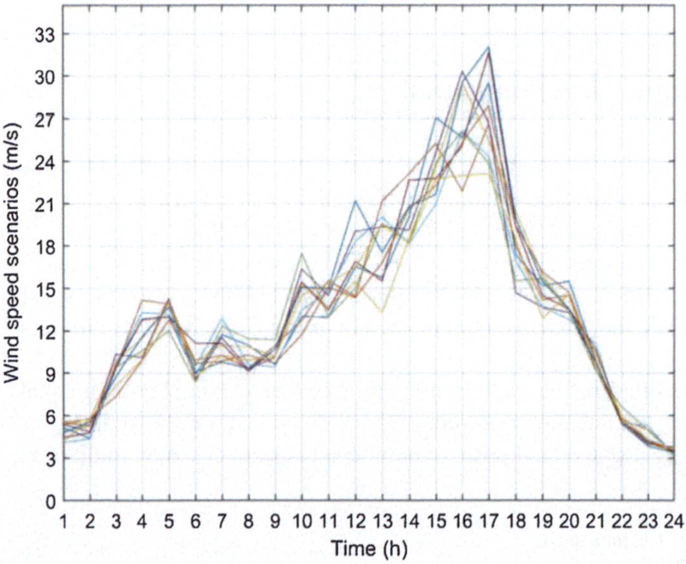

Fig. 12.3 Wind speed scenarios

is obvious that discharged power in scenario 9 is slightly higher than other scenario 9. On the other hand, in scenario 5, charged power is slightly higher in comparison with scenarios 6. As it has been expected, considering power price scenarios, the BSS is discharged during high price periods which are 6–7, 11–12, and 19–21. On

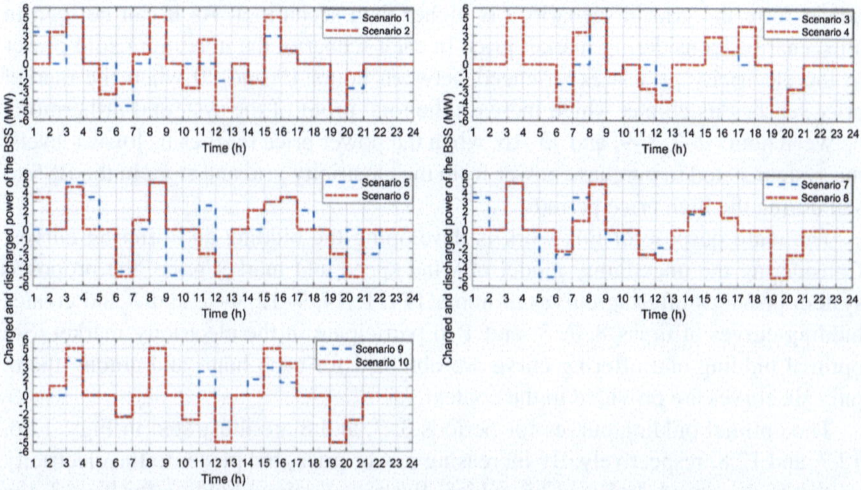

Fig. 12.4 Discharged and charged power of the BSS for different scenarios

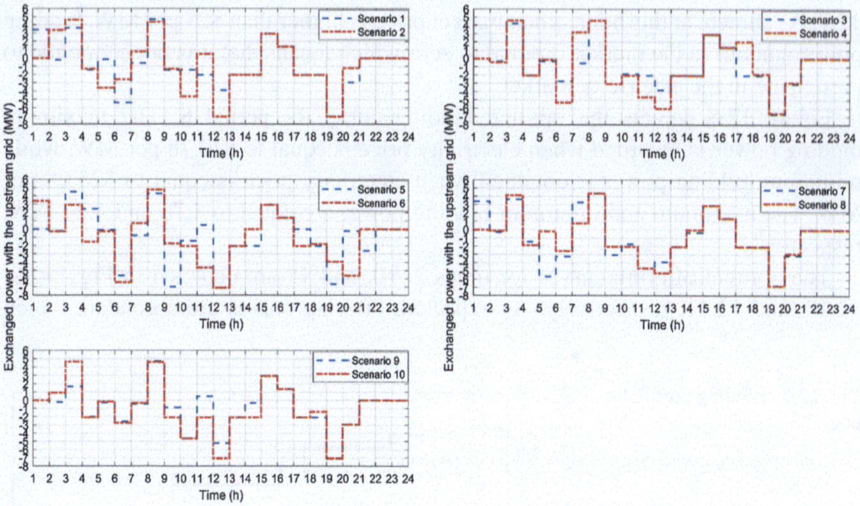

Fig. 12.5 Exchanged power with the upstream for different scenarios

the other hand, in the low price times which are experienced at hours 3–4, 8, 15–17, BSS is charged these periods.

Exchanged power with the electricity grid in each hour for all scenarios is shown in Fig. 12.5. Procured power from the electricity grid is shown by positive numbers while sold power to the electricity grid is shown by negative numbers. It should be denoted that sold power to the electricity grid is aggregation of WT output and discharged power of the BSS. By considering Fig. 12.5, it can be said that sold power to the electricity grid in scenario 5 is more than scenario 6 while procured power form the grid in scenario 2 is higher than scenario 1. As it can be seen in all scenarios, considering market price in each scenario, the maximum sold power to the electricity grid is experienced between hours 19 and 20 when the market price reaches its climax which increase the total profit of the system. Furthermore, between hours 2–3, 8–9, and 15–16, when the power price reaches its lowest levels, the system stated to purchase power from the electricity grid and store in the BSS to sell during the high price periods.

The main purpose of this work is developing the bidding and offering curves. Considering the uncertainty model of wind speed and market price, the proposed system provides offering curves at hours 17–21, 14, 9–12, 4–6, and 2 and creates bidding curves at times 8, 7, 3, and 1 to participate in the electricity market. So, optimal bidding and offering curve are obtained for each hour, and among them, only six curves are provided in the context.

The optimal bidding curves for periods 3, 7, and 8 are illustrated in Figs. 12.6, 12.7, and 12.8, respectively. By increasing power price, bid power to the electricity market is decreased. In Fig. 12.6, which illustrates optimal bidding for hour 3, the maximum bidding power is recorded equal to 4.66 MW when electricity price is equal to $17.92 per MW. The optimal bidding curve for period 7 is presented in Fig. 12.7. As shown, at this hour, when market price is higher than $25 per MW, bidding power amount to the market is equal to zero which means that it is not economic to participate in the electricity market.

Figure 12.8 depicts the optimal bidding curve for period 8. The maximum bidding power is recorded when electricity price is equal to $20.78 per MW, while minimum bidding power is obtained when electricity price is equal to $25.08 per MW. The maximum and minimum bidding power are equal to 4.76 and 4.27 MW, respectively.

The optimal offering curves for times 2, 10, and 18 are depicted in Figs. 12.9, 12.10, and 12.11, respectively. As expected, offering curves are ascendant which

Fig. 12.6 Bidding profile for hour 3

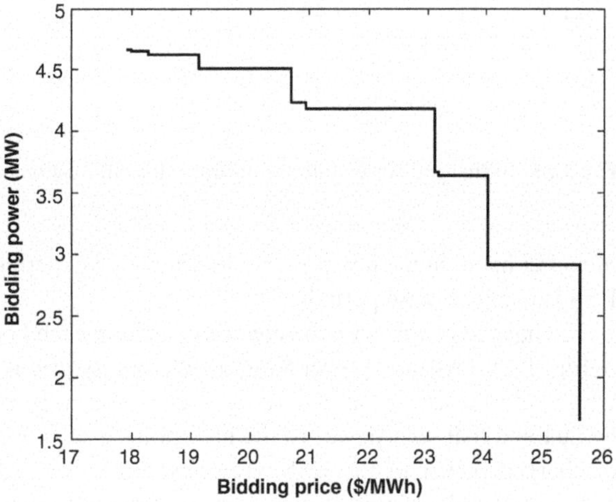

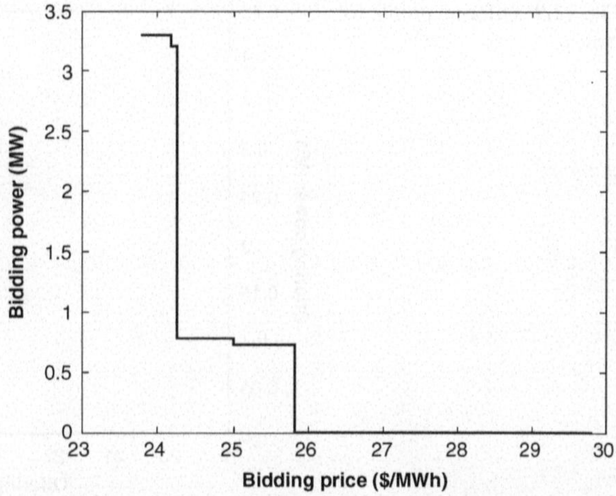

Fig. 12.7 Bidding profile for hour 7

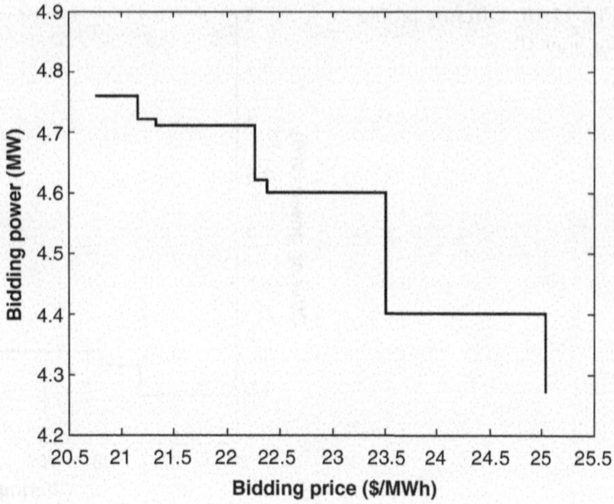

Fig. 12.8 Bidding profile for hour 8

is one the market's necessity. According to Fig. 12.9 which shows offering curve of hour 2, offering power is equal to zero when market price is less than $23.34 per MW. Figure 12.10 shows the offering profile for time 10. As seen in Fig. 12.10, maximum and minimum offering powers, which are equal to 4.65 and 1.43 MW, are recorded when market prices are equal to $43.29 and $36.06 per MW, respectively. The optimal offering profile for hour 18 is illustrated in Fig. 12.11 in which the maximum and minimum offering power are equal to 2.05 and 1.39 MW when power market prices are $59.25 and $47.31, respectively.

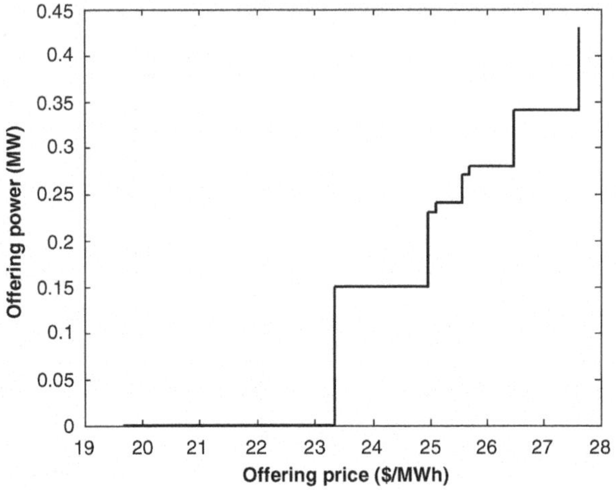

Fig. 12.9 Offering profile for hour 2

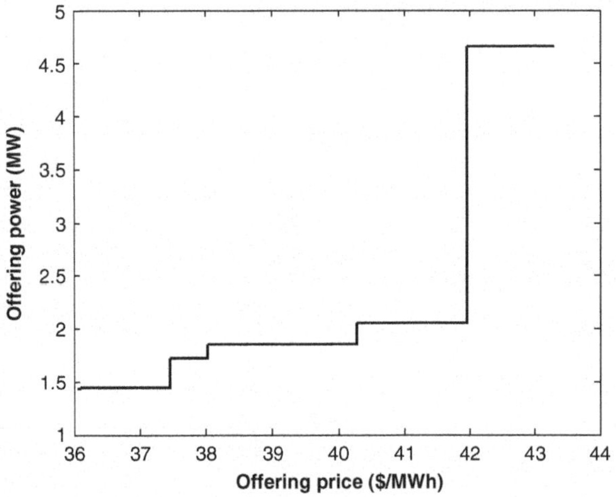

Fig. 12.10 Offering profile for hour 10

12.4 Conclusion

In this work, a novel model is proposed to integrate BSS and WT to maximize the profit. In the proposed model the BSS can be charged with WT or can procure power from market in low price periods. By selling stored power of the BSS in high price periods, the system can maker profit. On the other hand, by considering power market prices, output of WT can be directly injected to the grid or can be used to charge the BSS. To model uncertainty of electricity price, a set of ten discrete scenarios are used by applying a stochastic programming method. The Weibull distribution is applied to model the wind speed uncertainty and the normal distribution is applied to get power price scenarios. Obtained results conclude that total profit of the proposed model is equal to $1433. In addition, in order to

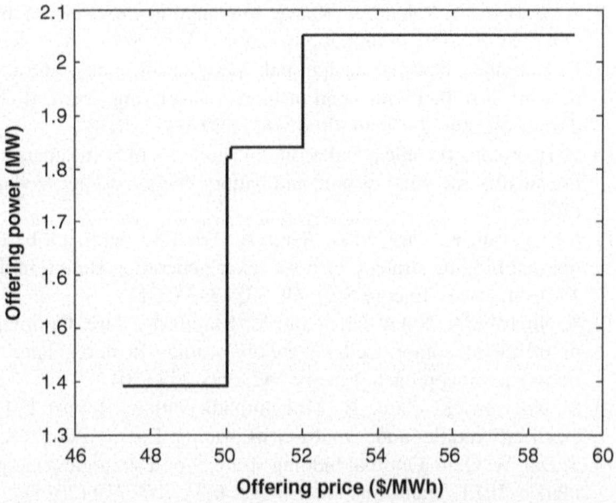

Fig. 12.11 Offering profile for hour 18

participate in the electricity market, the optimal bidding and offering strategies are developed based on an MIP method. Obtained results indicate that the proposed system intends to charge the produced power by the WT when the power price in the market is low. Also, during these periods, the system purchases energy to store in the BSS. Conversely, in the high price periods, the system sell produced power by the WT and stored energy in the BSS to the upstream grid. In this way, considerable profit can be achieved by the system.

References

1. S. Nojavan, M. Majidi, N.N. Esfetanaj, An efficient cost-reliability optimization model for optimal siting and sizing of energy storage system in a microgrid in the presence of responsible load management. Energy **139**, 89–97 (2017)
2. S. Weitemeyer, D. Kleinhans, L. Siemer, C. Agert, Optimal combination of energy storages for prospective power supply systems based on renewable energy sources. J. Energy Storage **20**, 581–589 (2018)
3. K. Saberi, H. Pashaei-Didani, R. Nourollahi, K. Zare, S. Nojavan, Optimal performance of CCHP based microgrid considering environmental issue in the presence of real time demand response. Sustain. Cities Soc. **45**, 596–606 (2019)
4. K. Johansen, J. Emborg, Wind farm acceptance for sale? Evidence from the Danish wind farm co-ownership scheme. Energy Policy **117**, 413–422 (2018)
5. A.A. Salimi, A. Karimi, Y. Noorizadeh, Simultaneous operation of wind and pumped storage hydropower plants in a linearized security-constrained unit commitment model for high wind energy penetration. J. Energy Storage **22**, 318–330 (2019)
6. K.P. Bassett, R. Carriveau, D.K. Ting, Integration of buoyancy-based energy storage with utility scale wind energy generation. J. Energy Storage **14**, 256–263 (2017)
7. V. Jani, H. Abdi, Optimal allocation of energy storage systems considering wind power uncertainty. J Energy Storage **20**, 244–253 (2018)

8. A. Soroudi, A. Rabiee, A. Keane, Information gap decision theory approach to deal with wind power uncertainty in unit commitment. Electr. Power Syst. Res. **145**, 137–148 (2017)
9. R. Hemmati, Optimal design and operation of energy storage systems and generators in the network installed with wind turbines considering practical characteristics of storage units as design variable. J. Clean. Prod. **185**, 680–693 (2018)
10. R. Hemmati, Technical and economic analysis of home energy management system incorporating small-scale wind turbine and battery energy storage system. J. Clean. Prod. **159**, 106–118 (2017)
11. S. Nojavan, K. Zare, M.A. Ashpazi, A hybrid approach based on IGDT–MPSO method for optimal bidding strategy of price-taker generation station in day-ahead electricity market. Int. J. Electr. Power Energy Syst. **69**, 335–343 (2015)
12. S. Nojavan, A. Najafi-Ghalelou, M. Majidi, K. Zare, Optimal bidding and offering strategies of merchant compressed air energy storage in deregulated electricity market using robust optimization approach. Energy **142**, 250–257 (2018)
13. S. Nojavan, K. Zare, B. Mohammadi-Ivatloo, Robust bidding and offering strategies of electricity retailer under multi-tariff pricing. Energy Econ. **68**, 359–372 (2017)
14. T. Dai, W. Qiao, Optimal bidding strategy of a strategic wind power producer in the short-term market. IEEE Trans. Sustain. Energy. **6**(3), 707–719 (2015)
15. A. Botterud, J. Wang, R.J. Bessa, H. Keko, V. Miranda, Risk management and optimal bidding for a wind power producer. In *Power and energy society general meeting, 2010 IEEE* (IEEE, 2010), pp. 1–8
16. G. He, Q. Chen, C. Kang, P. Pinson, Q. Xia, Optimal bidding strategy of battery storage in power markets considering performance-based regulation and battery cycle life. IEEE Trans. Smart Grid. **7**(5), 2359–2367 (2016)
17. D.T. Nguyen, L.B. Le, Optimal bidding strategy for microgrids considering renewable energy and building thermal dynamics. IEEE Trans. Smart Grid. **5**(4), 1608–1620 (2014)
18. G. Liu, Y. Xu, K. Tomsovic, Bidding strategy for microgrid in day-ahead market based on hybrid stochastic/robust optimization. IEEE Trans. Smart Grid. **7**(1), 227–237 (2016)
19. H. Wu, M. Shahidehpour, A. Alabdulwahab, A. Abusorrah, A game theoretic approach to risk-based optimal bidding strategies for electric vehicle aggregators in electricity markets with variable wind energy resources. IEEE Trans. Sustain. Energy **7**(1), 374–385 (2016)
20. M. Zugno, J.M. Morales, P. Pinson, H. Madsen, Pool strategy of a price-maker wind power producer. IEEE Trans. Power Syst. **28**(3), 3440–3450 (2013)
21. S. Nojavan, H. Ghesmati, K. Zare, Robust optimal offering strategy of large consumer using IGDT considering demand response programs. Electr. Power Syst. Res. **130**, 46–58 (2016)
22. M. Asensio, J. Contreras, Risk-constrained optimal bidding strategy for pairing of wind and demand response resources. IEEE Trans. Smart Grid. **8**(1), 200–208 (2017)
23. M. Kazemi, H. Zareipour, M. Ehsan, W.D. Rosehart, A robust linear approach for offering strategy of a hybrid electric energy company. IEEE Trans. Power Syst. **32**(3), 1949–1959 (2017)
24. S. Shafiee, H. Zareipour, A.M. Knight, Developing bidding and offering curves of a price-maker energy storage facility based on robust optimization. IEEE Trans. Smart Grid. **10**, 650–660 (2017)
25. A.V. Ntomaris, A.G. Bakirtzis, Optimal bidding for risk-averse hybrid power station producers in insular power systems: An MPEC approach. In *Innovative smart grid technologies conference Europe (ISGT-Europe), 2017 IEEE PES* (IEEE, 2017), pp. 1–6)
26. A. Baringo, L. Baringo, A stochastic adaptive robust optimization approach for the offering strategy of a virtual power plant. IEEE Trans. Power Syst. **32**(5), 3492–3504 (2017)
27. L. Baringo, R.S. Amaro, A stochastic robust optimization approach for the bidding strategy of an electric vehicle aggregator. Electr. Power Syst. Res. **146**, 362–370 (2017)
28. H. Morais, P. Kádár, P. Faria, Z.A. Vale, H.M. Khodr, Optimal scheduling of a renewable micro-grid in an isolated load area using mixed-integer linear programming. Renew. Energy **35**(1), 151–156 (2010)

29. M. Abbaspour, M. Satkin, B. Mohammadi-Ivatloo, F.H. Lotfi, Y. Noorollahi, Optimal operation scheduling of wind power integrated with compressed air energy storage (CAES). Renew. Energy **51**, 53–59 (2013)
30. The GAMS Software Website (2019). [Online]. http://www.gams.com/dd/docs/solvers/cplex.pdf
31. A. Brooke, D. Kendrick, A. Meeraus, *GAMS User's Guide* (The Scientific Press, Redwood City, CA, 1990). [Online]. http://www.gams.com/docs/gams/GAMSUsersGuide.pdf
32. H.A. Aalami, S. Nojavan, Energy storage system and demand response program effects on stochastic energy procurement of large consumers considering renewable generation. IET Gener. Trans. Distrib. **10**(1), 107–114 (2016)

Index

S. Nojavan, K. Zare (eds.), *Electricity Markets*,
https://doi.org/10.1007/978-3-030-36979-8

The manufacturer's authorised representative in the EU is Springer Nature Customer Service Centre GmbH, Europaplatz 3, 69115 Heidelberg, Germany. If you have any concerns regarding our products, please contact ProductSafety@springernature.com

Printed and bound by CPI Group (UK) Ltd, Croydon, CR0 4YY
30/09/2026
02225602-0007